MW01283347

Of Bicycles, Bakelites, and Bulbs

Inside Technology
edited by Wiebe E. Bijker, W. Bernard Carlson, and Trevor Pinch

Wiebe E. Bijker, *Of Bicycles, Bakelites, and Bulbs: Toward a Theory of Sociotechnical Change*

Wiebe E. Bijker and John Law, editors, *Shaping Technology/Building Society: Studies in Sociotechnical Change*

Stuart S. Blume, *Insight and Industry: On the Dynamics of Technological Change in Medicine*

Geoffrey C. Bowker, *Science on the Run: Information Management and Industrial Geophysics at Schlumberger, 1920–1940*

H. M. Collins, *Artificial Experts: Social Knowledge and Intelligent Machines*

Pamela E. Mack, *Viewing the Earth: The Social Construction of the Landsat Satellite System*

Donald MacKenzie, *Inventing Accuracy: A Historical Sociology of Nuclear Missile Guidance*

Of Bicycles, Bakelites, and Bulbs
Toward a Theory of Sociotechnical Change

Wiebe E. Bijker

The MIT Press
Cambridge, Massachusetts
London, England

This book was set in Baskerville by Asco Trade Typesetting Ltd., Hong Kong and was printed and bound in the United States of America.

Library of Congress Cataloging-in-Publication Data
Bijker, Wiebe E.
Of bicycles, bakelites, and bulbs : toward a theory of sociotechnical change / Wiebe E. Bijker.
p. cm.—(Inside technology)
Includes bibliographical references (p.) and index.
ISBN 0-262-02376-8
1. Technology—Social aspects. I. Title. II. Series.
T14.B54 1995
306.4′6—dc20 94-44413
CIP

For Tonny

Contents

Acknowledgments

One of the central themes of this book, the interlaced character of hitherto separate domains such as technology and society, equally applies to its own evolution. This book developed from an academic detour, as described in the first chapter. But in addition to this intellectual origin, its social roots were equally important, and these lie first in close work with many friends and colleagues.

At the University of Twente, the late Peter Boskma, Ellen van Oost, and Jürgen Bönig were my closest collaborators. Without them, this research would never have been started in the first place. A crucial influence was Trevor Pinch's semester as a visiting researcher; his friendship and scholarship have changed my professional life irreversibly. During this period we organized the "Twente 1" and "Twente 2" workshops, which turned out to be seminal in shaping the sociology of technology field. Editing the volumes that emerged from these workshops with Tom Hughes and John Law was an exciting experience. These books contained brief versions of the bicycle, Bakelite, and fluorescent lighting case studies, and the comments from Henk van den Belt, Robert Bud, Michel Callon, David Edge, Robert Frost, Ernst Homburg, Stephen Kline, Rachel Laudan, Donald MacKenzie, Simone Novaes, Jeffrey Sturchio, Sharon Traweek, and Steve Woolgar were much appreciated. Boelie Elzen, Frans-Bauke van der Meer, Arie Rip, Wim Smit, and Gerard de Vries offered valuable help in turning the first stage of the project into a dissertation.

Later, at the University of Limburg in Maastricht, the project benefited from further interweaving of teaching and research. The preparation of courses together with Karin Bijsterveld and Ger Wackers, as well as teaching various classes abroad, contributed much to my own understanding of the issues discussed in this book (and I can only hope that it had the same effect on students).

I also benefited from discussions with research groups in various places. Chapter 4 was largely written during my stay as visiting professor at the Technical University in Vienna. Chapter 5 received its penultimate and most radical revision during my visit to the Technical University of Denmark. The research group "Technological Culture" in Maastricht provided an important forum for discussing drafts of chapters 4 and 5. Tannelie Blom, Ton Nijhuis, Rein de Wilde (Maastricht), Rob Hagendijk (Amsterdam), Ulrik Jörgensen (Copenhagen), and Eduardo Aibar (Barcelona) have tried to help me avoid various pitfalls in the discussions of power in these last chapters (though I probably still fell into a good number of them). Ed Constant, Tom Misa, and Paul Rosen have provided stimulating discussion at various stages of writing.

Making a book, however, is not just a matter of academic research and teaching. In the final stage, the comments of two anonymous referees were stimulating and challenging. Bernie Carlson, Larry Cohen, and Trevor Pinch succeeded in critically following and shaping the project without jeopardizing our relationship as co-editors of the Inside Technology series. Melissa Vaughn did the crucial editing and production job of turning the manuscript into a book. Such were the professional ties, many of which have turned into friendships.

But the last—and in some respects most important—part of the weave has yet to be mentioned. This book could never have been written solely within the confines of academia. Liselotte, Else, and Sanne continually reminded me of this in their need for cooking and caring, and their claims to bicycle and football, to playing piano and cello. But mainly by just being there, three daughters provide a strong, continuous demonstration that life is more, and more complex, and more interesting, than the activities in the academic compartment of society. This book is dedicated to Tonny, who complements all those mentioned above as skeptical commentator, as supportive friend, as mother of the daughters, as love.

1

Introduction

The stories we tell about technology reflect and can also affect our understanding of the place of technology in our lives and our society. Such stories harbor theories. But stories can be misleading, especially if they aim for neatness and therefore keep to the surface of events. This book will be about both stories and theory. I will start with some of the stories:

• In 1898 a female cyclist was touring the English countryside. She was dressed in knickerbockers, which seemed the most practical and comfortable clothing for a woman on a safety bicycle. After a good lap, she spotted an inn and decided to take a bit of refreshment. To her surprise, the proprietor refused to seat her in the coffee room and insisted that, if she wanted service, she would have to go into the public bar. The innkeeper's objection centered on the cyclist's clothes; evidently she did not think it proper for a woman to appear in public in anything but a long skirt. The cyclist objected, of course, and eventually brought her grievance to court, which sided with the right of the innkeeper to refuse service. This was not the end of the story, though. This lost case had an important afterlife as a symbol in the battle for women's rights. Can we say, then, that the design of this technological artifact, the safety bicycle, which allowed our cyclist to travel on her own and to choose a more comfortable form of dress, played a role in challenging traditional gender roles and building modern society?[1]

• "God said, 'Let Baekeland be,' and all was plastic." Few individual inventors have had as great an impact on society as did Leo Baekeland. This brilliant inventor created the first truly synthetic material to replace natural and seminatural materials such as ivory and Celluloid, and developed many of the applications that led society into the era of plastics. At first glance, Baekeland seems an exemplar of the American scientist-entrepreneur. A poor Belgian immigrant to the United States, he worked his way up by cleverness and diligence, and by combining scientific discovery with commercial acumen. He became rich, served his new country during World War I in the Naval Consulting Board, and served humankind by giving it plastics. A longer look, however, shows that Baekeland was shaped not by a mythical act of creation but by several distinct sociotechnical traditions and cultures. It was only because he was enculturated in the technical and scientific practices of electrochemistry that he was able to escape the bondage of the Celluloid engineering tradition; but it was only because he was part of the Celluloid tradition that he undertook this research at all. Can one assert, then, that even cases of seemingly unique individual ingenuity and creativity are always linked to wider social interactions and cultural processes?

• When the General Electric Company tried to introduce the fluorescent lamp in 1938 as a source of color lighting, they quickly found themselves in a battle with the electric utilities, who feared that the lamp's high efficiency would jeopardize their electricity sales. Consumers and lighting engineers, however, were so eager to buy this new type of lighting that the utilities were forced to accept some form of the lamp. After a fierce confrontation that threatened the established power relations in the electric lighting business, an agreement was reached under which the lamp's design was substantially changed. This renewed cooperation did not escape the notice of the Antitrust Division of the federal government, which decided to sue General Electric and the utilities for forming a cartel. General Electric, in response, successfully lobbied the War Department into fending off this suit because, they argued, such litigation would endanger the war effort. The fluorescent lamp was thus the product of a complex economic power play in which General Electric, the electric utilities, the U.S. government, and consumers all played roles. Conversely, the power map of the electric manufacturing scene in the United States was substantially modified by the introduction of the new lamp. Can we then say that artifacts are not only shaped by the power strategies of social groups but also form part of the micropolitics of power, constituting power strategies and solidifying power relations?

These three stories highlight many of the issues that this book will address. For example, how can gender relations affect the design of a bicycle? Although it later became an instrument for women's emancipation, the first cycles in fact reinforced the existing "gender order"—women were only allowed to ride on tricycles, and preferably on two-

Figure 1.1
Women's emancipation: the wheel of the past and the wheel of the present (reprinted from Palmer (1958: 101).

seaters with a male as chaperon. It is therefore appropriate to ask: What impact did the evolution of bicycle design have on society? How did it shape social relations (see figure 1.1)? This is the companion issue of this book, for we shall explore both the social shaping of technology and the technical shaping of society.

Framing these issues in terms of "society" and "technology" should not obscure the fact that technology and society are both human constructs. Technology is created by engineers working alone or in groups, marketing people who make the world aware of new products and processes, and consumers who decide to buy or not to buy and who modify what they have bought in directions no engineer has imagined. Technology is thus shaped not only by societal structures and power relations, but also by the ingenuity and emotional commitment of individuals. The characteristics of these individuals, however, are also a product of social shaping. Values, skills, and goals are formed in local cultures, and we can therefore understand technological creativity by linking it to historical and sociological stories. This is the second set of central issues in this book: How can we link the interactions of individual actors such as engineers and users to societal processes? And how can we link the analysis of micro case studies to an understanding of macro processes of societal and technological change?

This linking of micro stories with macro structures involves questions about the internal structure of technology: about the nature of inventors' work, about the interaction of knowledge, skills, and machines, about the epistemology of technology. But it also involves the politics of technology. The quick summary of the story of the fluorescent lamp showed how it was shaped by the power relations of General Electric and the utilities and eventually helped shift those relations. How do artifacts become instruments of power? And conversely, how do power relations materialize in artifacts? Some artifacts are more obdurate, harder to get around and to change, than others. Who was in a position to modify the fluorescent lamp design that was proposed in 1938, and who was compelled to "take it or leave it" as it was? Exploring the obduracy of technology offers one way to gain understanding of the role of power in the mutual shaping of technology and science.

From Detour to Main Route

This book is the result of a personal detour that turned into a main route. My detour started from sociopolitical concerns about the role of technology in society and then carried me into academia. Like many Dutch engineering students in the 1970s, I was drawn to the science-technology-society (STS) movement, whose goal was to enrich the curricula of both universities and secondary schools by offering new ways to explore issues such as the risks of nuclear energy, the proliferation of nuclear arms and other new weapons systems, and environmental degradation. The movement was eventually quite successful, especially in the natural science and engineering faculties, where small groups were established to teach STS courses and some of the courses even became part of degree requirements. The secondary school science curriculum was also reformed to include STS issues, both optional and integrated into the regular physics program. At the same time, STS students and staff were among the central actors in the movement against the extension of nuclear power and the introduction of the "neutron bomb" and cruise missiles. After gaining access to the academy, however, and during our political struggles, we were increasingly confronted with the crudeness and inadequacy of our models of science and technology development. We were working in many instances on our gut feelings about technology, but were not able to back our positions with theoretical arguments. This is what spurred my detour into academia—a desire to see if I could help devise new ways to think about the development of technology and its relationship to society.

Many other researchers from the early STS ranks made similar detours. Now, two decades later, science and technology studies is a well-established discipline with chairs, journals, societies, and both undergraduate and graduate programs—everything that a respectable academic discipline requires.[2] But did this detour yield the politically relevant insights that we needed fifteen years ago? Or does our new discipline worry too much about its status in the academy? Have all our activists turned into scholars? The central argument of this book will be that STS can retain its edge even in the academy, that what started out as a detour can be turned into a main route without necessarily losing its societal relevance.

At the beginning of my detour I found at least three models open to me. First, there were those who looked down their noses at mere storytellers. These were the scholars, often with backgrounds in the social sciences, who advocated general typologies, precise conceptual definitions, and macrotheoretical schemes that could produce "real" insights and explanations. Second, there were those who poked fun at any theoretical generalization beyond the uniquely detailed story. These students, often of the historians' tribe, scorned the empty theoretical boxes and abstract schemata that did not display any familiarity with what "really" went on. Third, there were the political activists, who considered any detour into academia a betrayal of the immediate societal tasks that should be the constant overriding concern of critical intellectuals.

What finally changed my detour into a main route was the conclusion that all three approaches are equally necessary. I believe that effective societal action on issues of technology and science cannot do without scholarly support, while academic technology studies have much to gain from engagement with politically relevant issues. And only an integration of detailed empirical case studies with general conceptual frameworks can build this link between academia and politics. I have come to believe that an integration of case studies, theoretical generalizations, and political analyses is called for and possible, both *to understand* the relations between technology and society and *to act* on issues of sociotechnical change. This book will start with stories, then generate theoretical concepts, and finally argue for politics. The idea of a gap between the real world and academia, at the basis of the "detour" metaphor, proved misleading.

Guideposts

What guideposts can lead us as we embark on this journey to an integrated understanding of the STS problematic? The past decade has seen the emergence of a new research program in technology studies.[3] This program, commonly labeled "constructivist studies of technology," is based mainly on the combination of historical and sociological perspectives. Infusions from economics and philosophy have hitherto been quite small, although efforts are now being made to incorporate work from these disciplines into constructivist research.[4] A central adage for this research is that one should never take the meaning of a technical artifact or technological system as residing in the technology itself. Instead, one must study how technologies are shaped and acquire their meanings in the heterogeneity of social interactions. Another way of stating the same principle is to use the metaphor of the "seamless web" of science, technology, and society, which is meant to remind the researcher not to accept at face value the distinctions between, for example, the technical and the social as these present themselves in a given situation.

Within the constructivist research program we can distinguish three lines of work: the systems approach, the actor-network approach, and the social construction of technology (SCOT) approach. This book has developed in the main from SCOT studies, but I believe that the arguments are of general relevance for the whole spectrum of modern constructivist studies.

One pitfall that the newer research programs are designed to avoid is any implicit assumption of linear development. Such assumptions were often found in earlier technology studies, sometimes at the level of the singular invention (figure 1.2) and sometimes in the genealogy of related

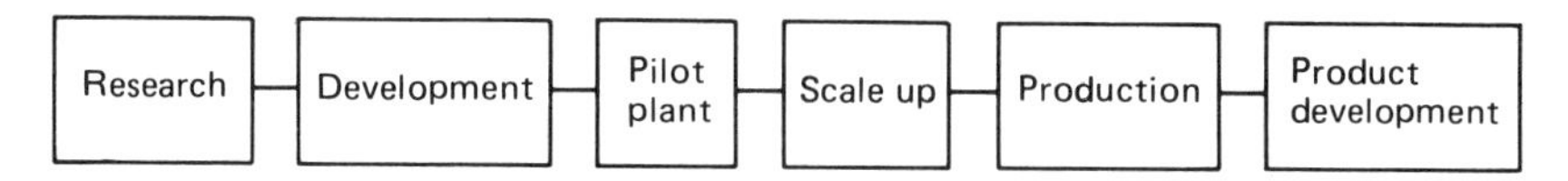

Figure 1.2
A six-stage model of the innovation process.

innovations (figure 1.3). The problem is that once students start expecting linearity, they blind themselves to the retrospective distortions that linear descriptions almost inevitably require. Too easily, linear models result in reading an implicit teleology into the material, suggesting that "the whole history of technological development had followed an orderly or rational path, as though today's world was the precise goal toward which all decisions, made since the beginning of history, were consciously directed" (Ferguson, 1974: 19). To name Lawson's bicycle "the first modern bicycle" is an example of such a false linearity, as I will show in the next chapter. This label seems appropriate at a surface level because this was the first bicycle with two relatively low wheels and a chain drive on the rear wheel. It was, however, at least in a commercial sense, a complete failure, and the relevance of the label "first" is therefore questionable. Bicycles such as the Star and the Geared Facile did much better commercially, but because they do not fit into a simple linear scheme, they are often written off into the margins of the story.

A second pitfall the constructionist programs are designed to avoid is what one might call the asymmetrical analysis of technology. Staudenmaier (1985) observed that in the first twenty-five volumes of the journal *Technology and Culture*, only nine articles were devoted to the analysis of failed technologies. The focus on successful innovations suggests an underlying assumption that it is precisely the success of an artifact that offers some explanatory ground for the dynamics of its development. Many histories of synthetic plastics, for example, start by describing the technically sweet characteristics of Bakelite. These features are then used implicitly to position Bakelite at the starting point of the glorious development of the synthetic plastics field, as in Kaufman's (1963: 61) quotation of God at the beginning of this chapter. However, a more detailed study of the developments of plastic and varnish chemistry following the publication of the Bakelite process in 1909 shows that Bakelite was at first hardly recognized as the marvelous synthetic resin it later proved to be.[5] A historical account founded upon the retrospective success of Bakelite (its "working") leaves much untold. More specifically, such an

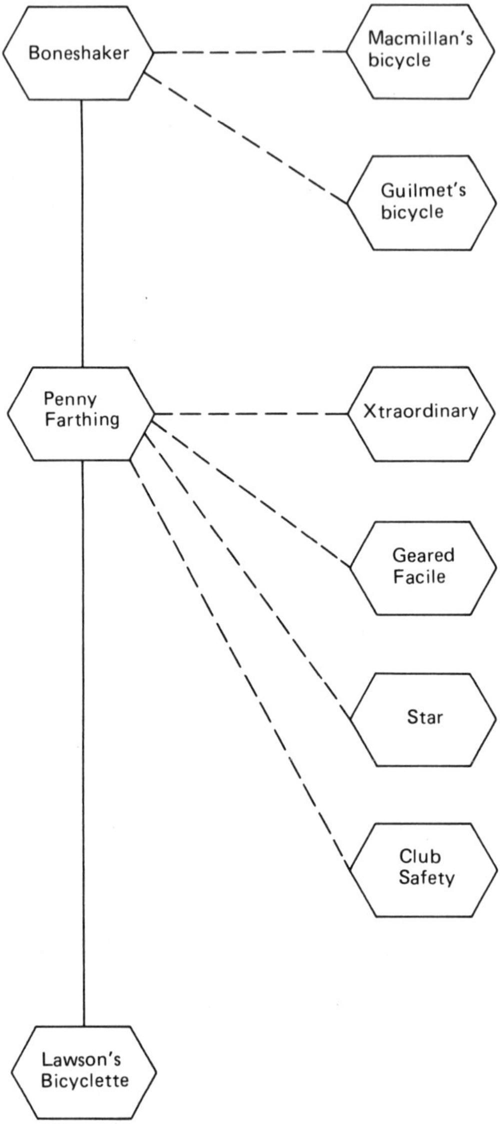

Boneshaker

Macmillan's bicycle

Guilmet's bicycle

Penny Farthing

Xtraordinary

Geared Facile

Star

Club Safety

Lawson's Bicycletté

Figure 1.3
The traditional quasi-linear view of the development of the high-wheeled Ordinary bicycle until Lawson's bicycle. The solid lines indicate successful development, while dashed lines indicate failure.

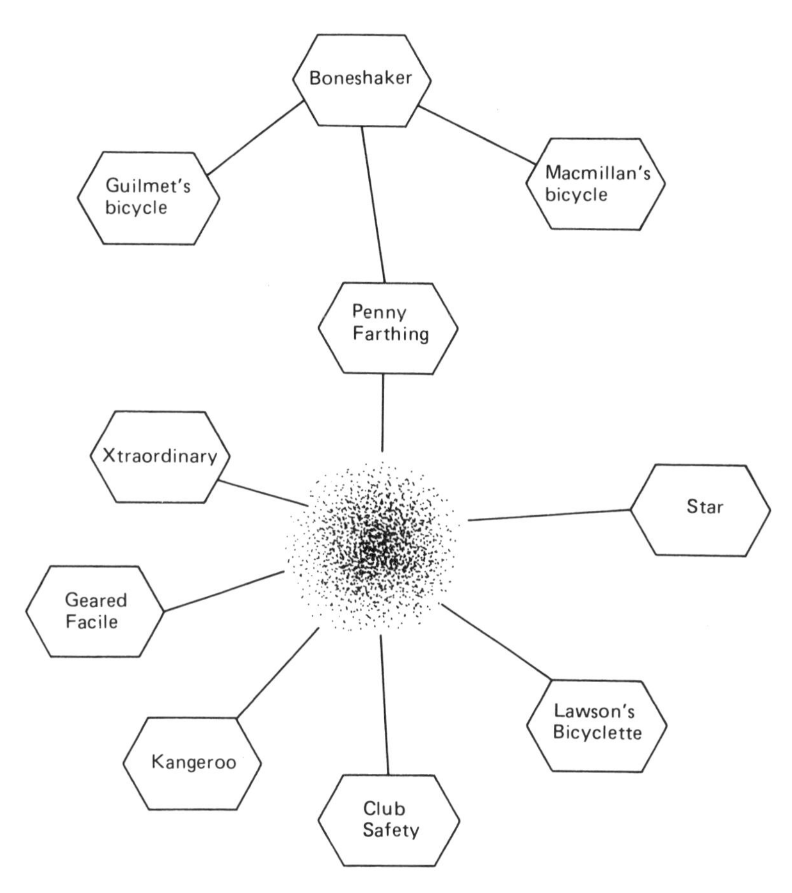

Figure 1.4
A nonlinear representation of the various bicycle designs since the high-wheeled Ordinary bicycle. The various designs are treated equally, without using hindsight knowledge about which design principles eventually would become most commonly used.

account misses the interesting question of how Bakelite *came to be seen as* a practical molding material; instead, in these asymmetrical accounts Bakelite simply *was so* all along. (Figure 1.4 shows a possible visualization of an alternative analysis for the bicycle example—symmetrical and without a linearity assumption.)

Other beacons to guide our journey come from the individual disciplines on which STS studies draw. For example, a key debate in the history of technology[6] has involved the primacy of internalist versus contextualist (or externalist) studies. Internalists maintain that we can understand the development of a technology only if we start with an understanding of the technology in all its minute details. Contextualists, by contrast, claim that the economic, social, political, and scientific context of a technology is as important to its development as are its technical design characteristics. I lean toward the contextualist side of this debate. To understand the development of bicycle designs, for example, I think it is important to know about the industrial development of Coventry, a visit to the Queen by the English "Father of the Bicycle," and the early professional bicycle races. At the same time, I believe that details are important, and I hope to demonstrate that it is only by going down to the "nuts and bolts" level of analysis that we can gain insight in the *design* development of technology.[7] By the end of the book I will also take one additional step outside the "pure" contextualist path, arguing that, rather than being satisfied with the distinction between technology and its context as the basic dimension for analysis, we must figure out a way to take the common evolution of technology and society as our unit of analysis.

Technological creativity has been another long-standing research topic from which we can draw guideposts. One key issue that absorbs many researchers is uncovering the "Mother of Invention." Is it "necessity," implying that an invention will sooner or later emerge out of felt needs, independent of individual creativity? Or is it the "act of ingenuity," without which needs might never be fulfilled (but perhaps not explicated either)? In arguments for the individual act, one still sees numerous references to the claim by Jewkes et al. (1958) that the majority of inventions in the twentieth century have resulted from individual work rather than large-scale organized research. Often, a stress on the role of the individual inventor is accompanied by a declaration that the topic is immune to research: "like a poet or an artist, therefore, the inventor participates in an act of creation, and no amount of theoretical construction can encompass the terms on which such creativity can be achieved."[8]

Nevertheless, two lines of research are now bearing fruit. The first focuses on inventors as system builders, thus combining analysis of individual creative actors with descriptions of their systemic constructs and contexts.[9] The second combines history of technology with the insights of psychology to explore acts of individual creativity.[10] A quite opposite approach is possible as well. Rather than taking individual ingenuity as given, this approach tries to describe the label "individual genius" as the result of a series of attribution processes by which one person eventually "wins all." My approach to the problem of creativity is probably closest to this last one. I tend to analyze the development of technology (including its invention) as a social rather than a psychological process. I will not argue, however, that individual engineers and their histories do not matter. For example, I found it quite illuminating to delve into Baekeland's early work in photographic chemistry as a source for his research gusto (he dissolved his father's watch when he needed more silver), and I found his work with the new electrochemical plants at Niagara Falls an aid to understanding his experience with upscaling chemical production. But I will also introduce a conceptual framework that will link these stories about individual inventors to a sociological analysis of their positions in a specific technological culture.

Yet another guidepost comes from political science. Power has always held a peculiar place in studies of technological development. It is hardly ever invoked by the historians as part of their explanations of events—mainly, I think, because their stories do a much better job of explanation than any crude "power" concept might. The older sociology of technology did not address either the question of technological development or that of the role of power in that process.[11] Recent sociohistorical studies of technology have also avoided the use of "power" as a central category, not because everyone is equal, or because there are no hierarchical relations between particular individuals and social groups, but because explanations in terms of power so easily result in begging what seem to be the most interesting questions. Thus it is just not very insightful to state that the introduction of the fluorescent lamp was held up because the electric utilities were more powerful than General Electric; nor is it illuminating to state that the fluorescent lamp finally appeared on the market because General Electric proved more powerful. Instead, I want to raise the question of which strategies the utilities and General Electric (and other companies, and the U.S. government, and all the other actors) employed to create a certain outcome—an outcome that can then be conveniently summarized by drawing a map of the power distribution. In this analysis of power strategies, I will especially focus on the role of artifacts.

Project Design

The core of this book is formed by three case studies: the safety bicycle, Bakelite, and the fluorescent lamp. In selecting these cases, I employed two crude criteria. The first was to focus on the actual design process of technology, on the details of the technical machines and processes. The second was to secure an empirical base broad enough to render generalizations interesting.

The first criterion suggested implicitly what in this study would constitute "technology." The aim was to select cases that would allow a focus on the "hard" contents of technology rather than its systemic aspects. I therefore decided to focus on "elementary innovations" rather than technological systems, and this led me to the bicycle rather than the automobile, Bakelite rather than synthetic materials in general, and the fluorescent lamp rather than electric lighting.

An intuitive and commonsense idea about what "technology" and "society" are, and what there is to be asked about their developmental process, further informed the selection. However, what starts out as an intuitive assumption about the object of research of this study will by the end of this book have become a key question: What constitutes "an artifact," "design," "technical change," "technology," "society"? The object of research will thus, in the course of the book, evolve from elementary technical artifacts to "sociotechnical ensembles."

My second criterion was founded on a desire to create a relatively broad empirical base for generalizations. Several dimensions were used to check the heterogeneity of alternative cases: the period in which the invention was made, the disciplinary background of the invention, the industrial context, the intended market, and the invention's process or product character. Thus I selected cases that, taken together, span most of the period after the second industrial revolution: the bicycle covers 1860–1890; Bakelite, 1880–1920; and the fluorescent lamp, 1930–1945. The cases are also varied in terms of their underlying engineering background: mechanical engineering (the bicycle), chemical engineering (Bakelite), and electrical engineering (the fluorescent lamp). With respect to industrial context, the cases move from a blacksmith's workshop (bicycle) to an early scientific laboratory (Bakelite) to a large industrial laboratory (fluorescent lamp). The bicycle was exclusively aimed at the consumer market, Bakelite as a molding material was aimed at the industrial market, and the fluorescent lamp has in this respect a hybrid character. In the patent literature a distinction is often made between product and process types of inventions. The bicycle and the fluorescent lamp are clearly both product inventions, while Bakelite is primarily a process invention.

Table 1.1
Requirements for a theory of technological development

1. Change/continuity	The conceptual framework should allow for an analysis of technical change as well as of technical continuity and stability.
2. Symmetry	The conceptual framework should take the "working" of an artifact as *explanandum*, rather than as *explanans*; the useful fuctioning of a machine is the result of sociotechnical development, not its cause.
3. Actor/structure	The conceptual framework should allow for an analysis of the actor-oriented and contingent aspects of technical change as well as of the structurally constrained aspects.
4. Seamless web	The conceptual framework should not make a priori distinctions among, for example, the social, the technical, the scientific, and the political.

Because I wanted to create a relatively broad base of data, I chose to present a larger number of cases using mainly published sources rather than one or two cases using unpublished archival material.[12] The studies are not intended primarily to unveil new historical facts, though they are presented in such detail that I hope readers will come away with new insights into the events they describe. I expect, however, that the primary benefit of the book will come from the generalizations made on the basis of the case studies–from the surplus value of the comparison of cases. It is to the requirements for this theoretical framework that I turn now (see table 1.1).

Requirements for a Theory of Sociotechnical Change

Elster (1983) has distinguished two approaches to the study of technical change. The first conceives of technical change as a rational, goal-directed activity. The second places emphasis on technical change as a process of trial and error, as a cumulative result of small and mostly random modifications. Two decades of studies in the sociology of scientific knowledge have stressed the contingent character of scientific development, and one of the basic assumptions of this book is that an analogous approach will be fruitful for studies of technical development.[13] This suggests that trial-and-error models, often cast in evolutionary terms, have specific advantages over models that stress the goal-oriented character of technological development. In the SCOT model that will be developed in this book I will also try to account for the contingent character of technical development, but will do so without employing a truly evolutionary framework.[14]

An emphasis on contingency seems to be the historian's delight as much as the sociologist's curse, offering no structuralist explanations for human action but free rein for individual actors. The other side of the coin, however, is that too much contingency would result in actors who have no meaningful history of their own: If there are no systematic, structural constraints, there are no limits to the spectrum of possibilities. There may be constraints, but they are contingent and unpredictable themselves. Therefore, evidently, one requirement for a theory of technical change is that it should be able to show how constancy and continuity exist in history, and under what conditions they exist. It should allow us to account not only for technical change but also for the stability of artifacts. If only rupture and revolution had a place in the analysis, while flow and evolution did not, the resulting framework would turn into (some) sociologist's delight and (some) historian's curse. Setting up such a truly dynamic conceptual framework is a notoriously difficult task. The typical way to tackle this problem is to give a static description and then add the time dimension—but to leave the concepts intrinsically static. Following this approach, one might try to explain the ability of a bicyclist to ride upright by drawing on a model of the bicycle as a pair of scales, with the bicyclist achieving balance by equating left- and right-hand forces.[15] The equilibrium of a rolling bicycle can, however, only be understood by using the intrinsically dynamic concept of "angular momentum." To meet the first requirement, our conceptual framework must have a similarly dynamic character.

Earlier in this chapter I discussed the idea of asymmetrical analysis—analysis in which the success and the failure of artifacts are explained in different terms. Using the "working" of an artifact as an *explanans* in the study of technology seems equivalent to using the "hidden hand of Nature" as an *explanans* in studies of science. That is to say, it was often assumed that scientific facts had to turn out the way they did because that is the way Nature dictated them to be. Recent science studies have called for an explanation of Nature as the result, instead of the cause, of scientific work. Similarly, for a theory of technology, "working" should be the *explanandum*, not the *explanans*. The "working" of a machine is not an intrinsic property of the artifact, explaining its success; rather, it should figure as a result of the machine's success. Thus, the success or failure of an artifact are to be explained symmetrically, by the same conceptual framework. An asymmetrical explanation might, for example, explain the commercial success of an artifact that we now consider to be working by referring to that "working," while the failure of that same artifact in another context might be explained by pointing at social factors. In a symmetrical explanation, "working" and "nonworking" will not figure as causes for a machine's success or failure. The claim is not that "working" is merely in the eye of the beholder, but that it is an achievement rather than a given. Understanding the construction of "working" and "nonworking" as nonintrinsic but contingent properties is the second requirement for the theory of technical change I shall try to develop.

The third requirement, pertaining to the actor/structure dimension, is closely related to the change/constancy requirement. The emphasis on the contingent character of technical change may seem to imply that anything is possible, that each configuration of artifacts and social groups can be built up or broken down at will. This, of course, cannot be: A theory of technology proposing such a view of our technological society clearly underestimates the solidity of society and the stability of technical artifacts. A theory of technical development should combine the contingency of technical development with the fact that it is structurally constrained; in other words, it must combine the strategies of actors with the structures by which they are bound.

The final basic assumption in my theoretical project is that modern society must be analyzed as a seamless web. The analyst should not assume a priori different scientific, technical, social, cultural, and economic factors. Rather, whatever creases we see are made by the actors and analysts themselves. Another way of expressing this idea is to recognize that a successful engineer is not purely a technical wizard, but an economic, political, and social one as well. A good technologist is typically a "heterogeneous engineer" (Law, 1987).

The metaphor of the seamless web has implications not only for empirical work but for our theoretical framework as well. I propose that we require our theoretical concepts to be as heterogeneous as the actors' activities. If we would do otherwise, the old a priori distinctions would return through the back door by the step of generalization, after having been kicked out through the front door by empirical research. The fourth requirement for a conceptual framework is thus not to compel ourselves to make any a priori choices as to the social or technical or scientific character of the specific patterns that we see by applying it.

To develop this empirically based theory of sociotechnical change, we need a descriptive model that will help us create a set of case studies that can be compared and combined in the process of developing generalizations. The descriptive model should allow analysts to get into the black boxes of the various cases, but also to get out again to compare case descriptions. Thus the model should strike a fine balance between describing the "nuts and bolts" and staying at a sufficient analytical distance to allow for cross-case comparisons. The development of such a descriptive model will be the main purpose of the second chapter, while in the third and fourth chapters the theoretical framework will be developed.

Book Design

The broad aim—to start from the more strictly disciplinary research questions of the history and sociology of technology, but to work toward an interdisciplinary result—implies certain constraints on form. This academic detour is built up from many smaller detours—sociological detours for historians and historical detours for sociologists. I can only hope that each detour will be attractive enough to be followed through by readers of whatever background, and that in the end the results will prove the detours worthwhile. These results should also yield the conclusion that what started out as a series of detours has turned out to be a new main route toward interdisciplinary STS studies.

At the outset, I have tried to use different styles of writing for the narratives and for the theoretical analyses and model building, highlighting the disciplinary character of both parts. Toward the end of the book the alternation between narrative and theoretical intermezzi fades away, and both blend into a single STS vocabulary. The narrative will present a thick description with many historical details. The cases per se merit such detailed descriptions—each offers a fascinating story of technical development and engineering life—but they also allow readers to check the interpretations I will propose during the theoretical detours. The theoretical framework, on the other hand, will be presented quite explicitly as a formal model. This is done for reasons of candor and clarity: It will allow more transparent discussions of the strengths and weaknesses of the theoretical framework. That theoretical framework does provide a coherent view of the joined development of society and technology; but it does not represent a closed world, outside of which no stories can be told. The aim, then, is not to make a model that pretends to explain all of technological development. The model will not be a set of narrowly defined concepts to be employed indiscriminately in empirical research. Rather, it will be a heuristic device, a set of sensitizing concepts that allow us to scope out relevant points, but one that will require adaptation and reformulation for use in new instances.

If this book has any usefulness as a teaching text, it will come from this combination of empirical case studies with theoretical modeling. I do not envisage the book as an instrument primarily meant to educate students about either the SCOT model or the details of the bicycle, Bakelite, or the fluorescent lamp. What I do hope is that the book will be useful in two other respects. First, it should make students think about the interplay of empirical research and theoretical modeling, about the relations between case studies and conceptual frameworks. Second, it should introduce students to recent constructivist perspectives in technology studies by putting one approach on the test bench. Finally, of course, I hope that the "detour-becomes-main-route" thesis will lead to high-spirited discussions about the relationships between society and technology, about the future of STS studies, and about one's societal roles and responsibilities as engineer, social scientist, or citizen.

In a way, this is an effort to write four books in one. The first three are the case studies, which focus on design, biography, and economics, respectively. The fourth is the combination of the first three, presenting a comparative analysis of changes in technology and society. All together I hope they will make the case for combining empirical case studies with theoretical analysis to strengthen the link between academic STS studies and politically relevant action.

2

King of the Road: The Social Construction of the Safety Bicycle

2.1 Introduction

Before the bicycle became "King of the Road,"[1] it was the "Prince of Parks." Aristocratic young men drove high-wheeled bicycles in Hyde Park to show off for their lady friends. The high-wheeled machine was not meant to provide ordinary road transportation, however, or to enable families to tour the countryside. These transportation and touring aims would be fulfilled by the safety bicycle—a low-wheeled vehicle with a diamond frame and a chain drive on the rear wheel—in the 1880s and 1890s. The process of emergence of this new bicycle will form the focus of this chapter.[2]

Why did the safety bicycle emerge only after the detour of the high-wheeled bicycle? A review of bicycle history shows an increase and subsequent decrease of front-wheel diameter, beginning and ending at about 22 inches, with a maximum of some 50 inches in between. The main difference between the first and last bicycles is the mechanical means of their propulsion: boots on the ground for the former versus a chain drive on the rear wheel for the latter. In retrospect, it seems that all the technical elements needed to modify the first bicycle (a "running machine") into the safety bicycle had been available since the time of Leonardo da Vinci. Why, then, did it take more than half a century for gears and a chain drive to appear on a working bicycle? What strange detour was this from the sure path of technical progress?

The high-wheeler has been described as a mechanical aberration, a freak. Its faults were its instability, the insane difficulty of getting on and off, and the fact that the large front wheel was driven and steered at the same time, which could be very tiring on the arms (Ritchie, 1975: 122). This will be the leading historical question of this chapter: How can we understand this detour as part of the construction of the safety bicycle?

The chapter has a theoretical as well as a historical goal. We will "extract" a descriptive model from the story of the bicycle, and then test this model by applying it to the other cases studies in the book.

I will start with an impressionistic sketch of early bicycle history, from the first machines up to the high-wheeled Ordinary bicycle. There follows a more detailed account of the specific groups involved in the transformation of the bicycle from high-wheeler to safety bicycle. I will then interrupt the flow to introduce the first element of the descriptive model: the idea of *relevant social groups*. A second methodological section will focus on problems involved in describing technical artifacts. The sixth section then shows how several solutions to the "problems" of the high-wheeler, especially the problem of safety, were designed in the form of alternative bicycles. This suggests the introduction of a crucial concept for our descriptive model: *interpretative flexibility*. The invention of the air tire—or rather its reinvention—is recounted in the next section. This proved to be a significant step in the formation of the safety bicycle and leads naturally to the introduction of the third and fourth elements of the descriptive model: *closure* and *stabilization*. The chapter closes by tracing in detail the stabilization process of the safety bicycle.

2.2 Prehistory of the Bicycle: From "Running Machine" to Ordinary

Leonardo da Vinci seems to have thought about the possibility of a humanly propelled vehicle that would be stable even though it had only two wheels (figure 2.1). The light-brown coloring of the drawing suggests that the machine was to be made of wood; it had wheels of equal size, a saddle supported by the rear axle, and a chain drive on the rear wheel. Da Vinci's role in this design has not been proved, although it is likely that the drawing was made in his atelier, thus suggesting his indirect involvement at least. The bicycle drawing in the *Codex Atlanticus* was found during a recent restoration. Data about da Vinci's pupil Salai, who is mentioned on the pages, suggest that the drawing was made around 1493 (Reti, 1974), at which time da Vinci was engaged in designing gears and chains, one of which looked much like that depicted in the sketch. There is no indication, however, that this vehicle was ever constructed.

The first vehicles with two wheels arranged in line were built at the end of the eighteenth century. Although there are some reports about machines of even earlier date (Minck, 1968; Daul, 1906), most accounts identify the Célerifère as the first such vehicle (figure 2.2). It had the

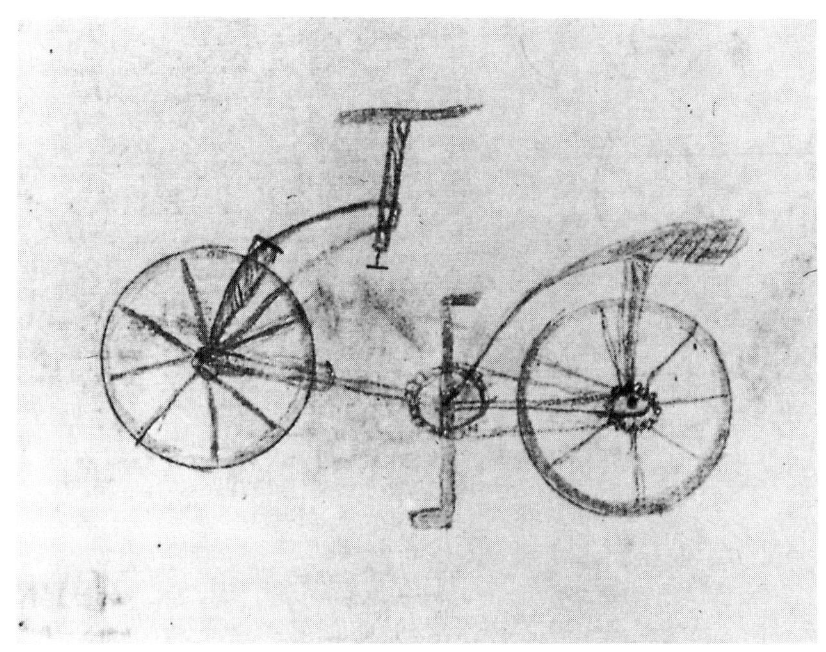

Figure 2.1
A bicycle-like machine, probably drawn by one of Leonardo da Vinci's pupils (*Codex Atlanticus*, page 133 verso; photograph courtesy the Biblioteca-Pinacoteca Ambrosiana in Milano).

form of a wooden horse with two wheels. Its maker is unknown. The Comte de Sivrac, a young man known for his eccentricities, is reported to have been seen riding it in 1791 in Parisian parks. While sitting on the "horse," he pushed the vehicle forward with his feet. Because there was no steering wheel, he had to go through a tedious procedure whenever he wanted to make a turn: stop, lift the machine, and then put it down facing the new direction. De Sivrac worked hard to earn the applause of the people walking in the Bois de Boulogne: "Il s'arrête de temps en temps, fort essouflé, fort fripé, mais toujours souriant."[3] Three years later the machine, renamed the *vélocifère,* had become a pastime for some of the more dashing young men of Paris, who showed their skills in the gardens of the Palais-Royale. Races were even held along the Champs Elysées. The initial enthusiasm faded quickly, however, after several riders strained themselves in lifting the heavy machines and others suffered rupture of the groin (Woodforde, 1970).

The turning problem was solved in 1817 by Karl Friedrich Christian Ludwig, Freiherr Drais von Sauerbronn in Mannheim. Karl von Drais,

Figure 2.2
The "Célerifère" of 1791, in 1793 renamed the "vélocifère."

as he is generally known, was employed by the Baden court as master forester and chamberlain. His true calling, however, was mechanical construction. He invented several machines, such as a meat chopper, a typewriter, and a periscope, which left no deep trace in history. In 1817, however, he constructed a *Laufmaschine*, a "running machine" that consisted of a wooden frame with two wooden wheels of equal size positioned in line; the front wheel was able to turn. Between the wheels, on the frame, a cushioned saddle was mounted (figure 2.3). In front of the saddle was a cushioned bar on which the underarms could be rested. In the first version, steering was done with this bar; later Drais provided a separate steering handle in front of the resting bar.[4] He moved his machine forward by pushing on the ground with his feet, which were suitably protected by iron toe caps worn on his shoes (Croon, 1939; McGonagle, 1968; Lessing, 1990).

On 12 January 1818 Drais acquired a Baden patent with a validity of ten years for his invention. He built and sold quite a number of "running machines." Unofficially his Draisienne, as he liked to call it after its demonstration in Paris, was recognized as a road vehicle: On Saxon road signs it was placed under the rubric of *Fuhrwerke* ("machine for moving"). Probably to demonstrate its military usefulness, Drais drove his Draisienne from Karlsruhe to the French border in the short time of four hours. In other races against the clock he showed that he could drive significantly faster than a stagecoach (Klinckowstroem, 1959; Croon, 1939).

Figure 2.3
The "running machine" or "Draisienne," constructed by Karl Drais von Sauerbronn in 1817. The photo shows a colored lithograph of the inventor on his machine, probably published in the *Weimarian Journal for Literature, Art and Fashion* in 1820, with the caption "Der Freiherr von Drais. Inventor of the fast-running machine. Known fast and sharp thinker." The technical details of the lithograph are correct in every aspect. The poplars in the background are reminiscent of those on the road to Schwetzingen, the hills belong to the Odenwalt mountains. (I am grateful to Prof. H. E. Lessing for offering me this picture as well as its interpretation. See Lessing (1990) for a richly illustrated history of von Drais's machine. Photograph courtesy of the Städt. Reiss-Museum, Mannheim.)

In the beginning, press comments were positive. The German post adopted a few machines for its postmen (Rauck et al., 1979). Drais tried to establish a manufacturing firm, but this venture did not take off. Then the auditor's office prevented the German post from buying more Draisiennes because of the wear on the postmen's shoes (Rauck et al., 1979). The Draisienne became an object of ridicule for caricaturists, pedestrians, and schoolboys. Drais himself, running into an English horseman who poked fun at the machine and its rider, started an argument that ended in a fight. By the end of the 1840s his situation was rapidly deteriorating, both socially and psychologically, probably because of inherited epilepsy (Lessing, 1990). It is reported that when he drove past the city hall in Karlsruhe, he was often invited by the sentry to drink a pint of beer; in return, he had to ride down the stairs in front of the hall on his Draisienne, which often resulted in a kind of "salto portale." Drais died, poor and disillusioned, in Karlsruhe on 10 December 1851 (Croon, 1939).

In other countries, notably England, the Draisienne had more success. Dineur in France, Johnson in England, and Clarkson in the United States had, in the name of Drais, taken out patents on the invention in 1818 and 1819. Denis Johnson in particular tried hard to stimulate the use of what he liked to call the "pedestrian curricle" in England. The machine became commonly known as the hobbyhorse or dandyhorse. He developed a version for women in 1819, and in 1820 he organized an experiment of employing hobbyhorses for postmen. In America as in England, several "riding schools" were established. Hundreds of hobbyhorses were produced and sold. But it appeared to be only a craze. The new sport seems to have been vaguely irritating to the general public, perhaps because the riders used the best footpaths, perhaps because they just looked silly. Going downhill was a thrill, but without brakes it was quite dangerous, and it was hard to be graceful when you had no place to rest your feet. A well-known joke was that users of the hobbyhorse could ride in their carriage and walk in the mud at the same time. Moreover, blacksmiths and veterinarians saw a direct economic threat in the vehicle. Blacksmiths are reported to have smashed hobbyhorses that passed through their villages. This horse, they pointed out, required no shoeing (Woodforde, 1970).

The lack of comfort posed another problem for many users. The wooden or iron-clad wheels, the rigid frame, and the potholed roads resulted in a rough ride. Moreover, the movement of the body, shifting over and bumping up and down in the saddle, caused strains and not a few hernias. Another problem was that steering—the key trick in the "craft" of riding a modern bicycle—could hardly be used to keep the rider's balance.[5] When one looks closely at the Draisienne, it becomes obvious that steering must have taken a lot of force: The friction of the crossbar sliding under the frame backbone was quite great because the turning point of the front fork was positioned relatively far forward. You had to use your feet to balance the vehicle while also using them to give the Draisienne its forward momentum.

The problems of the hobbyhorse were recognized by users at the time, but Drais had not wanted to revise his machine fundamentally once he had provided the extra steering handle. (Occasionally he did provide extras such as brakes and saddles whose height could be adjusted.) Others, however, did try to find solutions to the more fundamental problems and thus improve the hobbyhorse. Johnson, for example, constructed an iron version of the machine, and this enabled him to improve the bearing of the steering axis. With such a tube bearing, the axis of the steering front wheel could be positioned more precisely and the friction created by turning the wheel could be greatly reduced, so that the steering mechanism could be used to keep the vehicle upright. Indeed, this has remained the most effective way to do so ever since. In retrospect, we realize that this development raised in principle the possibility of getting one's feet up off the ground and keeping one's balance by steering. However, the problem of muddy feet stayed unresolved for some decades.

Several methods were tried to raise the feet off the ground. As early as 1839, Kirkpatrick MacMillan, blacksmith of Courthill, Dumfriesshire, Scotland, added cranks to the rear wheel of his hobbyhorse (see figure 2.4). These cranks were driven by a forward and backward motion of the feet on two long treadles. The machine seems to have functioned quite well, although MacMillan is said to have caused the first bicycle road accident in 1842 by knocking over a child in the crowd cheering his entry into Glasgow; he was arrested and fined five shillings. He had designed the treadles so that they could be adapted to the leg length of various riders. Nevertheless, there is no record of his selling this hobbyhorse (Robertson, 1974).

Another revision of the "running machine" took the form of cranks attached to the front wheel. These cranks were usually pushed by the feet, thus enabling the rider to sit in his carriage without walking in the mud. Several people made this addition, probably independently of one another: for example, Gottlieb Mylius in Themar (Sachsen-Meiningen,

Figure 2.4
Kirckpatrick Macmillan constructed cranks with treadles to drive his "hobbyhorse" (1839). The feet still made a walking movement. Photograph courtesy of the Trustees of the Science Museum, London.

Germany) in 1845, Philipp Moritz Fischer in Oberndorf (Germany) in 1853, Joseph Baader in Munich (Germany) in 1862. Lewis Gompertz in Britain constructed cranks for the front wheel that had to be moved by the rider's hands; the "feet in the mud" problem of course remained, but he may still have thought that his feet were needed for balancing (Croon, 1939; Feldhaus, 1914; Klinckowstroem, 1959; Rauck et al., 1979). In the late 1860s, as I will show, several inventors constructed rear-driven velocipedes as well—most of them probably not knowing about MacMillan's hobbyhorse. In summary, the 1860s seem to have been filled with numerous, and widely varying, designs of improved Draisiennes. Only one of them, built by Pierre Michaux, became a commercial success.

In 1861 Michaux, a coach builder in Paris, was asked to repair a Draisienne. One story is that his son Ernest, after testing it, complained about the great effort required to ride the machine and that, subsequently, he and his father designed the front-driven velocipede. The other story is that Pierre Lallement, employed in the Michaux workshop, first constructed such a front-driven Draisienne; he then went to America and left the honor to Michaux. In any case, Michaux continued to improve this velocipede and on 24 April 1868, a French patent was issued to him. The prototypes were made of wood, but by 1866 he had started

Figure 2.5
The vélocipède constructed by Pierre Michaux in about 1865. Photograph courtesy of the Trustees of the Science Museum, London.

to use iron. His machines were made with front wheels of various diameters (80, 90, and 100 cm) and a smaller rear wheel (see figure 2.5). The cranks had slotted ends so that their radius might be adapted to the length of the rider's legs. The frame was a solid wrought-iron bar with a fork for the rear wheel. A socket at its front end embraced the head of the driving wheel fork, to the top of which the steering handle was fitted. A brake block acting on the rear wheel could be applied by tightening a cord tied around the handlebar. He had also found a solution for the vibration problem: by making the rear wheel smaller, he obtained enough space to position the saddle on a spring brace. The saddle could be moved forward and backward along that spring to adjust to the rider's height. Leg rests were provided for coasting and a step for mounting (Caunter, 1958).

In the meantime, Pierre Lallement had received an American patent on his machine in 1866 and founded a business, but he could not cope with the rapidly increasing competition. The Hanlon brothers, a popular acrobat duo in New York City, were granted a patent on 7 July 1868, in which they suggested the use of rubber rings around the wheels to make them noiseless and to prevent slipping. The Hanlon brothers patented several other small improvements, most of which could be found on the Michaux velocipedes as well. Immediately after the Lallement patent, Americans did not pay much attention to the velocipede, but the Hanlon brothers' activities aroused much interest. December 1868 is identified as the moment at which there began a sudden wild enthusiasm for the Boneshaker, as the velocipede came to be known. Carriage makers commenced to produce the Boneshaker, which became very popular, especially among Harvard and Yale students. Riding schools with such names as "Amphicyclotheatrus" and "Gymnocyclidium" were established. Initially, the Boneshakers were priced at around $125, but soon models could be bought for around $75. The craze died as suddenly as it had started: in August 1869 the machines could be bought for some $12. There was one obvious problem related to the construction of the velocipede: the tendency to push one's body backward and away from the pedals when the going became heavy and more force was needed. The vibration problem also became serious, especially when cities began to pass ordinances against riding on the (smooth) pedestrian walks, thereby condemning the velocipede to the rough road—reminding its users of the origin of the name Boneshaker (Oliver and Berkebile 1974). Lallement returned to France.

In France, Michaux's business was prospering. Already in 1865 his workshop produced 400 velocipedes a year. During the 1867 World's Fair in Paris, he was so effective in promoting his machine that in the months afterward he could not respond in time to all orders he received. The firm decided to deliver velocipedes to the most prominent customers first. This in turn had quite a promotional effect; when the Imperial Prince Louis Napoleon and his friend the Duke of Alba were seen riding Michaux velocipedes, this provided one of the best and surely the cheapest promotion one could imagine. In 1869 the Michaux assembly moved to a new plant, where 500 workers were employed and about 200 velocipedes were produced each day. In England and Germany, the Michaux velocipede was not noticed until about 1867, when it was exhibited at the Paris World's Fair. In 1869 the first English and German designs were marketed. The Franco-German War of 1870–71 halted further development of the velocipede in France and Germany, and the lead was passed to the English (Rauck et al., 1979).

Side slipping, which was not so prominent with the hobbyhorse, was one of the major problems of the velocipede. It is difficult to imagine the skill involved in riding the velocipede: one had to continually adjust one's hold on the handlebar against the tendency of the front wheel to change direction with each thrust on the pedals (Minck, 1968; Woodforde, 1970). Those thrusts, in combination with the turning of the front wheel, made the velocipede frequently subject to side slipping because of its broad, flat, iron-shod wheels.

Before I move on to discuss further developments on the other side of the channel, it is worth noting that only by using commercial criteria can we attribute to Michaux the kind of prominence in the history of the bicycle that he has garnered. Application of either the "who was first" or the "who made the best" criterior would yield different answers. Other inventors were either earlier with their advances or closer to what would later become the bicycle design now considered the "best working machine." Mylius, Fischer, and Baader, who all constructed velocipedes with front-wheel drive, have been mentioned already. More interesting still, several other designs incorporated rear-wheel drive. Because these necessarily involved some mechanical means of transmitting the movement of the feet to the wheel, whether using gears, cranks, or treadles, most of them made it possible to incorporate some "amplification factor" in this movement. This applies to MacMillan's hobbyhorse and to Thomas McCall's similar machine; but also to the machine that was supposedly built in 1869 by André Guilmet and Meyer & Cie. Such an amplification factor would, if fully realized, have made the detour of the high-wheeler unnecessary.

Michaux had continued to modify his velocipede models. The last models, exhibited at the World's Fair, were distinctly lighter and had higher front wheels than the earliest models. The back wheel was kept relatively small. The handlebar was broader to help in controlling the side-to-side movement of the front wheel (Woodforde, 1970). The trend of enlarging the front wheel continued after the center of innovation had moved to England. This trend was further enhanced by the increasing focus on sports and racing as a context for riding the velocipede. One of the first velocipede races was held in May 1868 in St. Cloud, over 1,200 meters. In November 1869 an eighty-three-mile race from Paris to Rouen was held with two hundred participants, including five women. In England the sporting context was further emphasized, which had implications for the design of the bicycle. Because the pedals were fixed to the front wheel without any gearing system, the only way to realize a greater translational speed over the ground while maintaining the same angular velocity of your feet, rotating the wheel, was to increase the diameter of the front wheel. And this is exactly what happened.

At the end of the 1860s, the scene shifted to England completely. The hobbyhorse craze had not lasted, and people had almost forgotten about riding on two wheels when young Rowley Turner brought a Michaux velocipede back to Coventry after visiting the World's Fair in Paris in 1867. Turner, the Paris agent of the Coventry Sewing Machine Cy. Ltd., convinced his uncle Josiah Turner, manager of the company, to accept an order for manufacturing 400 velocipedes for export to France. However, the France-German War of 1870–71 made business with the continent difficult, and the order could not be filled. So, more was needed to get the bicycle industry going. Rowley Turner is reported to have escaped from the besieged city of Paris on his velocipede, after the last train had left (Williamson, 1966: 48). Safely back in England, he was quite energetic in promoting the velocipede, and the sewing machine company trimmed its sails to the new wind. Thus Coventry became one of the centers of the British cycle industry (Grew, 1921; Woodforde, 1970). In the next section I will follow more closely the shaping of this social group of manufacturers.

An important step toward increasing the wheel diameter was the application of wire spokes under tension instead of rigid spokes acting as struts. This enabled the manufacturers to keep the wheels relatively light while making them bigger. This improvement was patented in 1869 by W. F. Reynolds and J. A. Mays in their "Phantom" bicycle (Caunter, 1958). In the same year, the term "bicycle" was introduced in a British patent granted to J. I. Stassen, and thereafter it quickly replaced all other names (Palmer, 1958). In 1870 the bicycle "Ariel" was patented by James Starley and William Hillman (see also figure 2.6). The difference between this vehicle and the Michaux velocipede is striking: where the two wheels of the velocipede were indeed a little different in size, on the Ariel a man was "hurtling through space on one high wheel with another tiny wheel wobbling helplessly behind" (Thompson, 1941: 18). Generally speaking, this was the first lightweight all-metal bicycle, setting the stage for what would become known as the "high-wheeled Ordinary bicycle," or "Ordinary" for short.[6]

2.3 Social Groups and the Development of the Ordinary

The high-wheeled bicycle did not have one unambiguous meaning, but was evaluated in varied ways by different social groups. To describe

Figure 2.6
The "Ariel," patented in 1870 by J. Starley and W. Hillman, is generally considered to be the first high-wheeled "Ordinary bicycle." The lever to rotate the hub with respect to the rim and thereby increasing the tension of the spokes can be clearly seen. Photograph courtesy of the Trustees of the Science Museum, London.

its development, I will concentrate in this section on the various social groups involved—in its production and use, as well as in criticizing and fighting it. These groups will be described in some detail, and at the same time I will further trace the development of the high-wheeled bicycle. Let us first return to the story of Rowley Turner and the Coventry Sewing Machine Cy. Ltd., and examine the social group of producers.

The Bicycle Producers

The Coventry Sewing Machine Cy. Ltd. changed its name to Coventry Machinists Co., Ltd. in 1869 when it embarked, as Rowley Turner had suggested, on the manufacturing of velocipedes. Such a change of production was quite common in those days, in part because the Franco-German War had a destabilizing effect on British industry. As export opportunities grew scarce, several machine manufacturers started looking for other trades. Weapon makers, sewing machine manufacturers, and agricultural machine producers were only too happy to shift their production to bicycles (see figure 2.7). It is significant that at this stage of velocipede development the machine industry enters the story. Until the late 1860s, the basic skills needed to make a velocipede were those of the carriage builder: working with cast iron, making long bow springs for saddles, bending steel rims, and constructing wooden wheels—this was all well within his trade. But tubular backbones, wire spokes, more sophisticated bearings, special stampings and castings—that was quite another business (Grew, 1921: 27).

Figure 2.7
Not only sewing machine manufacturers but even weapons makers turned to cycle production (part of advertisement reprinted in Grew (1921))

To understand this development better, I will briefly go back in history to sketch the founding of the Coventry Machinists Co. and the role of James Starley, often called in England the "Father of the Cycle Industry." Starley ran away from his Sussex home because he hated farming and wanted to be a mechanical inventor. He was subsequently employed as gardener in a large household. During this period he made several successful contraptions—for example, for use in the garden. One of his more colorful inventions, a "self-rocking basinette," was dropped when the prototype made a young child violently ill by rocking a bit too effectively (Williamson, 1966: 27). Starley repaired watches and clocks in the evening and thus educated himself about the basics of fine machine construction. Then one day he was asked to repair the sewing machine of the lady of the house. At that time a sewing machine was an expensive novelty that not many could afford, and it represented the most complicated mechanism that Starley had ever handled. He took the risk of stripping down the entire machine. He spotted the trouble (a tiny screw had worked loose), reconstructed the machine, and made it run better than ever. This impressed starley's employer so that he persuaded his friend Josiah Turner, manager of the company that was the actual maker of this particular machine, to take on Starley as an employee of the London factory of Nelson, Wilson & Co. (Williamson, 1966: 33).

Turner quickly identified Starley as "a sort of mechanical genius." He helped him to take out a patent on a treadle arrangement that kept the sewing machine running while its operator's hands were free to guide the cloth. By this time Turner had such faith in Starley's technical capabilities that he proposed that they leave the London firm and start a new company together to exploit this invention. They did so, moved to Coventry, and founded the Coventry Sewing Machine Cy. in 1861 (Williamson, 1966: 36–37). Turner recruited other technicians from the London region as well: Thomas Bayliss, William Hillman, and George Singer, to name a few (Grew, 1921: 2; Williamson, 1966: 41). In Coventry they found a receptive atmosphere. To highlight the particular combination of unemployment in technically skilled and unskilled labor, I shall briefly review the economic circumstances of this county.

The Warwickshire city of Coventry was economically and socially in bad shape. The weaving industry had been weakened by a decade of social conflicts between workers and employers. The long conflict, instigated partly by the tariff policies of the national government and partly be class struggle, almost ruined the ribbon weaving industry. Of the original eighty weaving masters existing befare 1855, only twenty remained by 1865; there had been at least fifty bankruptcies. Unemployment was very high in Coventry. Poverty spread, and so many families were threatened by starvation that a national appeal was launched in the early 1860s (Williamson, 1966: 38–40). From the census reports of 1861 and 1871, a remarkable decrease of the population of Coventry can be traced, especially when these figures are seen in the perspective of the population growth in other towns in the Midlands.[7] The watchmaking industry, which had been expanding between 1830 and 1860, had declined as well, although for other reasons. Coventry watchmakers did not have factories, and the individual masters in their isolated workshops were not able to compete with the cheaper machine-produced products imported from America and Switzerland (Williamson, 1966: 40). Despite the displacement of people from Coventry, the new sewing machine company still found many skilled workers. For Coventry this meant the beginning of its development into an engineering city. The watch trade provided the nucleus of skilled labor, and the ribbon trade the pool of unskilled labor, with which Coventry would graduate from the sewing machine and the bicycle to the motor bicycle and the motorcar, climbing back toward prosperity as the nineteenth century drew to a close (Prest, 1960: x).

The Coventry Sewing Machine Cy. prospered to such an extent that larger premises were necessary after seven years. The company had continuously improved its sewing machines by adding innovations and turning out new models with names such as "The European," "Godiva," "Express," and "Swiftsure." When Rowley Turner convinced his uncle Josiah to start making velocipedes, the new product was approached in the same innovative spirit (Williamson, 1966: 41). Starley's immediate reaction when confronted with the new machine was to lift the velocipede and criticize it for being weighty and cumbersome (Williamson, 1966: 48). Starley learned to ride the machine, however, and he quickly thought of a series of small but important modifications. For example, he fitted a small step to the hub of the rear wheel to enable the rider to simply step on. The usual way of mounting a velocipede was to take a short run and leap into the saddle. Some of these modifications probably have been incorporated in the first velocipedes produced by the Coventry Machinists Co., but there are no records of these early products.

Starley and Hillman then concentrated on designing a new, light velocipede. As sewing machine constructors rather than carriage builders, they employed quite different techniques than had Michaux. For one thing, there were no wooden parts on their machine. They followed Reynolds and Mays by using wire spokes under tension to make the wheels without heavy struts (made of wood or, later, hollow steel tubes) loaded by pressure forces. But added to this was a mechanism to tighten these radially positioned spokes and thus stiffen the wheels, which in Reynolds's and Mays's case still lacked rigidity. This was done by fixing two levers to the middle of the hub; the levers were connected by wires to opposite positions on the rim. By tightening these wires, one could make the rim turn relative to the hub until the spokes had the required tension (Caunter, 1958: 6). Finally, they followed the trend of enlarging the front wheel. Thus Starley and Hillman patented the Ariel on 11 August 1870 (see figure 2.6). They had such a confidence in their new product that they left the Coventry Machinists Co. and started a new business (Williamson, 1966: 49). Almost at the same time, W. H. J. Grout took out a patent on his "Grout Tension Bicycle" (Grout, 1870). This patent added some further basic elements to the scheme of a high-wheeled bicycle, notably the hollow front fork that further reduced the frame's weight, massive rubber tires, and a new means of mounting the spokes. Grout's radial spokes were threaded into nipples loosely riveted into the rim, which could be used to adjust the tension of the spokes and thus to true the wheel by screwing them on and off the spokes. These two patents can be said to have laid the basic pattern of the high-wheeled bicycle in the early 1870s.

Of course Turner, Starley, and the other Coventry Machinists Co. men were not alone in identifying the velocipede as an attractive new line of manufacture. The city of Coventry soon saw a variety of former watchmakers, ships' engineers, cutlery shop workers, and gun makers starting small workshops in which to build velocipedes.[8] In other towns, such as Leicester and Liverpool, velocipede makers were commencing business as well. Coventry was not a manufacturing town, however, as was Birmingham; thus in search of suitable materials, the Coventry engineers had to turn elsewhere. For example, Sheffield provided bar steel for bearings and wire for spokes; Walsall supplied saddles; springs came from Redditch and Sheffield; Birmingham firms provided the drawn steel tubes crucial to making those light metal frames, and it supplied the steel balls for bearings (Grew, 1921: 27). The assistance of Birmingham was not without risk for Coventry. Although at first the firms in Birmingham produced only half-finished materials and velocipede parts, they started to look around for outlets for their production in slack times.

For example, Perry & Co., pen makers, and the Birmingham Small Arms Co. (B.S.A.) began to supply sets of fittings and parts for small workshops, which were in this way able to build velocipedes without requiring the more expensive tools and machinery (Grew, 1921: 29–30). However, making good parts is not the same as making good bicycles, and Coventry remained the center of the British cycle industry for a long time. In the 1870s and 1880s the industry spread all over the Midlands, Yorkshire, and part of London.

Starley and Hillman did not immediately market their Ariel, but first produced and sold velocipedes in which they incorporated many of Starley's improvements. Hillman suggested that the launching of the high-wheeled bicycle had to be marked by a spectacular promotional feat. They decided to set up a kind of unusual test: completing the ride from London to Coventry in one day. And they did, probably in 1871.[9] Both gentlemen took their bicycles to Euston Station on the train, spent the night in the station hotel and got up before daylight. They had a light breakfast and started out along the cobbled roads of London. Once outside the city the roads became better, and at about 8:30 A.M. they reached St. Albans, where they stopped to have an ample breakfast. The next stretch ran over the Chiltern Hills. On some of the steeper hills they had to walk, but compensation came on the long downhill portions where speeds of some twelve miles an hour were attained. "Disaster might have overtaken the gentlemen who wished to take full advantage of the hills, had it not been for Mr. Starley's ingenious brake." By one o'clock the riders had covered about half the distance, and they enjoyed dinner and an hour of rest near Bletchley. Mounted again, they were cheered or by the inhabitants of towns and villages, few of whom had seen a bicycle before. Only one mishap befell them. "Mr. Hillman was thrown from his machine when the rubber tyre of his front wheel came off but escaped with nothing worse than a grazed hand. He was able to bind the tyre on again and proceed without further trouble." The last miles from Daventry to Coventry were hard. The men were tired and the darkness made it difficult to avoid stones and holes in the road. But just when the clock of St. Michael's struck midnight, it is said, they reached Starley's residence in Coventry. The ninety-six miles had been completed within one day and with the bicycles still in almost perfect condition. The contemporary account finishes by stating that "the bicycle that has been developed by Messrs. Starley and Hillman from the velocipede is a most efficient form of human transport. It may be recorded that the two intrepid gentlemen, though tired, and stiff after their long ride, were no worse for their adventure." However, an intimate footnote was added in the margin of this account that for both riders the experience was painful enough to oblige them to remain in their beds "for two or three days." When, subsequently, the Ariel was marketed in September 1871, it was priced at £8 (Williamson, 1966: 54).

So they were quite active, these bicycle producers. But for whom were they producing? For whom was the Ariel's spectacular promotion intended? The demand increased. By the end of the 1870s clubs and associations for cyclists had been established in most countries. To continue the story of the high-wheeled Ordinary bicycle, we will now shift our focus to its users.

The Ordinary Users

The memorable ride of Starley and Hillman enhanced the image of the new high-wheeled bicycle as a sport machine, and records were set and contested on all classic roads of England. For example, the Brighton Road is associated with the earliest bicycle performances, as is Watling Street, on which Starley and Hillman crossed the Chiltern Hills. Especially on the Brighton Road, relay rides were often held against the four-horse coach (Grew, 1921: 78). Track racing started soon as well. Probably the first was in 1869 in Crystal Palace, London (Woodforde, 1970: 161), but other tracks sprang up in Birmingham, Wolverhampton, and Leicester (Grew, 1921: 67–68). After the Franco-German War, racing on Ordinaries began on the continent as well.[10] Because the German local ordinances were rather limiting—for example, restricting racing on public roads to the early morning and late evening hours—in Germany the bicycle clubs started to build separate racing courses.[11] We will come back to bicycle racing below, for often there was more at stake than just a medal and a small cash prize.

Whereas skiing began as a way of getting about and evolved into a sport, bicycling began as a sport activity and evolved into a means of transport. Even when the rider of a high-wheeled bicycle was not actually racing, he viewed his activity primarily as an athletic pastime. It was not easy to mount the high-wheeled bicycle, even with the provision of a step directly above the trail wheel. Uwe Timm (1984: 17–22) gives a convincing and colorful description of his uncle Franz Schröder's efforts to learn how to ride high-wheeled bicycle: "Schröder experienced this afternoon the large and fundamental difference between theory and practice. He mounted and fell down. The crowd of spectators was standing there and kept silent. He stood up again and fell off again."[12] He repeated this motion several times, to increasingly enthusiastic clapping and cheering: "Hopf, hopf, hopf, immer aufem Kopf!"[13] By the end of the afternoon he had learned how to mount and ride in a straight line; making a curve and dismounting were not yet in his repertoire, so each little ride ended in a fall. However, after another week of trying (in which he lost two finger tips between the spokes of the front wheel), he had mastered the art of riding a high-wheeled bicycle. No wonder bicyclists wore an anxious air. "Bicyclist's face," this expression was called, and newspapers predicted a generation with hunchbacks and tortured faces as a result of the bicycle craze (Thompson, 1941: 18). Going head-over-heels was quite common, as we will see shortly. Partly for that reason, and because there was no freewheel mechanism—which implied that the cranks were permanently turning around when riding—a special mode of riding was practiced when moving downhill. This "coasting" again required some athletic ability: when the bicycle was moving fast, the legs were thrown over the handlebar to the front (see figure 2.8).

Learning to ride a bicycle became a serious business in the 1870s. In some European cities, bicyclists had to pass an examination to prove their proficiency (Woodforde, 1970: 120). Bicycle schools existed in most towns of some importance. Partly this was made necessary by the maturity of the riders, none of whom had learned cycling as a child, which is now the usual way, at least in bicycling countries.[14] On the other hand, the bicycles of those days were definitely more difficult to ride than are modern versions. Even walking a bicycle could result in a bruised leg when the novice had not yet learned how to keep free of the revolving pedals.

Charles Spencer, owner of the London gymnasium bicycling school, described in his instruction book how to mount a high-wheeled Ordinary:

> Hold the handle with the left hand and place the other on the seat. Now take a few running steps, and when the right foot is on the ground give a hop with that foot, and at the same time place the left foot on the step, throwing your right leg over on to the seat. Nothing but a good running hop will give you time to adjust your toe on the step as it is moving. It requires, I need not say, a certain amount of strength and agility.[15]

The cycling schools and instruction books tried to make the art of bicycling as explicit as possible. For example, what

> each learner must remember is simply to turn the handles in the direction in which he is falling. Having drummed this into his head, the rest is easy. He will

Figure 2.8
The first rider has thrown his legs over the handlebar when coasting downhill.

soon discover that there is a happy medium and that the bars require only to be turned slightly, and instantly brought back to the straight as soon as the machine has resumed the perpendicular.[16]

It is unlikely that a modern bicyclist would be able to describe so adequately what exactly she is doing when keeping her balance. Her craft of riding a bicycle is almost completely "tacit knowledge." However, riding the high-wheeler could be just as pleasant and comfortable as it was dangerous. Having mounted an ordinary bicycle—by this time implicitly meaning a high-wheeler—one would immediately feel its easy-rolling, billowy motion as very different from the bone-shaking effect of the velocipede.[17] Moreover, the pedals almost directly beneath the saddle enabled one to sit comfortably upright, with the bar in one's lap; on the velocipede, there had always been the pushing forward of the legs and the pulling on the handlebar to compensate for that pushing. There was a direct advantage of being so high above the ground: the roads had worsened since the railways eclipsed the horse coach, and the large wheel could keep its rider well above the water-filled holes and mud, while dealing effectively with the bumps.

Few men over middle age, and even fewer women, attempted to ride the high-wheeled bicycle. The typical bicyclist—by this time meaning an Ordinary rider—had to be young, athletic, and well-to-do. Accordingly, bicycling still had, as in the early days of the hobbyhorse, an element of showing off:

> Bicycle riding, like skating, combines the pleasure of personal display with the luxury of swift motion through the air. The pursuit admits, too, of ostentation, as the machine can be adorned with almost any degree of visible luxury; and differences of price, and, so to speak, of caste in the vehicle, can be made as apparent as in a carriage. It is not wonderful, therefore, that idle men sprang to the idea.[18]

Generally, bicycling was associated with progress and modern times. This was sometimes voiced in grandiose terms:

> The bicycle: the awakening of a new era. The town comes into the village, the village comes into the town, the separation comes to an end, town and village merge more and more. Cyclisation: the era of the bicycle, that is the new time with richer, broader and more mobile civilisation, a back to nature which however keeps all advantages of culture.[19]

But cycling was also linked with new social movements in more concrete ways. The first meeting of the bicycle society of the town of Coburg was observed by a local police officer, who had to ensure that this society was not an undercover meeting of the forbidden social democratic party. Schröder's wife Anna was pointed out for committing subversive actions that were intuitively understood as revolutionary and the first exemplification of the women's movement in Coburg. "*Petroleuse* on a highwheeler" read the headline in the local newspaper, thus associating female bicyclists with *petroleuses* of the 1871 *Commune*.[20] And especially in the days of the low-wheeler, after the high-wheeled bicycle had become obsolete, cycling was explicitly linked to feminism. I shall return to this point. For an instrument of the liberation of the proletariat, the bicycle was too expensive. The laborer who would have liked to use the machine for his transportation to work could not afford one, until a second-hand market had developed.[21] Indeed, many workers were still riding their high-wheeler after 1900; by that time it had been nicknamed "Penny-farthing" because it was not "ordinary" any more. In Ashford, Kent, a gas lamp lighter still used it in 1914, finding it useful in his work (Woodforde, 1970: 49).

The Nonusers of the Ordinary

With only the group of "young men of means and nerve" riding the Ordinary, there were many more people not using it. Some of them wanted to ride a bicycle but could not afford one, or were not physically able to mount the high-wheeler, while others actively opposed the machine.

There were several reasons for the antagonism against bicyclists. One was irritation caused by the evident satisfaction with which the riders of the high-wheeler elevated themselves above their fellow citizens. This irritation gave rise to derisive cheers such as "Monkey on a gridiron!" (Wells, 1896: 24) or the loudly hailed pronouncement that "your wheel is going round!" (Woodforde, 1970: 50). Jokes like this inflicted no injury, "but when to words are added deeds, and stones are thrown, sticks thrust into the wheels, or caps hurled into the machinery, the picture has a different aspect."[22] The touring clergyman who made this observation added, "All the above in certain districts are of common occurrence, and have all happened to me, especially when passing through a village just after school is closed. The playful children just let loose from school are generally at this time in an excitable state of mind."[23]

Another reason for the antagonism was the threat posed by the bicyclists to those who were walking.

> Pedestrians backed almost into the hedges when they met one of them, for was there not almost every week in the Sunday newspaper the story of some one being knocked down and killed by a bicycle, and letters from readers saying cyclists ought not to be allowed to use the roads, which, as everybody knew, were provided for people to walk on or to drive on behind horses. "Bicyclists ought to have roads to themselves, like railway trains" was the general opinion. (Thompson, 1941: 18)

Police and magistrates supported this view. Local ordinances posed various restrictions on bicycling, often widely different in different towns. A German cantonal judge observed that these local ordinances stipulated many obligations for the cyclists, but hardly any rights.[24] Elaborating on these rights, he remarked that the offense bicyclists suffered from most frequently was defamation. Carriage drivers being overtaken by a bicycle, pedestrians having to wait a few seconds before crossing a street—

they all would shout insults at the cyclist. The judge described the various forms of defamation recognized in German law and added that the so-called *einfache Beleidigung* (simple slander), which could be exerted by words, gestures, or pawing, was most common. An enthusiastic bicyclist himself, he used to write down all insulting words shouted at him; he was amazed by the public's creativity. Newspaper reports about fights between bicyclists and pedestrians or coach drivers were quite common. A particularly flagrant attack, Woodforde reports, happened on 26 August 1876, when a coach driver lashed an overtaking bicyclist with his whip and the coach guard actually threw an iron ball, which he had secured to a rope, between the spokes of the wheel (Woodforde, 1970: 52). An offense with which bicyclists were frequently charged was "riding furiously," especially on roads with excellent wood paving such as the high road between Kensington and Hammersmith in London. The antagonism of the general public can be sensed through the following excerpt from a court hearing transcript, concerning four men charged with furious riding: "Police constable ZYX 4002 deposed that he was on duty the previous evening, and saw the defendants riding at a rate of forty miles an hour; he walked after them and overtook them . . . taking them to the station handcuffed."[25] If we can assume that this speed of forty miles an hour was a gross overstatement, the acceptance of such a statement suggests a generally negative opinion about bicycling in those days.

There were also people who wanted to ride a bicycle but could not do so. One reason has been mentioned already: the price of the Ordinaries prevented middle-class and working-class people from buying a new machine. The other main reason was the problem of safety. This problem made older men and women reluctant to mount the high-wheeler. For women there was an additional problem, and I will turn to that first.

In 1900 it was still possible to find newspaper articles such as the following, reporting on the observation of a two-seater bicycle with a man and woman on it:

> The numerous public that was walking in the Maximilian-strasse, yesterday at noon, witnessed an irritating spectacle that gave rise to much indignation. . . . Unashamed, proud like an Amazon, the graceful lady displayed herself to men's eyes. We ask: Is this the newest form of bicycle sport? Is it possible that in this manner common decency is being hit in the face without punishment? Finally: is this the newest form of advertising for certain female persons? Where is the police?[26]

This must have been a rather exceptional outcry in 1900, especially as it concerned merely a low-wheeled two-seater. But it makes one think about the sentiments expressed two decades earlier against women wanting or actually trying to ride on high-wheeled bicycles. The whole weight of Victorian prudery set itself against women taking such a masculine and, on the high-wheeler, revealing posture.[27] Some bicycle producers tried to find a solution for what was euphemistically called "the dress problem." In 1874 Starley and Hillman pursued the idea of S. W. Thomas, patented in 1870, of having two pedals on one side of the velocipede, thus enabling it to be "side-ridden."[28] This is what Starley and Hillman did with their successful Ariel high-wheeler. The rider sat in a sidesaddle position, the handlebars being shortened on one side and lengthened on the other. The rear wheel was mounted on an overhung axle, and the front wheel was offset from the track of the rear wheel to counteract the bias of the sidesaddle posture (see figure 2.9). It all seems rather complicated, and the machine must have been quite difficult to master. This technical solution to the dress problem did not become a success, and few sidesaddle bicycles were sold.

However, solutions other than purely technical innovations were tried—and indeed, were more successful. First, Victorian morals could occasionally be a little more flexible than one might assume. For example, a young lady who wrote to a magazine in 1885 about having used a bicycle (which at that date must have been a high-wheeled Ordinary) was reassured in the reply: "The mere act of riding a bicycle is not in itself sinful, and if it is the only means of reaching the church on a Sunday, it may be excusable."[29] Another solution to the dress problem posed by the Ordinary was to modify the designs of women clothing and, accordingly, to set new standards of fashion. A third way for women to ride cycles while avoiding the Ordinary was to use tricycles.

The "safety problem" was pressing for many nonusers of the Ordinary. As mentioned, the Ordinary rider was liable to go head over heels when encountering a small obstacle like a stone, a hole in the road, or an animal wandering about. The trend of enlarging the front wheel of the velocipede had continued once speed had become so important, and this made it necessary to move the saddle forward in order to keep pedals within reach of the feet. This implied a reduction of the rear wheel's diameter—partly because otherwise the machine could not be mounted at all, partly to reduce the bicycle is weight, and partly for aesthetic reasons (it set off the grandeur of the high wheel). But these two developments moved the center of gravity of the bicycle and rider far forward, to

Figure 2.9
The ladies' model "Ariel" (to the right), designed in 1874 by J. Starley and W. Hillman. The pedals do not drive the cranks directly, but are placed at the ends of levers, pivoted some distance in front of and slightly above the front wheel axle on the left side of the bicycle. About halfway along these levers, short connecting-rods communicate the motion of the pedals to the overhung crankshaft. The axle forming the pivot for the pedal levers is supported on the inside by an arm attached to the front fork and on the outside by a stay that joins the lower crosspiece of the steering head.

The lever to rotate the hub with respect to the rim and thereby increasing the tension of the spokes can also be seen in both bicycles. Photograph courtesy of the Trustees of the Science Museum, London.

a position almost directly above the turning point of the system. Thus only a very small counter force—for example, from the bumpiness of the road, but also from the sudden application of the brake—would topple the whole thing. Another serious and frequent cause of falls was getting a foot caught between the spokes, for example when feeling for the step before dismounting. Different ways of falling forward even got their own labels (as in present-day wind surfing), so that an experienced Ordinary rider remarked, "The manoeuvre is so common, that the peculiar form of tumble that ensues is known by the distinctive name of 'the cropper' or 'Imperial crowner.'"[30] Falls were such an accepted part of bicycling that producers advertised their bicycles' ability to withstand falls, rather than claiming that they did not fall at all. In the *Humber Bicycle Catalogue* of 1873, a letter from a customer is reproduced, saying that although his Humber bicycle "on several occasions [had] been engaged in universal spills and collisions, it is now almost as sound as when first despatched from your works."[31] This, however, was to change within a few years, when manufacturers began to regard women and older men as potential bicycle buyers.

2.4 Relevant Social Groups

In this section, the flow of the historical case study is halted for the first methodological intermezzo. The concept "relevant social group" will be introduced.

I have described the development of the Ordinary bicycle by tracing what various groups thought of it. I used these perspectives to avoid the pitfall of retrospective distortion. If we are to find out how the so-called detour of the high-wheeled bicycle came about, it seems wise to stick as closely as possible to the relevant actors, rather than bringing our own evaluations to bear on the story. Thus we may be able to show that what in a Whiggish account of bicycle history seemed a strange and ineffective detour was indeed quite straightforward when viewed from the actors' perspective. ("Whiggish" is an account that presents history as uninterrupted progress, implying that the present state of affairs follows necessarily from the previous.)

But there is another reason to focus on social groups than merely the desire to avoid retrospective distortion. One of the central claims in this book will be that such social groups are relevant for understanding the development of technology. I will first show how empirical research can identify the social groups that are *relevant for the actors*. Then I will argue that these social groups are also theoretically *relevant for the analyst* when he or she sets out to explain the development of technical change.

Empirical Research to Identify Relevant Social Groups

Relevant social groups may be identified and described by following two rules: "roll a snowball" and "follow the actors." The snowball method is used in contemporaneous sociological research, and I will use a study of a scientific controversy to illustrate this method.[32] Typically one starts by interviewing a limited number of actors (identified by reading the relevant literature) and asks them, at the end of each interview, who else should be interviewed to get a complete picture. In doing this with each interviewee, the number of new actors at first increases rapidly like a snowball, but after some time no new names will be mentioned—you have the complete set of actors involved in the controversy.[33] This is a neat methodological solution to the problem of how to delineate the group involved in a scientific controversy, at least when interviewing is a possible technique.

The same method is applicable in historical research. Just as we can find relevant actor by noting who is mentioned by other actors, we can identify what social groups are relevant with respect to a specific artifact by noting all social groups mentioned in relation to that artifact in historical documents (see figure 2.10). When after some time the researcher does not find reference to new groups, it is clear that all relevant social groups have been identified.

By using the snowball technique, a first list of relevant social groups can be made. Using this as a starting point, the researcher can then "follow the actors" to learn about the relevant social groups in more detail.[34] This can be quite a straightforward process: because these social groups are relevant for the actors themselves, they typically have described and delineated the groups adequately. Thus marketing people will identify user groups and describe them as far as is relevant; producers thus had identified rich, young, athletic men as bicyclists; and anticyclists had identified tricyclists and bicyclists. Thus after the first step of identifying the relevant social group, two subsequent steps were taken: a second to describe the relevant social groups in more detail, and a third to further identify the relevant social group by delineating it from other relevant social groups. In practice, these description steps are of course interdependent, and it is not practical to carry them out completely separately.

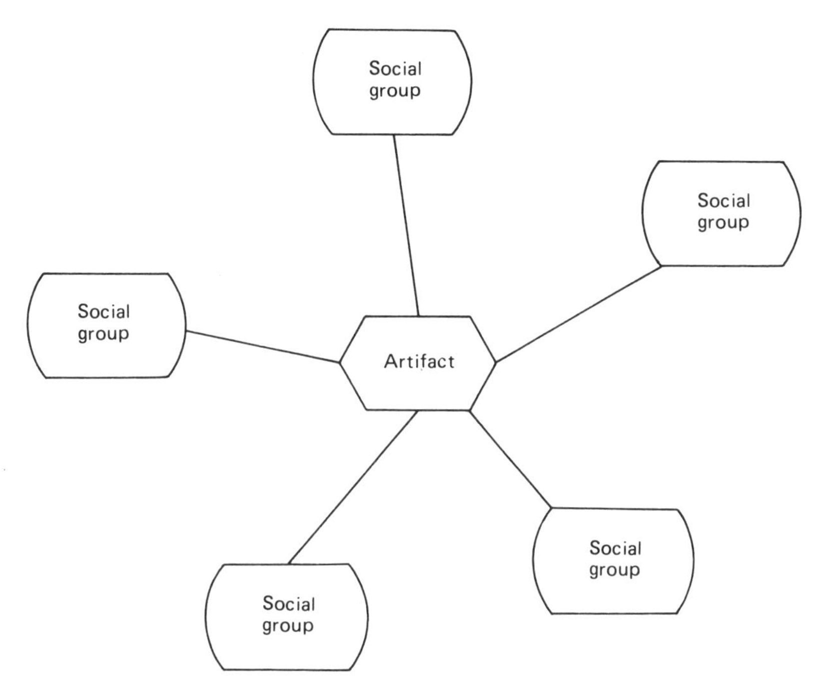

Figure 2.10
Related to an artifact, the relevant social groups are identified.

For example, the relevant social group of Ordinary users was characterized, in the first descriptive step, as being constituted of people who saw the Ordinary as a sporting machine that was rather hazardous to ride. In the second step this relevant social group was further described as consisting of young athletic men, distinctly upper and upper-middle class. A brief reference to road maintenance in the new railway era hinted at the wider socioeconomic context. The description of relevant social groups is as important as the detailed description of artifacts in standard technical histories. So I will, when turning to a discussion of tricycles, devote substantial space to women, postmen, and queens as well as to differential gears, big wheels, and brakes.

Then, for the third step, the relevant social groups' boundaries, intuitively assumed at the outset, are traced more precisely. Again, the actors can be followed. In the turmoil of technical development actors, to make sense of their world, will identify new relevant social groups or forget about others. Thus the boundaries of social groups, although once clear-cut, may become fuzzy; new groups may split off and old groups may merge into new ones. Actors thus "simplify" and reorder their world by forgetting about obsolete distinctions or by drawing new boundaries.[35] As I will show, at some point bicycle producers concluded, for example, that within the relevant social group of nonusers, women should be separated out as an important relevant social group. Similarly, the relevant social group of Ordinary users did not remain unchanged. At first it coincided completely with the group of cyclists. With the coming of the low-wheeled bicycle, some parts of the relevant social group of nonusers became users of the safety bicycle, and the relevant social group of Ordinary users changed accordingly. Its boundaries changed—some categories of cyclists switched from the high-wheeler to the safety. But its key characteristics changed as well: in the beginning its members could be labeled "young men of means and nerve," and Franz Schröder, a typical Ordinary rider, successively passed through the stages of being associated with social democracy and other "revolutionary" movements to being simply the laughingstock of town.

Relevant Social Groups: Also Relevant for the Analyst

The concept of "relevant social group" is an actor's category. Although actors generally do not use these words, they actively employ the concept to order their world. A crucial claim in the development of a social constructivist model of technology is, however, that "relevant social group" is also an important analyst's category. It will help us to describe the development of technical artifacts in terms that meet the requirements set out in the first chapter.

Technological development should be viewed as a social process, not an autonomous occurrence. In other words, relevant social groups will be the carriers of that process. Hence the world as it exists for these relevant social groups is a good place for the analyst to begin his or her research. Thus the analyst would be content to use "cyclists" as a relevant social group, but introduce separate "bicyclists" and "tricyclists" only when the actors themselves do so. The basic rationale for this strategy is that only when a social group is explicitly on the map somewhere does it make sense for the analyst to take it into account.

There seems to be one obvious problem with this argument, which has two important aspects, the political and the epistemological. The political aspect arises out of recognition that powerless social groups—those that do not have the ability to speak up and let themselves be found by the analyst—will thus be missing in the account. The epistemological aspect of the problem concerns the suggested identity between actors' and analysts' categories. The first formulation of the problem is relevant for the practical and political relevance of technology studies, to which I will return in the last chapter. The second formulation addresses a classic debate in the philosophy of the social sciences. This problem, however, does not need to exist, in either of the two formulations.

The problem of the "missing groups" does not exist if the conceptual framework I am developing is taken in the right spirit—as a collection of sensitizing concepts that aims to provide the researcher with a set of heuristics with which to study technological development. Another slightly rhetorical way of making the same point is to emphasize that the goal is to develop a framework for *scientific research*, not a computer program for an expert system to carry out social studies of technology. Let me give one example, comparing my treatment of the bicycle and the fluorescent lamp case studies. In the bicycle case I followed the lead of the actors and included the relevant social group of women in my description. In the fluorescent lamp case I did not, as will become clear in the fourth chapter. An expert system would have done so, however, because a General Electric manager once mentioned "the housewife" as a relevant social group. Another human researcher might have decided differently, and have included the relevant social group of housewives in her description—there is no way of deciding "mechanically" which of the two accounts would be best. This is where the approach outlined in this book draws most heavily on methods of interpretative research.

Similarly, no simple identity between actors' and researchers' categories is advocated. I am proposing the combined method of "snowballing" and "following the actors" as heuristics—a negative heuristic to avoid a facile projection of the analyst's own categories, which might lead to retrospective distortion and Whiggish accounts; and a positive heuristic to help identify relevant social group that do not figure in the standard histories of the specific technology. In the next chapter some concepts will be introduced that are exclusively analysts' categories.

What I have been arguing here about the identification, delineation, and description of relevant social groups also applies to the characterization of artifacts. If we want to understand the development of technology as a social process, it is crucial to take the artifacts as they are viewed by the relevant social groups. If we do otherwise, the technology again takes on an autonomous life of its own. Thus in this descriptive model the meanings attributed to the artifact by the different relevant social groups constitute the artifact. I described, for example, the artifact Ordinary bicycle "through the eyes" of members of the relevant social groups of women, older men, and Ordinary users. The definition of the Ordinary as a hazardous bicycle (for the relevant social groups of women and elderly men) was supplemented by listing specific ways of using the artifact, such as track and road racing, touring, and showing off in parks (for the Ordinary users). The risky aspects of riding the Ordinary were explicated by describing in some detail the techniques involved in mounting the machine and in coasting downhill. Also the pleasure and comfort of riding the Ordinary were described and contrasted with the bone-shaking experience of riding bicycles with smaller wheels.

As an aid in describing the meanings attributed by the relevant social groups to an artifact, I will now focus on the problems and solutions as seen by these relevant social groups.

2.5 Focus on Problems and Solutions

When my daughters want to find out about a ball, they do not sit down and stare at it. They pick it up, throw it against the wall, kick it, or play catch. When a physicist wants to study an atom, she excites it and studies the emission spectrum when electrons fall back into their lower energy states. When you want to study a social system—for example, the relationship of a married couple—not much would be learned by looking at it in steady state. Rather, it would help if you could induce a change—for example, by sending in a newborn child. Then insight might be gained about the hidden properties and processes that keep the social system together, or not.[36] This principle of focusing on disturbances when studying a system can be usefully employed when describing the meanings attributed by relevant social groups to an artifact.

Therefore in describing the artifacts I have tried to avoid the uninformative states of equilibrium and stability. Instead the focus was on the problems as seen by the various relevant social groups (see figure 2.11). Linked to each perceived problem is a smaller or larger set of possible solutions (see figure 2.12).

What kind of model is emerging? First, focusing on the different relevant social groups seems to be an effective way of guarding against the kind of implicit assumptions of linearity that I have criticized in the first chapter. From a traditional, quasi-linear view, the bicycle's history was depicted as a simple genealogy extending from Boneshaker to velocipede to high-wheeled Ordinary to Lawson's Bicyclette, the last labeled "the

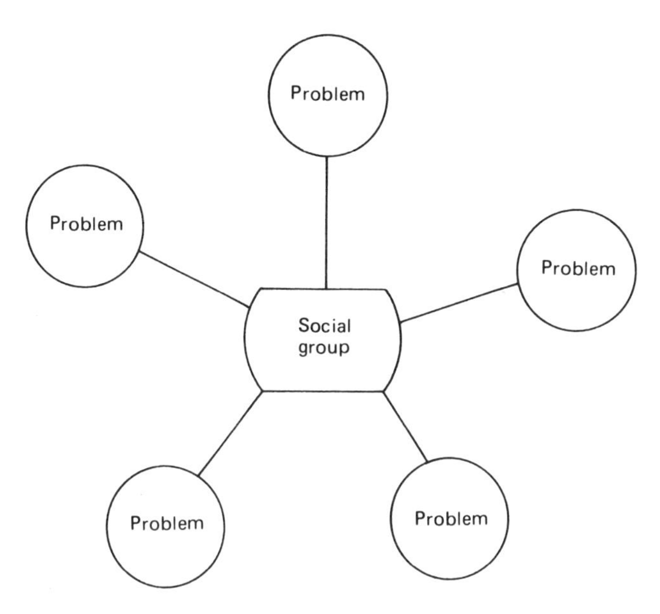

Figure 2.11
Artifacts are described by focusing on the problems perceived by the relevant social groups.

first modern bicycle" (see figure 1.3). All other bicycles were pushed to the margins of history because they are, retrospectively, seen to have failed. If, on the contrary, the various alternatives to the Ordinary are initially[37] put on an equal footing and considered as variants from which the next stable artifact had to be selected (as in figure 1.4), this view could be helpful in avoiding an implicit assumption of linearity.

Second, parts of the descriptive model can effectively be cast in evolutionary terms. A variety of problems are seen by the relevant social groups; some of these problems are selected for further attention; a variety of solutions are then generated; some of these solutions are selected and yield new artifacts. Such an evolutionary representation would thus not exclusively deal with artifacts, but would consist of three layers: variation and selection of (1) problems, (2) solutions, and (3) the resulting artifacts. Thus the results of variation and selection on the level of problems is fed into a further evolutionary process of variation and selection of solutions, which subsequently generate the artifacts (see figure 2.13).

One could try to summarize the narrative of the case study in one enormous drawing, compiling artifacts, relevant social groups, problems,

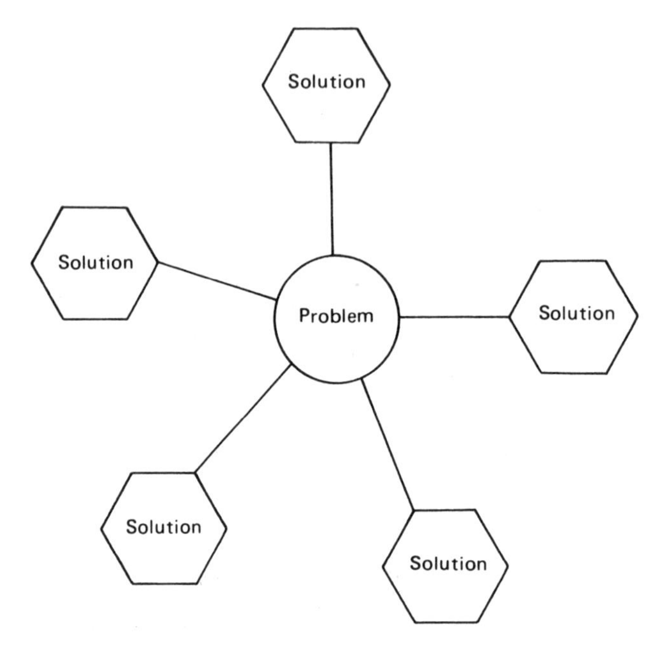

Figure 2.12
Finally, the solutions are described that are seen as available to each of the perceived problems.

solutions, and the subsequently modified artifacts. There are, however, two related problems lurking behind such an evolutionary representation. The first is practical and becomes obvious should the reader accept the challenge and try to make such a drawing of the case study presented in this chapter—it simply cannot be done because of the immense complexity. The other problem is that, if such a multilayered representation of problems/solutions/artifacts is not completely adequate, one almost inevitably ends up with the assumption that an artifact is a constant, fixed entity—to be generated in the variation process and then ushered through the selection processes.[38] In the remainder of this chapter we will find that this is not the case. Rather, an artifact has a fluid and ever-changing character. Each problem and each solution, as soon as they are perceived by a relevant social group, changes the artifact's meaning, whether the solution is implemented or not.

In the next section the history of the bicycle is followed further, with a focus on the solutions that various relevant social groups saw to the safety problem of the Ordinary.

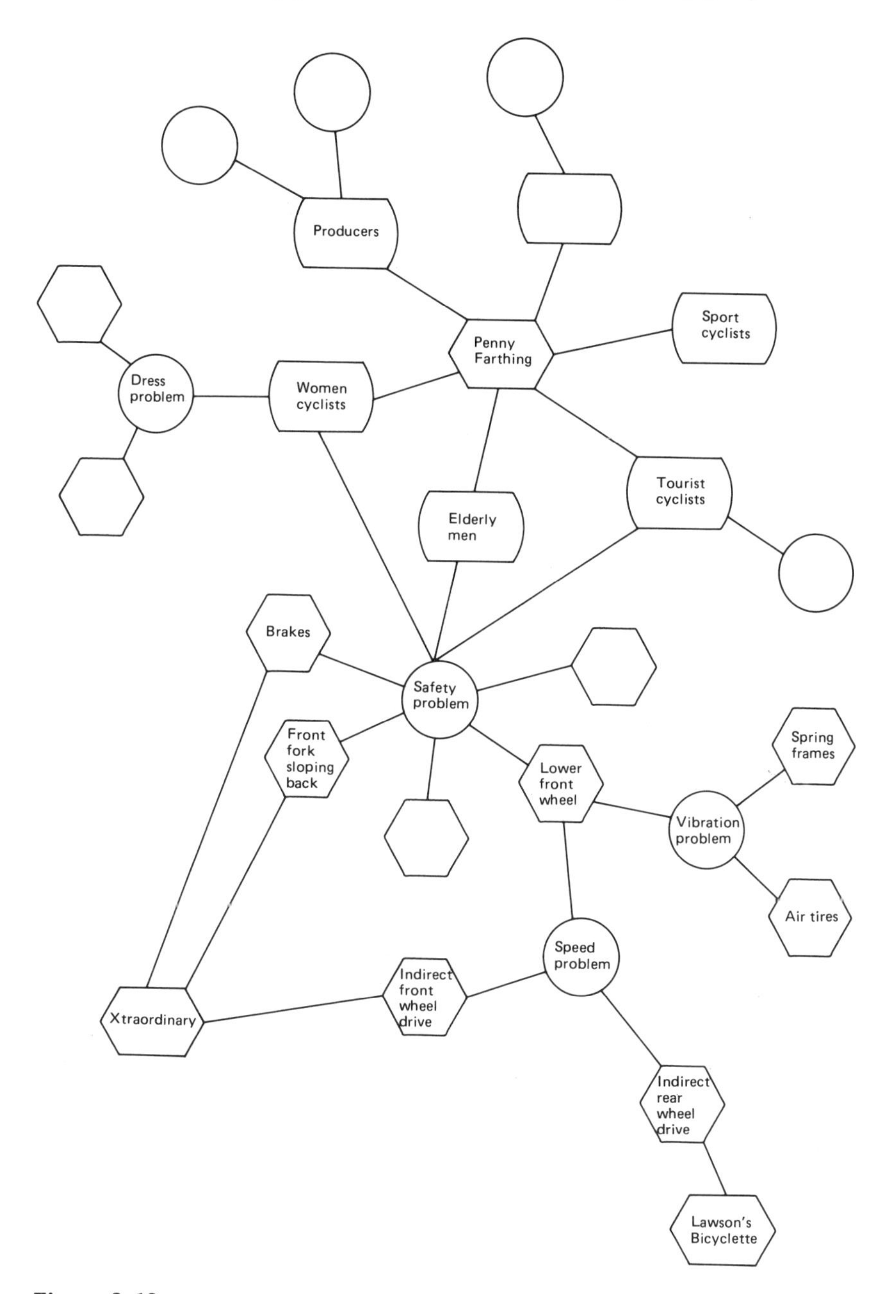

Figure 2.13
Three levels of evolutionary processes may be compiled by superposing figures such as 2.10, 2.11, and 2.12, onto figure 1.4

2.6 Solutions to the Safety Problem of the Ordinary

A great variety of solutions were tried to tackle the safety problem, once it was recognized by manufacturers and inventors. I will discuss them under three rubrics: three-wheeled cycles, modifications of the basic scheme of the Ordinary, and some more radical departures from that basic scheme.

Tricycles

It would be only partly true to describe the tricycle as a solution to the safety problem of the Ordinary, because parallel to the development I have described, from the Draisienne to the bicycle, another genealogy could be drawn of human-driven vehicles with three and four wheels, starting for example with the inventions of Demetrios of Phaleron (308 B.C.) and including the machines of Drais von Sauerbronn (1814).[39] None of these machines, however, reached a stage of commercial viability, and it is unlikely that these designs ever led to more than one prototype. Given the quality of the roads, it is not surprising that these vehicles could not compete with horse-pulled carriages. Moreover, those who could afford such a machine normally would prefer not to propel the "muscle-power cart" themselves—although some of these machines were big enough for servants to ride along and propel, leaving ample space for the owner to enjoy the ride at ease.[40]

A few inventors tried to overcome the disadvantages of the hobby-horse by developing heavy machines with three and four wheels, but none of these went beyond the prototype or toy stage (Woodforde, 1970: 13–15). But in the early 1870s the situation changed. The bicycle had created a market for human-propelled transport vehicles and when the safety problem of the Ordinary had been identified, the tricycle was reinvented as a solution. Moreover, the tricycle promised to solve the problem of staying upright, which many new and less athletic riders found difficult. One of the first successful tricycles was designed by James Starley. The "Coventry Lever Tricycle," patented in 1876, was a two-track machine, originally driven by a lever gear. The lever mechanism was soon changed to a chain drive (Caunter, 1958: 8). This machine became quite popular, especially among lady cyclists. Other designs followed, using all possible schemes to combine the three wheels (see figure 2.14). Propelling the tricycle on the pair of parallel wheels, as for example on the "Coventry Front-Driving Tricycle," caused problems. Because the two wheels were fixed on the axle, the smallest turn caused

Figure 2.14
The Doubleday and Humber Tricycle was a great success on the racing track, but because of its tendency to swerve on passing over a stone it was not much used as a roadster. The front wheels, also used for steering, are driven by a chain; the right wheel is mounted solidly on the axle, while the left wheel has a nonrigid connection to the axle to allow for different rotation speeds when turning a corner. Photograph courtesy of the Trustees of the Science Museum, London.

tension in the axle, then friction between the two wheels and road, thus making the tricycle swerve. This would happen when meeting a stone, let alone when the machine turned a corner, and the parallel wheels followed circles with different radiuses. The first solution was to let one of the parallel wheels run loose or with a friction coupling on the axle. James Starley developed a solution that is still used in all modern motor vehicles. He is reported to have made such a swerving maneuver one day, when he was cycling with his son William. They were riding a strange contraption called the "Honeymoon Sociable," consisting of two high-wheeled Ordinaries with axels fixed rigidly together to form a four-wheeled two-seater. After their unhappy landing, when sitting in the bed of sticking nettles at the roadside and applying a dock leaf to his hands, Starley got the idea of the differential gear: don't connect the two axles rigidly, but use two bevel wheels in the middle. Immediately after returning home, he started to make a model of this device and the next day he left for the patent office in London.[41] Starley applied this differential gear to a new tricycle, the Salvo Quad, with one steering wheel in front and a fourth small trailing wheel for extra stability in the rear (hence the "quad").

Tricycles were advertised as being adapted to the requirements of women and elderly men. Their novelty gave them—just as had happened to the early velocipedes—a social cachet too. And indeed the aristocracy did go for it, especially after the invention received Queen Victoria's blessing. Williamson (1966) described how this came about. During one of her regular tours on the Isle of Wight in June 1881, Queen Victoria had seen what seemed to be a young woman amid a flashing mass of spinning spokes. Her horse carriage was unable to catch up with this amazing sight, and the Queen could not inspect more closely. Servants were sent out to track down the young woman and to summon her to the royal residency at Osborne House. The girl was found to be Miss Roach, daughter of the local Starley agent who encouraged his daughter, for promotional reasons, to ride the new Salvo Quad as much as possible. She came to Osborne House and demonstrated the tricycle to the Queen, who "must have been gratified to see that her performance was really very graceful and one which by no stretch of the imagination could be termed 'unladylike'" (Williamson, 1966: 76). Queen Victoria was interested enough to order two tricycles immediately, and a royal command was added that the inventor should be present on delivery. Thus, a few weeks later, James Starley, very nervous and with a brand new silk hat, traveled to Osborne House where the Salvo Quads were delivered through the local agent. The Queen was sitting on the lawn on a small garden chair, reading papers with a secretary; Prince Leopold, then about twenty-seven years old, was examining one Salvo Quad the stood under a tree. Starley was presented to the Queen, who said some pleasant words to him and gave him a leather case containing a silver watch as a memento of his visit. Then, Starley wrote in a letter to his wife,

> I was quite overcome and bowed so low that I nearly toppled over as I said I am very honoured, Ma'am. Then the gentleman led me away and I was surprised and pleased when the Prince came along and asked me to explain the working of the tricycle to him. A servant was wheeling it behind. We found a nice level drive where I got on and was soon rolling along in fine style. He seemed very pleased with it and thanked me very kindly.[42]

The Salvo Quad was immediately renamed into Royal Salvo, and within a few years the tricycle had become fashionable among the elite. Lord Albemarle wrote that there was not a crowned head who had not a fleet of tricycles, both within and outside Europe:

> I have seen a picture in which the Maharajah of an Indian state, together with the British resident at his court and all the great officers of the durbah, are seated on tricycles at the gate of the palace, and gaze at the lens of the camera with the breathless attention usual on such occasions.[43]

It was no surprise then that the Tricyclists' Association sought special privileges in the London parks because tricyclists were supposed to be better bred than bicyclists.[44] And in 1882 a Tricycle Union was founded, because many tricyclists wanted to distance themselves from the bicyclists, "who are a disgrace to the pastime, while Tricycling includes Princes, Princesses, Dukes, Earls, etc." (Ritchie, 1975: 113).

It is difficult to appreciate the importance of the tricycle in hindsight, as most of us associate the tricycle with children. However, the tricycle was a very viable alternative to the bicycle in the 1880s and 1890s and not some "historical mistake," as it may seem now. Most cycle manufacturers produced bicycles as well as tricycles.[45] For example, Messrs. Singer & Co manufactured a tricycle in 1888 that embodied the same state-of-the-art technology employed in their 1890 bicycles. A 1886 catalogue of all British cycles described 89 different bicycles and 106 tricycles.[46] Thus, notwithstanding the initial efforts to restrict the use of tricycles to the upper class, the tricycle gained widespread acceptance. In 1883, the "Bicycle Touring Club," founded in 1878, changed its name to the "Cyclists Touring Club." Many people were convinced that it would just be a matter of time before the tricycle was the only commercially available cycle (Rauck et al., 1979: 60). Especially as a vehicle for business purposes, it seemed to have a splendid future. The *Evening Standard* was distributed by means of the Singer tricycle called "Carrier," and the Post Office employed scarlet tricycles for delivering parcels. A specific illustration of the impact the tricycle had on designers of the time features the Otto Dicycle (see figure 2.15). This was a bicycle in the sense that it had but two wheels, but they were arranged in a "tricycle way": the dicycle had two large parallel wheels, and the rider was seated in between. The machine enjoyed some popularity; the Birmingham Small Arms Co. built 1,000 machines of this design. The "Ottoists" claimed that their dicycle was especially effective when riding against the wind,

Figure 2.15
The Otto Dicycle was patented between 1879 and 1881 by E.C.F. Otto. From the back of the frame, supported by the axle, projects a small rubber-tired roller, which prevents the frame and rider from swinging too far backward. This roller can be used as emergency brake by leaning back; normally it will be well off the ground. The wheels turn loose on the axle and are driven by two rubber-sheathed pulleys. Handles on each side of the rider allowed the pulleys to be slackened selectively, so that one wheel could turn faster than the other and the machine could make a turn. Although keeping one's balance was said to be rather easy, steering downhill took longer to master. Photograph courtesy of the Trustees of the Science Museum, London.

because by leaning forward the riders could get out over the pedals, thus bringing all their weight to bear directly on them.

As mentioned, the tricycle played an important role in providing an opportunity for women to cycle. The acceptability of women riding tricycles was linked to the association of tricycling with the upper classes: "Tricyclists will generally be of a better class than bicyclists, and will seldom consist of mere beardless youths, but men of position and experience, and above all, by the fair sex."[47] Tricycling made it possible for young ladies of good breeding to get out of their stuffy Victorian homes. The tricycle (as some time later the bicycle) was not so much used by women to go somewhere, but rather to get away. And thus it showed the way to a loosening of customs, for example in the domain of dress. The Cyclists Touring Club seriously discussed the dress that should be worn by lady tricyclists (Woodforde, 1970: 123). The crucial element was assuring propriety by wearing knickerbockers or trousers beneath a full-length skirt. Still, this English "C.T.C. uniform" was a long way from the American "Bloomers," which will be discussed later. "One reason for the protection which ladies undoubtedly find in the C.T.C. uniform lies in the fact that it is so little remarkable, and so closely resembles that ordinarily worn by the wife of the parson or doctor."[48] Tricycling, too, was an activity during which a woman should not display herself too freely. But even so, tricycling engaged women in cycling and thus paved the way for women's participation in bicycling. Because the tricycle was appreciated for solving the safety problem of the high-wheeled Ordinary and thereby allowing women and elderly men to engage in cycling, it made bicycle producers acutely aware of these groups as potential markets for bicycle sales. This was further stimulated by recognizing that the tricycle was not without problems itself.

Surely it was more easy to keep one's balance on a tricycle than on a bicycle, and making a "header" was less likely too. But the tricycle appeared to have safety problems of its own. Most tricycles had three tracks, where the bicycle had only one when riding straight on. This made the tricycle more subject to the perils of the roads, for it was more difficult to avoid stones and holes. On the roads of the 1890s, this was a considerable drawback. Another circumstance that caused tricycles to be involved in accidents was that most of them did not have effective brakes. The rider had to "reverse the action of his machine" by trying to backpedal. And this could be difficult. Especially when going downhill, it was crucial not to let your feet slip off the pedals. When trying to regain control over those more and more quickly revolving pedals, many tricyclists were lifted from their seats. As a passing cyclist commented when helping a tricyclist after such a downhill accident, "You lost control. Should never do that, you know. Might have ruined your machine."[49]

Further, sitting between the two large wheels, as required by most tricycle configurations, was a safe and stable position as long as you were rolling along smoothly, but it became a very hazardous place to be when taking a spill. In such a case, it was almost impossible not to get entangled in the spokes of the large wheels. In 1883, tricycle accidents seem to have outnumbered accidents with bicycles, and the *Times* of that year reported a death caused by a fall from a tricycle (Woodforde, 1970: 67).

Thus the tricycle offered a partial solution to the safety problems of the Ordinary, and therefore it was a substantial commercial success. By the 1920s new tricycles were still used and sold, although few large cycle manufacturers were producing them. Instead, local assemblers were the typical producers of these custom-designed machines (Grew, 1921: 22). But since these machines posed some new problems of their own, the success was not complete and there was room for alternative solutions to the Ordinary's safety problem.

Safety Ordinaries

Another class of attempts to solve the high-wheeler's safety problem was based on modifying the basic scheme of the Ordinary bicycle. Moving the saddle backward was an obvious way to reduce the problem; without further modifications, however, this would bring the rider's weight above the small rear wheel and thus make its vibration more manifest. The only way to cope with this vibration problem was to enlarge the rear wheel. An additional advantage was that once the rear wheel was of significant size, the rider was positioned between the two wheels, rather than above one; this would also reduce vibration.[50] But this alteration made the bicycle heavier and thus more difficult to handle. Moreover, such an enlarged rear wheel was out of syne with the aesthetic norms of the community of high-wheel bicyclists, where the smallness of the rear wheel emphasized the loftiness of the rider. However, because the goal of making the Ordinary safer was already out of sync with the high-wheelers' norms, bicycle designers were probably prepared to put up with this drawback, expecting the relatively bigger rear wheel to be acceptable to potential buyers of these new machines. This new class of bicycles was soon to be called safety Ordinaries.

Another disadvantage of moving the saddle backward was that treading the pedals became less comfortable: because he was now behind the pedals rather than almost directly above them, the bicyclist, as in the case of the velocipede, would push himself backward with his legs, and counteract that force by pulling forward on the handlebar. One way to tackle this problem was to replace the pedals with some lever mechanism extending backward. John Beale had already patented such a mechanism in 1869, but its application in a commercial bicycle had to wait until about 1874, when the Facile bicycle was produced by Ellis & Co in London. The front wheel was reduced in size to 44 inches, the saddle was placed farther back, and the pedals were lowered by placing them on the rear ends of levers mounted below the axle. These levers were pivoted to forward extensions of the fork and their midpoints were connected to the cranks with short links (Caunter, 1958: 8).

On the Facile, the rider's feet made an up-and-down movement, rather than a rotary action. This was claimed to be very effective, especially when climbing a hill (see figure 2.16). The question of which of the two types of motion was best for cycling was hotly debated at the time. As was so often the case in bicycle history, enthusiasts tried to settle this issue by testing the bicycles in races and record-breaking efforts. Significantly, the Facile was not used for high-speed racing and sprinting, but primarily for hill climbing and long-distance riding (Griffin, 1886: 32). The rotary motion was generally preferred for sprinting (Ritchie, 1975: 126).

In 1878, G. Singer patented a device similar to the Facile (Singer, 1878). In this design, named the Xtraordinary, the backward position of the saddle was realized by tilting the front fork backward (see figure 2.17). However, such a sloping front fork, without further modification, would have made steering quite difficult: the center of the wheel—and thus the point of action of the bicycle's weight—was forward of the point at which the wheel had contact with the ground, and so the wheel tended to veer sharply and needed to be kept straight by continuous application of force. This problem was solved by the idea (also included in this patent) of giving the front fork such a form that the center line of the steering head met the ground at the point of contact between wheel and ground. The pedals of the Xtraordinary were brought backward by mounting them on levers that moved the crank pins. The upper end of each lever, attached by a link to a point near the top of the front fork, moved in an elliptical arc while the pedals made their "normal" rotary movement (Caunter, 1958: 9).

Although in the case of the Xtraordinary, the rotational speed of pedal and crank was still the same, one could choose different force-movement

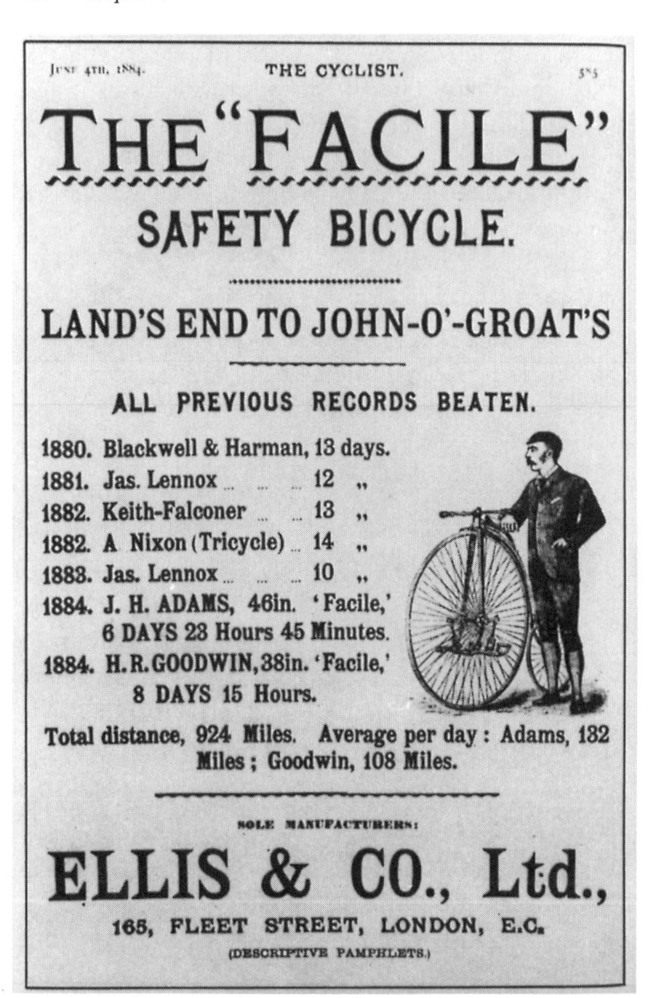

Figure 2.16
The "Facile" was advertised by referring to the records set in long-distance racing. The riders performing these feats were often paid by the manufacturing firm: the first "professional" bicyclists. (From an advertisement in *The Cyclist*, June 4, 1884: 585; reprinted from Ritchie (1975).)

Figure 2.17
The "Xtraordinary," or "Xtra" for short, was produced by Messrs. Singer & Co., Coventry, in 1878. The levers allow the rider a downward push, although the saddle is moved backward. Photograph courtesy of the Trustees of the Science Museum, London.

ratios by varying the length of the levers and the position of the linking point of the crank pins. In principal this was no different from choosing specific lengths of cranks as possible in earlier models, but here the lever mechanism was mode more flexible (Griffin, 1886: 11). The levers of the Facile and the Xtraordinary perhaps point to the designers' growing awareness of the gearing possibilities of using "intermediate" driving mechanisms. In any case, several designs were tried to solve the Ordinary's safety problem, primarily by lowering the front wheel, and as an intrinsic part of that modification, to incorporate an accelerating mechanism to compensate for the resulting lower top speed.

Complicated combinations of levers and gears were employed in the Sun and Planet, the Devon Safety, the Dutton Safety, and the Raccoon Safety.[51] These machines still had the upright front fork and the forward-positioned saddle of the Ordinary, but their front wheels were significantly lower. None of these became a commercial success.

In Marseille, Rousseau was the first to add a chain drive to an Ordinary. He designed a bicycle called Sûr in 1877 that had a front wheel with a radius two-thirds that of an Ordinary. The wheel was driven by a gears-and-chain mechanism with a gear ratio of 2 : 3, exactly compensating for the smaller wheel radius. The Sûr, however, was not successful either, although a very similar design by E. C. F. Otto and J. Wallis did become a commercial success in Britain. Their Kangaroo had a front wheel of 36 inches, which was geared up to 54 inches (see figure 2.18). One problem with their mechanism was the arrangement with two independent chains: each pedal had to be raised by the "slack" side of its chain, which caused, unless it was kept carefully tightened, two shocks per revolution, jarring the gear (Caunter, 1958: 9–10). The Kangaroo

Figure 2.18
The "Kangeroo" safety Ordinary, patented in 1878 by E.C.F. Otto and J. Wallis, was built by several well-known manufacturers. Photograph courtesy of the Trustees of the Science Museum, London.

was manufactured by the firm of Hillman, Herbert, and Cooper. They publicly launched the Kangaroo in 1885 by organizing a race, which the professional cyclist G. Smith won on a Kangaroo. The average speed he obtained (14 miles per hour; 22.4 kilometers per hour) was more than twice the speed Hillman and Starley had achieved in their historic ride from London to Coventry. The Kangaroo scheme was taken up by several designers, and in the 1886 catalogue of bicycles some ten different makes of chain-driven Ordinaries were described (Griffin, 1886).

These safeties were claimed to be safer than the Ordinaries: the Facile, for example, was hailed by its makers as "Easy to learn. Easy to ride. Easy to mount. Easy to dismount. Safe from side-falls. Safe from headers."[52] And not only the manufacturers were enthusiastic. New bicycles were routinely tested and reviewed in the various journals, and most of these new machines were well received. For example, the Kangaroo was said to be

> a thoroughly sound and reliable little mount, likely to win its way more and more into popular favour, particularly among those who value their necks too highly to risk them upon the ordinary bicycle, or who are occasionally apt to characterize the propulsion of a heavy three-wheeler—as Dickens' friend did the turning of the mangle—as "a demm'd horrid grind."[53]

But from the advice given by *Cycling* in 1887 about coasting on a Kangaroo, the conclusion can be drawn that there was still considerable chance of being sent flying over the handlebar. A Kangaroo rider was cautioned to "throw his body as far back as possible" and to apply the brake very gradually. Thus the greater safety of the safety Ordinaries seems to boil down to falling less hard rather than less often.

One rather colorful solution left the height of the Ordinary unmodified, and only sought to enable its rider to land on his feet in case of a header. Franz Schröder constructed a safety handlebar, or "Non-Header" or "Non-Cropper" as he proposed to call it. When colliding head-on with some obstacle, the rider would be projected forward along with the handlebar, which disconnected automatically from the front fork (Rauck et al., 1979: 51). Schröder arranged a demonstration for the director and chief engineer of the bicycle manufacturing firm Frankenburger & Ottenstein in Neurenberg. When Schröder ran into the large stone that he had brought to the demonstration, he landed squarely in front of the Ordinary, on his feet. The director was wildly enthusiastic but the engineer less so, wondering what would happen to a somewhat less athletic rider. They summoned a worker to test the device, and he indeed landed on his bottom. This was nonetheless considered an improvement over landing on his head, and the "Non-Header" entered into production. The invention, however, did not fundamentally change the course of events by solving the Ordinary's safety problem in the eyes of all. A bitter patent fight evolved between Schröder and a Czech named Havlik who had simultaneously patented a similar device. In addition to challenging the fact that Schröder's invention had predated his own, Havlik also disputed its effectiveness, claiming that it would castrate its rider because a part of the steering tube was left on top of the front fork. Moreover, consumers were sending angry letters to Frankenburger & Ottenstein, complaining that the handlebar disconnected after only a minor push during regular cycling; frantically trying to keep control of the bicycle by holding on to the front fork, hapless riders inevitably landed on the ground entangled in their Ordinary (Timm, 1984: 194–202).

Reordering the Basic Scheme

All bicycles inspired by the safety problem of the high-wheeler were developed in roughly the same period and to some extent in parallel. Ordinaries, tricycles, safety Ordinaries, and the machines to be discussed in this section were all striving for the cyclists' favor. Considering the uneven quality of the historical material and the inevitable overlap of the various designs, it is hazardous to lend much weight to the chronological order as distilled from available sources. For example, the bicycles described in this section cannot be considered to form a logical follow-up to the safety Ordinaries of the previous section. Moreover, they do not bear much relation to one another. These designs deserve a separate discussion, because they differed more radically from the basic scheme of the Ordinary than the machines previously described.

One radical solution to the Ordinary's safety problem was to reverse the order of the big and small wheels. One bicycle that can be viewed as the outgrowth of this idea was patented by Henry J. Lawson and J. Likeman in 1878. Others, discussed below, were primarily of American origin. The Lawson and Likeman bicycle bore a close resemblance to MacMillan's machine. But when its frame is closely analyzed, the Lawson and Likeman bicycle clearly reveals its origin as an Ordinary. However, it is driven backward. To bring the rider within reach of the handlebar, which was of the normal Ordinary construction but now mounted on the small wheel, the saddle had to be moved to a position between the wheels. This low positioning of the saddle enabled the cyclist to reach the ground with his feet while staying on the bicycle (Caunter, 1958: 8). Indeed, this machine could have been called a "safety."[54] However, it seems not to have been much of a commercial success.

Another family of bicycles, designed according to the same basic idea, did meet with success. Viewed from a distance, the obvious difference with the previous design was the position of the rider, who sat much more on top of the large wheel, which had consequences for the steering mechanism. Several patents were taken out on designs according to this scheme.[55] The available sources are ambiguous about the construction date of the first successful bicycle of this principle. One of the first of these machines was probably produced by the H. B. Smith Machine Company of Smithville, New Jersey, and publicly exhibited at the meeting of the League of American Wheelmen in Boston, on November 23, 1981.[56]

The Star bicycle, as the Smith machine was called, had its saddle forward of the big rear wheel and thus needed a lever-type of driving mechanism to bring the pedals forward to the position of the rider's feet (see figure 2.19). Two drums were attached to the ends of the rear axle.

Figure 2.19
The American "Star" bicycle, first manufactured around 1881. Photograph courtesy of the Trustees of the Science Museum, London.

A leather strap was wound around each drum several times, one end of the strap being attached to the drum and the other end to the lever on that side. As a lever was pushed down, its strap was pulled, which made the drum turn. The drum was attached to the axle by a ratchet mechanism and thus forced the wheel to turn as well. At the end of a stroke, the foot pressure was released and a spring within the drum wound the strap back, bringing the lever to its original position. The effective attachment point of the straps on the levers could be adjusted, thus providing a kind of "gear shift," as two different driving ratios were possible.[57] Normally the levers would be pushed down alternatingly, but because they worked independently of each other, they could be pressed down together in one big stroke. This was considered an advantage for racing purposes, to obtain a quick start or produce a spurt (Caunter, 1958: 14).

Although the small front wheel of the Star suggests difficulty in steering as well as in coping with rough ground, this seems to have been compensated for by its safety and the advantages of the driving mechanism.[58] The Star had considerable success in the United States. But although it was sold in Europe, it did not acquire a significant share of the market in Britain or on the continent.[59] Perhaps the Star was evaluated in comparison with the safety Ordinaries and not found a very credible competitor. In turn, the British safety Ordinaries did not obtain a foothold in the United States.

Another way of reordering the basic scheme of the Ordinary was to move the drive to the rear wheel. In 1879, H. J. Lawson, by that point manager of the Tangent and Coventry Tricycle Company, took out a patent on a design of a bicycle that had a chain drive on the rear wheel (see figure 2.20).[60] The diameters of the wheels revealed its origin: the Ordinary. Now the only function of the relatively large front wheel was to offer a comfortable ride, but the comfort was reduced by the still quite small rear wheel. Because the saddle was mounted on a spring, the result may have been acceptable, though. The front wheel was 40 inches and the rear wheel 24 inches, but geared up to 40 inches as well (Caunter, 1958: 10–11). Lawson called his machine a "Bicyclette."

Whereas the latest types of Ordinary were considered slim and graceful, the aesthetic aspect of the Bicyclette was not much appreciated. Both the public and the trade just could not swallow the grotesque form of the Bicyclette, which was compared to a crocodile because of its elongated frame.[61] A small number of Bicyclettes were manufactured, but they proved to be a commercial failure even though they were rather exten-

Figure 2.20
Lawson's "Bicyclette," patented in 1879. Photograph courtesy of the Trustees of the Science Museum, London.

sively advertised and exhibited. In many bicycle histories the Bicyclette is said to be "ahead of its time."

Although Lawson's Bicyclette was not successful, in the early 1880s there was enough awareness of the safety problem of the Ordinary to stimulate further attempts along these lines. A real boom of different bicycle designs occurred between 1884 and 1888: besides new Ordinaries, safety Ordinaries, and tricycles, lower-wheeled bicycles proliferated. Frequently the words "dwarf" and "safety" were combined in their names. John Kemp Starley, a nephew of James Starley, had started a partnership with William Sutton. After manufacturing Ordinaries and tricycles for some years, they presented in 1884 a new design, comprising a 36 inches front wheel, coupling-rod steering, a chain drive to the rear wheel, and a frame that seemed, when viewed from a distance, to have a diamond shape (see figure 2.21). This machine, named the Rover, is a curious mixture of elements found in the other bicycles described in this section. The relative size of the wheels, the chain drive, and the steering mechanism are very similar to Lawson's Bicyclette, while the main tube of the frame is formed like that of the Star (and the Ordinary). New is the extra support of the saddle by a fork on the rear wheel. This extra support would soon develop into a true part of the

Figure 2.21
"The Rover," designed by J. K. Starley and W. Sutton in 1884, was the first dwarf safety with a diamondlike frame. Photograph courtesy of the Trustees of the Science Museum, London.

frame, thus resulting in the diamond-shaped frame that most (male) bicycles would have up to today. Effectively, however, the Rover still had a single backbone frame.

The Rover was presented to the public during the Stanley Exhibition, a large annual event held in London. All important cycle manufacturers sent their latest models for display, and all members of the trade came to keep abreast of recent developments and to place orders for the coming year. Starley and Sutton's new Rover must have seemed a pygmy among all the lofty Ordinaries. Some traders approved its small wheels and assumed it to be suited for nervous or less athletic cyclists. Others were inclined to scoff and nicknamed it "Beetle" or "Crawler" (Williamson, 1966: 103). Starley and Sutton started to organize races in which they had professional racers riding their Rover. For the first race Starley and Sutton used the same route Hillman had used for the launching race of the Kangaroo in 1885. They could be satisfied by this promotional move: the record set on a Kangaroo was beaten by the same George Smith, but this time on a Rover.[62] As the sales started to increase, Starley continued to revise his Rover, turning out a second and third model with only a few months in between. I will return to these revisions shortly.

Figure 2.22
From 1884 to 1886 several new designs were developed, in various aspects widely different from the basic scheme of the Ordinary bicycle. The Humber "Dwarf Safety Roadster," also designed in 1884, had a trapezial frame. In their model of 1886, Humber introduced a front fork with the forward bending of the Singer 1878 bicycle. Photograph courtesy of the Trustees of the Science Museum, London.

The new designs from the mid-1880s clearly show that all elements of the basic scheme of the Ordinary had been called into question (see figure 2.22). The Birmingham Small Arms Company, for example, made a bicycle of which the cross frame was radically different from all previous frames based on the single backbone of the Ordinary. This bicycle had a large chain-driven rear wheel and a small indirectly steered front wheel. The new frame consisted of an almost straight tube between the axle of the traveling wheel and the bracket for the steering front fork, and a second rod, perpendicular to the first, which supported the saddle, the indirect steering mechanism with handlebar, and the crank bracket with sprocket wheel. This invention had for its object "to give greater rigidity to the framing of the bicycle so that the seat and steering mechanism may be free from the unsteadiness of those parts in bicycles of the ordinary kind."[63]

Lawson played an indirect role in stimulating the design of this bicycle. He had approached B.S.A. with a proposal to manufacture his new design for a lady's safety bicycle. This machine had a large chain-driven rear wheel, a smaller front wheel, and a single-tube frame that bent upward and forward to support the saddle (Lawson, 1884). B.S.A. declined the offer but agreed to make two prototypes for Lawson. While doing that, they decided to design their own safety bicycle and assembled this machine as much as possible from their standard tricycle parts (Caunter, 1955: 35).

Another design that departed from the old frame scheme was patented by the gun maker H. Wallis in 1884 and produced by Messrs. Humber & Co (Wallis, 1884). Its frame had a trapezial form, which proved stiffer and more compact than the single backbone frame of the Ordinary bicycle. Another important feature, in comparison with competing 1884 designs, was its direct steering. The small front wheel proved to be problematic on rough roads, however.

J. McCammon patented a dwarf bicycle suitable for ladies. The single backbone of the frame dropped deep between the wheels, so as to allow women to mount the machine more easily (Caunter, 1958: 14). This McCammon bicycle had the same steering wheel scheme as the Humber, but also featured the slightly backward bending of the front fork that was used in the B.S.A. machine.

Starley and Sutton further revised their Rover by making the steering direct, giving the front fork much rake and changing the frame from the original single backbone to the beginning of a diamondlike shape. The front fork is straight, so it is not surprising that "at first the steering feels rather difficult, as the pilot wheel has no automatic assistance or fly-back spring to keep it straight" (Griffin, 1886: 44). It is unclear why they did not use Singer's idea, described previously, of bending the front fork to counter this effect. Nevertheless, consumer tester Griffin concluded his report on this Rover on a positive note, predicting a successful future for it.

In spite of the emergence of quite a number of dwarf safeties, many people still were convinced that the high-wheeled Ordinary bicycle would never be superseded by those geared-up small-wheelers. In a report on the yearly Stanley Exhibition of Cycles, it was observed that

> No radical changes have been made in the construction of cycles during the past year, and the tendency is to settle down to three types of machines—the ordinary bicycle, the rear-driven safety bicycle, and the direct front-steering tricycle, whether single or tandem. (Engineer, 1888a: 118)

Besides mud splashing on the rider's feet[64] and the power wasted by the chain drive, the most prominent problem was the vibration of the low-wheeler (Woodforde, 1970: 87).

At the 1888 exhibition most safeties were equipped with some sort of antivibration gear. Many frames were constructed with several hinges instead of rigid connections. Springs were mounted between the wheel axles and the frame, between handlebar and front fork, between saddle and frame, and between crank bracket and frame (Engineer, 1888a: 118). The awareness of the vibration problem seems to have increased in subsequent years. In the 1889 cycle show it was clear that

> With the introduction of the rear-driving safety bicycle has arisen a demand for antivibration devices, as the small wheels of these machines are conducive to considerable vibration, even on the best of roads. Nearly every exhibitor of this type of machine has some appliance to suppress vibration. (Engineer, 1889: 158)

In the report on the 1890 show, the situation is even more pronounced (Engineer, 1890b: 138). One of the spring frames that was most successful at the 1890 cycle show had been patented in 1885 by O. Macarthy. The machine was manufactured by Messrs. C. A. Linley and J. Biggs (see figure 2.23). The sloping backbone that joined the rear axle with the steering head and front fork was connected to the rest of the bicycle by springs and hinges. Thus all bicycle parts with which the rider had contact (the saddle, the handlebar, the cranks) had an elastic connection to the rest of the machine (Caunter, 1958: 14–15 and Caunter, 1955: 35–36). However, not many of the antivibration devices were strong and durable: "Of those exhibited for the first time too many are conspicuous by their complication; we should imagine that their designers were in many cases ignorant of the first principles of mechanics" (Engineer, 1889: 158). And of course, even the successful "Whippet" with its many movable parts needed more attention than an ordinary bicycle.

It is not surprising then that the safety bicycle was not more than one of the three alternative types of cycle, without threatening the market share of the other two, the Ordinary bicycle and the tricycle. This changed when the air tire was made available for bicycles.

2.7 Interpretative Flexibility

The bicycle story will be interrupted again to discuss another issue related to the descriptive model I am developing. In the previous sections I have described the various artifacts through the eyes of relevant

Figure 2.23
The "Whippet" safety bicycle was patented and built in 1885. The relative positions of saddle, handlebar, and cranks were fixed, since these three formed a rigid triangle that was isolated from the main backbone of the frame by a strong coil spring, a movable shackle in the steering rod, and a hinged tube between backbone and steering pillar. Photograph courtesy of the Trustees of the Science Museum, London.

social groups. Where the differences between the various social groups were taken seriously, quite different descriptions did result. Until this point, however, this was left rather implicit. I shall now discuss more explicitly the consequences of those differences in the meanings attributed to an artifact by various relevant social groups.

For example, for the social group of Ordinary nonusers an important aspect of the high-wheeled Ordinary was that it could easily topple over, resulting in a hard fall; the machine was difficult to mount, risky to ride, and not easy to dismount. It was, in short, an *Unsafe Bicycle*. For another relevant social group, the users of the Ordinary, the machine was also seen as risky, but rather than being considered a problem, this was one of its attractive features. Young and often upper-class men could display their athletic skills and daring by showing off in the London parks. To impress the riders' lady friends, the risky nature of the Ordinary was essential. Thus the meanings attributed to the machine by the group of Ordinary users made it a *Macho Bicycle.* This Macho Bicycle was, I will argue, radically different from the Unsafe Bicycle—it was designed to meet different criteria; it was sold, bought, and used for different purposes; it was evaluated to different standards, it was considered a machine that worked whereas the Unsafe Bicycle was a *nonworking machine.*[65]

Deconstructing the Ordinary bicycle into two different artifacts allows us to explain its "working" or "nonworking." There is no universal time- and culture-independent criterion with which to judge whether the high-wheeled bicycle was working or not. Is the Ordinary a nonworking machine because it was highly dangerous and very difficult to master? Or was it a well-working device because it displayed so nicely the athletic skills of the young upper class and because it dealt so effectively with bumps and mud puddles in the road? Only by reversing the question—that is, by asking under what conditions the high-wheeled Ordinary constituted a well-working machine and under what other conditions it was utterly nonworking—can we hope to begin to understand technical development.

In terms of the descriptive model, this implies the following. The artifact Ordinary is deconstructed into two different artifacts. Each of these artifacts, the "Unsafe," and the "Macho" are described as constituted by a relevant social group, and this description also includes a specification of what counts as "working" for that machine, for that group. In this way, the "working" and "nonworking" are now being treated as *explanandum*, rather than used as *explanans* for the development of technical artifacts. The "working" and "nonworking" of an artifact are socially constructed assessments, rather than intrinsic properties of the artifact. One artifact (in the old sense) comprises different socially constructed artifacts, some of which may be "working" while others are "nonworking." I am not primarily making a metaphysical claim here—I am stressing this point because in this way the descriptive model will allow for a symmetrical analysis of technology, as called for in chapter 1. This is analogous to arguing that "Nature" should not play a role as *explanans*, as David Bloor (1973, 1976) did in his strong programme.[66] "Nature" should not be invoked to explain the truth of scientific beliefs; and neither should specific sociological circumstances—such as the scientist being excessively ambitious, having a bad marriage, or living under a totalitarian regime—be used exclusively to explain the falsity of scientific beliefs. This "symmetry principle" calls for sociologists analyzing scientific development to be impartial with respect to the truth or falsity of scientific beliefs. They should explain truth and falsity symmetrically, that is, by using the same conceptual framework.

Thus I want to argue that the account of bicycle development can be adequately rephrased by distinguishing two separate artifacts: the Unsafe Bicycle and the Macho Bicycle. Although these two artifacts were hidden within one contraption of metal, wood, and rubber (the so-called Ordinary), they were not less real for that. This can be seen from the different designs spectrums they to which they belonged. The Unsafe Bicycle gave rise to a range of new designs that sought to solve the safety problem. Many of these efforts were described in the previous section: moving the saddle backward (Facile, Xtraordinary), adding auxiliaries (the "Non-Header"), reversing of the positioning of small and large wheels (Star), or making other radical changes to the basic scheme (Lawson's bicycle). The Macho Bicycle developed in the opposite direction: the front wheel was made as large as possible. This design trend produced important and lasting effects in bicycle technology, even though the high-wheeled Penny-farthing became obsolète in the end. The making of higher wheels, for example, necessitated the development of better (and specifically, stiffer) spoked wheels. To distinguish two different artifacts in this way is more straightforward than trying to cope with the wide spectrum of different designs, even though one needs some imagination to see them within that one Ordinary.

I will call this sociological deconstruction of the Ordinary into an Unsafe Bicycle and a Macho Bicycle "demonstrating the interpretative flexibility of the Ordinary." The possibility of demonstrating the interpretative flexibility of an artifact by deconstruction implies that there is an immediate entrance point for a sociological explanation of the development of technical artifacts. If no interpretative flexibility could be demonstrated, all properties of an artifact could be argued to be immanent after all. Then there would be no social dimension to *design*: only application and diffusion—or *context*, for short—would form the social dimensions of technical development. But demonstrating the interpretative flexibility of an artifact sets the agenda for a social analysis of the design of technology as formulated in the "working as result" requirement for a framework.

Indeed, demonstrating the interpretative flexibility of an artifact can only be the first stage in a social analysis of technical design. The sociological deconstruction of an artifact leaves the sociologist's desk full of pieces that have to be put together again. After all, the analyst may deconstruct the Ordinary into two artifacts, but that does not change the fact that late nineteenth-century English society eventually did construct the high-wheeled bicycle into *one* black box, eliminating the Macho Bicycle and focusing development efforts on the Unsafe Bicycle, thereby preparing the way for the Safety Bicycle. Thus the demonstration of the interpretative flexibility sets the agenda for a sociological analysis of technical development. Once an artifact has been deconstructed into different artifacts,[67] it is clear what has to be explained: how these different artifacts develop; whether, for example, one of them peters out while the other becomes dominant. In the bicycle case, the "Macho," although dominant in the beginning, was in the end superseded by the "Unsafe," and the Ordinary thus developed from a working into a nonworking machine. In the next sections I will follow this process in the case of the safety bicycle.

Relevant social groups do not simply see different aspects of one artifact. The meanings given by a relevant social group actually *constitute* the artifact. There are as many artifacts as there are relevant social groups; there is no artifact not constituted by a relevant social group. The implications of this radically social constructivist view of technology will be addressed in the remainder of this chapter, especially in the third intermezzo, section 2.9. Then I will discuss how the "pluralism of artifacts," brought to the fore by demonstrating the interpretative flexibility, will eventually be reduced again, when one of the artifacts stabilizes.

2.8 The Air Tire

In 1845 William Thomson, a civil engineer of Adelphi, Middlesex, had already found what we now view to be the solution of the vibration problem (see figure 2.24). He patented "elastic bearings round the tires of the wheels of carriages, for the purpose of lessening the power required to draw the carriages, rendering their motion easier, and diminishing the noise they make when in motion," (Thomson, 1845: 2). He did so by using "a hollow belt composed of some air and water tight material, such as caoutchouc or gutta percha, and inflating it with air, whereby the wheels will in every part of their revolution present a cushion of air to the ground or trail or track on which they run" (ibid.). One of the ways, specified by Thomson in his patent, to make this elastic belt was to cement together a number of folds of canvas with india rubber, then render the belt more pliable by immersing it in melted sulphur, and finally, provide a strong outer casing by sewing circular segments of leather around the tire. Pipes were provided, passing through the wheel

Figure 2.24
The first air-filled elastic belt was patented by R. W. Thomson in 1845. Photograph courtesy of the Trustees of the Science Museum, London.

and fitted with an airtight screw cap, to inflate the elastic belt. Thomson made wide claims in his patent and specified the application of his elastic belt to horse carriage wheels, to railway wheels running on timber rails, and to such objects as bath chairs and rocking chairs.

Thomson's belts were tried on horse carriages and found very useful. One journey of more than 1,200 miles is reported to have taken place without any damage to the belts.[68] A carriage-making firm, Whitehurst & Co. was licensed to produce the belts and started carrying out some promotional activities. In 1847 Thomson described some tests to compare the traction forces required by a carriage with elastic belts and one with iron wheels. Here he also presented an air pump very similar to the modern bicycle hand pump (Thomson, 1847). But soon this enterprise came to an end. Part of the explanation for this failure is probably that the belts must have been rather expensive. Moreover, in carriage construction other antivibration devices were feasible, such as large leaf springs and luxuriously cushioned seats. Apparently the small and light Safety bicycle was needed to create a market for this type of tire.

From the early 1870s onward, noninflated rubber tires represented the state of the art in bicycle construction. They came in various forms: solid rubber tires, cushion tires (containing air under atmospheric pressure), and rubber tires with some solid filling as reinforcement. By 1885 several adequate techniques for mounting the tire on the wheel were practiced and the kind of accident Hillman experienced during the famous ride with James Starley from London to Coventry rarely happened anymore. But the small wheels of the Safety bicycles caused too much vibration for these tires to handle. Thomson's patent, his article, and the constructions were completely forgotten in twenty years' time. When John Boyd Dunlop started to consider the idea of air-filled rubber tubes, he did the work all over again.

Dunlop, born in Scotland on 5 February 1840, was a veterinary surgeon in Belfast. He invented various special surgical instruments and was especially experienced in making rubber appliances (Caunter, 1958: 44; Du Cros, 1938: 33). What prompted Dunlop to start working on an air-filled bicycle tire is unclear. One story suggests that a doctor had advised him that cycling would be healthy for his son, but that it would be even more beneficial if the jarring could be reduced (Du Cros, 1938: 39). Another story depicts Dunlop as having long been interested in road transportation and purposefully searching for a means to reduce vibration. In the course of these efforts he constructed various spring wheels and flexible rims, and at last turned to rubber pipes.[69] There is agreement on the secret test ride that Dunlop's son Johnny made on the night of 28 February 1888, using his tricycle equipped with two new air tires on its rear wheels. After this test, which went well, Dunlop asked the bicycle manufacturing firm Edlin & Sinclair of Belfast to make him a new tricycle on which he mounted his air tires himself. When the tires passed this trial as well, he applied for a patent in June 1888.

In his patent Dunlop specified the use of his tires for "all cases where elasticity is requisite and immunity from vibration is desired to be secured," but he also mentioned the "increased speed in travelling owing to the resilient properties" of his tires (Dunlop, 1888: 1). Neither in later parts of this patent, nor in the next patents (Dunlop, 1889a,b) did he comment further on that second objective, and we will see that all early reports about the pneumatic tire focused on its value as an antivibration device. It is unsure how important the speed-increasing potential of his tires was for Dunlop, but judging by Thomson's experiments on traction forces, the latter seems to have been more aware of the friction-reducing possibilities than was Dunlop. This aspect of the air tire will play an important role.

Dunlop employed a hollow tube of india rubber, surrounded with cloth canvas or other material adapted to withstand the pressure of the air. This canvas or cloth was again covered with rubber or other suitable material to protect it from wear on the road (Dunlop, 1888: 1). The tire was provided with a nonreturn valve. In a separate patent he further developed this valve for specific application to bicycle tires (Dunlop, 1889a). Dunlop also mentioned in the patent "any ordinary forcing pump" to inflate the tire; according to his daughter, he had used Johnny's football pump for this purpose. In a third patent Dunlop specified a means for mounting the tires on the wheel rim (Dunlop, 1889b).

The first Dunlop tires were made by the Pneumatic Tyre and Booth's Cycle Agency in Dublin (Du Cros, 1938: 84). Two exhibition rides played an important role in the early days of promoting the air tires. The first was undertaken by R. J. Mecredy, the editor of an Irish cycling paper and a renowned cyclist, who rode from Dublin to Coventry on a tricycle fitted with the new tires. In Coventry he aroused great interest: "The tyres were quite unknown, and when the tricycle was left outside a hotel (not in the centre of the city) for ten minutes, a crowd of 400 or 500 people were found pushing each other to obtain a sight of it."[70] Within a few months everybody interested in bicycles knew all about the new tires. The second feat was a bicycle race held in Dublin. On 18 May 1889 all four races of the Queen's College Sporting Games were won by W. Hume on a "pneu bicycle." This proved to be important because the well-known Du Cros brothers appeared to be among the defeated cyclists. Their father, Harvey Du Cros, was impressed enough to buy Dunlop's patent rights and to found the Pneumatic Tyre Company in Belfast (Rauck et al., 1979: 108). This was the beginning of commercial production of the air tire.

The first tyres were very expensive: about £5 a pair, whereas a complete Ordinary or safety bicycle, fitted with solid rubber tires, cost only around £20. Apart from this, technical difficulties induced much skepticism within the trade, as is obvious from a report on the newly exhibited pneumatic tires at the Stanley Exhibition of Cycles in 1890:

> Not having had the opportunity of testing these tires, we are unable to speak of them from practical experience; but looking at them from a theoretical point of view, we opine that considerable difficulty will be experienced in keeping the

tires thoroughly inflated. Air under pressure is a troublesome thing to deal with. (Engineer, 1890a: 107)

Besides this technical skepticism, the reporter also had arguments pertaining to road behavior and aesthetic complaints:

From the reports of those who have used these tires, it seems that they are prone to slip on muddy roads. If this is so, we fear their use on rear-driving safeties—which are all more or less addicted to side-slipping—is out of the question, as any improvement in this line should be to prevent side slip and not to increase it. Apart from these defects, the appearance of the tires destroys the symmetry and graceful appearance of a cycle and this alone is, we think, sufficient to prevent their coming into general use. (Engineer, 1890a: 107)

Arthur du Cros's memories of his very first ride through London on an air-tired bicycle underline the same aesthetic disapproval:

Omnibus and hansom drivers, making the most of a heaven-sent opportunity, had the time of their lives; messenger boys guffawed at the sausage tyre, factory ladies simply squirmed with merriment, while even sober citizens were sadly moved to mirth at the comicality which was obviously designed solely to lighten the gloom of their daily routine. (Du Cros, 1938: 54–55)

Another problem was that the tires were easily punctured. Repairing an original Dunlop tire was a job that the average bicyclist undertook only with fear and trepidation. Because the tires were cemented to the wheel, you had to peel back the solutioned tread of the rubber cover, then slit the canvas across and withdraw the air tube. Then you had to find the puncture, fix the patch, replace the tube, stitch up the canvas with needle and thread, refix the tread, and reinflate the tire. This task required more skill than many cyclists possessed, and wheels could be seen turning around with huge blobs on the tires where the amateur sewing and repairs were too weakly done to prevent the air tube from bulging out of the canvas and cover. The comfortable ride for which the air tires had been bought was rather spoiled by this. The blobs would hit the forks time after time while the wheel revolved, until the friction wore away the cover and bang went the tube again—but this time condemning the rider to the railways or a long walk (Grew, 1921: 54). Moreover, you needed to bring along a box of tools and materials to accomplish repairs.

However, like so often in the cycle history, it was on the racing track that the air tire's fate changed radically. Here it won its first and probably most important battle against the solid rubber tire. Hume's victory in Dublin in May 1889 had already led to the participation of Du Cros in Dunlop's enterprise. But further participation in races quickly spread the news. Spectators confronted with a "pneu bicycle" for the first time invariably hailed it with derisive laughter, but, as the racing cyclist Arthur du Cros (1938: 51) remembers, "to the stupefaction of the onlookers the ugly interloper outpaced all rivals so decisively that their derision was turned to hysterical applause." Within a year no serious racing man bothered to compete on anything else than air tires.

As to the specific way of mounting the tires, cycle makers had to send their wheels to the Pneumatic Tyre Company in Belfast where the tires were fitted. Obviously this presented a serious barrier to further increasing the sales of air tires. Du Cros decided to move his business to Coventry, the heart of the British cycle industry. Soon the Dunlop carts could be seen careering about Coventry, collecting the tire wheels and delivering them, fitted with air tires, back to the various factories (Grew, 1921: 54).

But then, in the autumn of 1890, Dunlop was officially informed that his patent was invalid. Thomson's patent had turned up and all Dunlop could claim was the application of an existing invention to cycles—and that was not patentable. Anyone was allowed to make and sell air tires. This seemed disastrous for the Pneumatic Tyre Company, which was built solely on the strength of Dunlop's patents. But they found a narrow escape. Many inventors had been attracted by the Dunlop air tire as the basis for patent modifications. C. K. Welch patented a new way of mounting the rubber cover of the inner air tube to the wheel rim. He specified what became known as a "wired on" cover. The cover had inextensible wires molded into its edges, which slipped over the wheel rim and were then pressed into ledges at the inner side of the wheel rim by the inflated air tube (Welch, 1890). Thus no cementing was needed and the air tire was readily detachable. Two other patents, by W. E. Bartlett, came close to Welch's patent. They applied to mounting solid tires molded to fit into clinches on the wheel's rim (Bartlett, 1890a,b). Du Cros bought these three patents and the new mounting his engineers made on the basis of these modifications subsequently proved to represent substantial improvements. Now the tires could be mounted by the bicycle producers themselves, and the expensive transportation of wheels was no longer needed. Moreover, it was rather easy for the cyclist to detach and refit the tire, which was especially important for repairing punctures. It was on the basis of these three patents that the Pneumatic Tyre Company prospered after all. Numerous infringements had to be fought by Du Cros and Dunlop, but gradually all competitors were either forced by legal action to stop their business or their companies died a natural death (Grew, 1921: 56). In 1895 the Dunlop Pneumatic Tyre Co. Ltd. was founded with capital of £5,000,000 (Doorman, 1947: 506). The histories of Thomson and Dunlop thus provide, paradoxically, an illustration of the adage "few patentees grow wealthy." Thomson did not see any commercial return from his patent, although it effectively covered the most crucial aspects of the air tire. Dunlop's air tire enterprise prospered, but without his own original patents having been granted.

In France the firm E. E. Michelin held a patent on an air tire with no outer cover, which could be mounted rather easily. However, in Europe this tire lost popularity to Dunlop's "wired-on" tires. Only in the United States did the Michelin tire hold out for some years.[71]

The competition between the air tire companies, prior to the settlement of the patent controversies, was fierce and greatly stimulated the popularity and promotional importance of racing. Previously the bicycle maker bought his solid or cushion rubber tires from manufacturers who limited their advertising to occasional announcements in the trade journals and did not approach the general public directly. Not so the pneumatic tire producers, who started to announce loudly every win on a machine fitted with their product. Now men were racing to advertise tires as much as to promote the bicycles themselves.

Although the air tire was taking over quite effectively (see table 2.1), the vibration problem had not completely been solved. In 1896, for example, new spring frames were still being exhibited. The "New Whippet" bicycle, incorporating a spring frame, was produced until the late 1890s. That this bicycle was not designed by an eccentric engineer, loving old-fashioned technology and fearing progress, is demonstrated by the fact that this machine also introduced some of the most innovative

Table 2.1
Percentages of exhibits of three types of tires in Britain, from 1890 to 1894

	1890	1891 (begin)	1891 (end)	1892	1893	1894
Solid	98.6	29.1	16.6	4.0	3.1	0.4
Cushion	0.06	54.2	32.2	14.9	14.7	3.3
Pneumatic	1.2	14.0	39.7	65.5	69.3	89.5

Source: *Encyclopedia Britannica.*

cycle accessories of the time, such as the free wheel, rim brakes, and a four-speed changeable gear (Caunter, 1958: 19). But the spring frame was becoming more and more obsolete and soon did not form a normal part of the pattern of the low-wheeled Safety bicycle, which was slowly becoming dominant.

2.9 Closure and Stabilization

In this third methodological intermezzo, I will discuss the last elements of the social constructivist descriptive model: the concepts of "closure" and "stabilization" of an artifact. It is with these concepts that we will clean up the sociologist's desk, littered with artifacts after the sociological demonstration of an artifact's interpretative flexibility. It is with these concepts that, after having carried out the sociological deconstruction, we will now trace its social construction.

For Dunlop and the other developers of the air tire, the tire originally had the meaning of a solution to the vibration problem, in other words, the air tire was an antivibration device. In the first advertisement, which appeared in a weekly cycle journal in Dublin in December 1888, the only claim made for the new pneumatic tire was that it made "vibration impossible."[72] However, the group of sporting cyclists riding their high-wheelers did not consider vibration to be a problem at all. Vibration presented a problem to the (potential) users of the low-wheeled bicycle only. Three important social groups were therefore opposed to the air tire; for these relevant social groups, the air tire did not work. But then the air tire was mounted on a racing bicycle, and another artifact was constructed. When the tire was used at the racing track for the first time, its entry was met with laughter. As I have described, this derision was quickly silenced by the sweeping victory realized on the new tire. Very soon handicappers had to give cyclists on high-wheelers a considerable start if riders on air-tired low-wheelers entered the race. After a short period no racer with any ambition bothered to compete on anything else. What had happened? By two important groups, the sporting cyclists and the general public, another artifact had been constructed: a high-speed tire.[73]

We thus have deconstructed the air tire into an antivibration tire and a high-speed tire, and demonstrated its interpretative flexibility. Now the question is: How did these two artifacts develop further? The tire company spared no efforts to develop the high-speed tire. They sponsored cycle racing, arranged training facilities under a competent trainer, and organized a regiment of professional racing teams with multiple machines.[74] Thus they succeeded in redefining the key problem for which the artifact was meant to provide a solution—now it solved the low-speed problem, rather than the vibration problem.[75] It is by no means self-evident that this should have been the outcome of the trial; enabling high speed is not an unambiguous, intrinsic property of the air tire that could dictate the course of events. On the contrary, taking for a moment the ahistorical viewpoint of an engineer, I find it very unlikely that it was the pneumatic tire that tipped the scales for the Du Cros brothers. Probably more influential were other differences between the high-wheeled Ordinaries and the low-wheeled bicycles with air tires: the chain drive on the latter and the high wind resistance on the former.[76] Thus the artifact "high-speed air tire" was socially constructed.

This social construction of an artifact is the outcome of two combined processes, closure and stabilization. These actually are two aspects of the same process, but for analytical purposes I will present them separately and only at the end of this section indicate that they are two sides of the same coin. The concept of closure relates to the interpretative flexibility argument, and is analogous to the discussion of closure of scientific controversies in recent social studies of science. The concept of stabilization is grounded in a critical evaluation of the naive invention-as-an-act-of-genius approach to the study of technology and draws on work in linguistics and recent laboratory studies in the sociology of science. Stabilization can most easily be introduced by analyzing the intragroup development of artifacts, while closure is primarily relevant to an intergroup analysis. If the closure concept has a primarily social interactionist origin, the stabilization concept is colored more by semiotics.[77]

Let me start a discussion of closure by briefly reviewing the analogous issue in controversy studies in the sociology of scientific knowledge.[78] When a scientific controversy is closed by the participants reaching consensus, scientific facts are created. This consensus means that the interpretative flexibility of, for example, an observation statement disappears, and from then on only one interpretation is accepted by all. Such a closure is not gratuitous, but has far-reaching consequences: it restructures the participants' world. History is rewritten after such a closure, and it is difficult to recapture the factual flexibility as it existed prior to the ending of the controversy. On the other hand, it is in principle always possible—although in practice very difficult—to reopen up a controversy once closure is reached.

Several closure mechanisms have been identified. For example, in the case of the "rhetorical closure mechanism," there is a "crucial experiment" or a "knock-down argument" that has the effect of closing a controversy without being completely convincing to the core set of scientists; its effect is based on the appeal that the experimental results or arguments have on a wider and less expert audience.[79] The case of the air tire is an example of the "redefinition of problem" closure mechanism (Pinch and Bijker, 1984). Additional closure mechanisms have been identified in other case studies (Beder, 1991; Misa, 1992).

Let us return to the analysis of technology and the case of the Macho Bicycle versus the Unsafe Bicycle. There, an example of rhetorical closure could almost be seen—if it had succeeded. One producer tried to make the case for his Macho Bicycle by claiming its "absolute safety": "Bicyclists! Why risk your limbs and lives on high Machines when for road work a 40 inch or 42 inch 'Facile' gives all the advantages of the other, together with almost absolute safety."[80] If this producer had succeeded, the Unsafe Bicycle would have become obsolete. This is an example of (failed) "rhetorical closure," for the engineers still considered the height of the bicycle and the forward position of the rider a safety problem.

Closure, in the analysis of technology, means that the interpretative flexibility of an artifact diminishes. Consensus among the different relevant social groups about the dominant meaning of an artifact emerges and the "pluralism of artifacts" decreases.

I will now turn to the concept of stabilization, which underscores the observation that technical change cannot be the result of a momentous act of the heroic inventor. Here the focus will be on the development of an artifact within one relevant social group. Employing the descriptive model, we should be able to trace growing and diminishing degrees of stabilization of the different artifacts. In principle the degree of stabilization will be different in different social groups.

This process of increasing or decreasing stabilization can be traced by using an established type of rhetorical analysis first employed in science studies by Latour and Woolgar (1979). They showed that in the construction of scientific facts "modalities" are attached or withdrawn from statements about facts, thus connoting the degree of stabilization of that fact. Thus the statements: "The experimenters claim to show the existence of *X*," "The experiments show the existence of *X*" and "*X* exists" exhibit progressively fewer modalities and thereby show progressively greater degrees of stabilization of *X*.[81] In the study of technical cases, similar varieties can be seen in the number of definitions, specifications, and elucidations attached to statements about the artifact. Of course, as Latour and Woolgar also observed, there is a methodological problem in this use of language as a medium through which to trace stabilization. The need to add definitions and elucidations to be able to communicate about an artifact depends on more than the degree of stabilization of that artifact in that social group; it will at least depend as well on the context in which the statement is used (e.g., a research paper, a patent, or a handbook). When, however, we take one relatively stable social group and analyze a communication situation that is relatively constant over time, this problem seems negligible. To trace the Safety bicycle's stabilization process, I will sketch how the bicycle was described within the social group of cycle engineers in one specific communication channel, their journal *The Engineer*. Thus the "invention" of the Safety bicycle will be depicted not as an isolated event (for example, in 1884), but as an eighteen-year process (1879–1897).

How are the processes of closure and stabilization related? In my analysis of the concept of closure I implicitly focused on the meanings attributed by different relevant social groups to an artifact. In contrast, for the analysis of stabilization, the focus was on the development of the artifact itself within one relevant social group, in terms of the modalities used in its descriptions.

Closure leads to a decrease of interpretative flexibility—to one artifact becoming dominant and others ceasing to exist. As part of the same movement, the dominant artifact will develop an increasing degree of stabilization within one (and possibly more) relevant social groups.

It is important to recognize that the process of closure is almost irreversible—almost, but not completely. It is now difficult to see the Ordinary bicycle as anything other than a very unsafe machine that is extremely difficult to ride. Trying to envisage the Ordinary as the artifact it was for its contemporaneous users seems to require not only the mental gymnastics of interpretative flexibility, but physical skills as well.[82] Thus it could be said in 1889 that "it must be understood that the safety bicycle is far more difficult to ride than the one of the ordinary type."[83] Evidently closure involves more than a psychological gestalt switch. The irreversibility aspect of closure may seem to induce a static element in the description of technical change. This is not necessary, however, because we have the stabilization process to highlight the continuous character of technical change. The combination of stabilization and closure processes makes it understandable that technical change is a continuous process, although not one that occurs at equal rates at every point in time; it is more like a punctuated evolution.[84] In this way the concepts "closure" and "stabilization" are especially important in making the SCOT framework meet the change/continuity requirement introduced in the first chapter.

2.10 The Safety Bicycle

The pneumatic tire made the scales tip in favor of the safety bicycle. As Gwen Raverat, a granddaughter of Charles Darwin, remembered in her autobiographical sketches:

> Then, one day after lunch, my father said he had just seen a new kind of tyre, filled up with air, and he thought it might be a success. And soon after that everyone had bicycles, ladies and all. (Raverat, 1952: 238)

The high-wheeled bicycle was less frequently called Ordinary, and received instead the nickname of Penny-farthing. One of the last Penny-farthings was a racing machine designed by Rudge-Whitworth in 1892. It had air tires, which shows to how large an extent the air tire was identified with making high speed possible, instead of (only) with serving as an antivibration device. The chain drive of the low-wheeler was thought to waste power; hence the designers' efforts to further develop the scheme of the high-wheeled bicycle (Caunter, 1958: 18).

The victory of the safety over the Penny-farthing did not come about without active opposition of the proponents of the high-wheeler. The tale of Franz Schröder's fight against the low-wheeler is only an anecdote, but it typifies the views held by many confirmed Ordinary riders. When the first low-wheeled Rover appeared in the small town of Coburg, Schröder planned a demonstration ride together with his wife on two high-wheelers. When a woman could ride an Ordinary bicycle, he reasoned, there would be no reason to opt for the low-wheeler. By this time, Schröder also had a commercial interest, as he had acquired the dealership of Bayliss, Thomas & Co. and was now trying to sell Ordinaries. He "publicly" announced his Sunday tour through the barber shop conversation network, and when the Schröders departed, many faces peeped through the curtains, more horrified than impressed by such an indecent sight (Timm, 1984: 120–122). The result was disappointing, and Schröder's competitor sold his Safeties. Then it was tried, also in Coburg, to end the controversy between proponents of the low-wheeled and the high-wheeled bicycles through races. In the first race, for which the bets were 1 : 14 in favor of Schröder on his high-wheeler, he won convincingly. The next week he sold three high-wheelers. Then he started to write a book in which he argued that cycling led the way to "Cyclisation": the high-wheeler obliged its rider to watch the road carefully; all brooding disappeared and instead the senses of sight, hearing, and touch were sharpened. Only riding an Ordinary bicycle really was bicycling. The Ordinary was a "sense-sharpening machine" that constituted the essential feeling of life: moving, with great pleasure, but continuously in danger of falling. And finally, it was the ultimate aesthetic experience: "still und bewegt" (andante con moto) (Timm, 1984: 147).

The Ordinary riders were not considered conservative. On the contrary, when the bicycle society in Coburg split into a high-wheeler and a low-wheeler society, the adepts of the Ordinary were reproached for being Jakobiner and revolutionary anarchists. This splitting of the Coburg bicycle society indicated the beginning of the end for the high-wheeler in that small town. As the quote from Schröder's book notes indicated, he did not value speed most prominently. Nevertheless, the rhetorical power of the speed argument was effective for others, and the final blow against the high-wheeled bicycle in Coburg was delivered in a second race, for which the Safety dealer had hired a semiprofessional racer and a lightweight Peugeot racing bicycle on pneumatic tires. Now the bets were 1 : 12 against Schröder. When Schröder finally made it to the finish line, the public had left and the rainy Schlossplatz was empty but for his wife and his competitor, who had arrived hours earlier. No Penny-farthing was ever sold in Coburg after that.

Development of the Design of the Safety Bicycle

With the air tire, the low-wheeled bicycle was gaining a decisive advantage over the high-wheeled Penny-farthing. But this does not mean that a basic design scheme of the low-wheeler had been settled. Developments in designing the driving mechanism and the bicycle frame indicate how such a basic scheme gradually emerged. I will follow this in some detail, partly because of its interest for bicycle history per se, but also to give a background for better understanding of the safety's stabilization.

Once the scheme of the Ordinary with its direct drive was abandoned, a great variety of other driving mechanisms were designed. Forms of transmission included the lever (Coventry tricycle), epicycle gears,[85] independent ratchet (Star), pivoted lever (Facile), steel band (Otto dicycle), and front and rear chain drives. One of the debates centered on the best kind of movement for the rider's feet: up-and-down versus rotary action. However, low-wheeled bicycles had never been equipped with lever mechanisms, and after the Safety Ordinaries became obsolete, rotary action became standard. Within the domain of rotary-action drives, two different mechanisms continue to exist: the chain drive and the shaft drive (the latter now used only in motorcycles).

The chain drive has a long history. First in the form of a simple pin or stud chain, then as a roller chain, it was extensively used in driving textile machinery, for example.[86] In 1880 the bush roller chain was patented by Hans Renold, and was further improved by patents in 1891 and 1899. Here, the hollow bushes spread the load over the entire length of the rollers. A point that required particular attention was a device for adjusting the tension of the chain. At the Stanley Exhibition of 1890, the washer-and-threaded-fang device, still used today, was presented for the first time (Engineer, 1890b: 139–140). The chain drive had several problems. One was that it could not be applied to the middle of the axles, so that there was an asymmetric pull on the machine. A possible solution was to use two chains, one on each side of the bicycle, but that would amplify other problems. A second problem was that chains were inevitably affected by the dust and mud splashing up from the road. This created extra wear if they were not cleaned and lubricated frequently. Third, the chain tended to damage clothes, as "there is a growing tendency, especially among the more fashionable devotees, to ride in ordinary walking costume, and this involves damage to the nether garments by reason of their entanglement with the chain" (Engineer, 1897c: 569). One way to try to solve the last two problems was to design effective protective chain casings. A solution to all three problems was the shaft drive.

Already in 1882 a shaft and bevel gear drive designed by S. Miller (1882) was used on a tricycle and on several bicycles such as the Humber & Goddard, the Columbia, and the Acatène shaft-driven bicycles.[87] One of the disadvantages of the shaft drive was that it had relatively high friction. Various designers sought to overcome this problem through improved gearing.[88] The chainless drives were valued for their noiseless running, but it required more accurate workmanship to manufacture them than the chain drive, and consequently their price was higher (Engineer, 1898: 514). With the development of an adequate chain casing, two important benefits of the shaft-driven bicycle over the chain-driven machine had disappeared, and after 1900 almost all bicycles were driven by chain.

Some of the basic frame forms have been described already. Of these, the cross frame and the diamond frame were preponderant. In 1888 a synthesis of the two schemes was proposed, as reported by *The Engineer* (1888a: 118): "Of the two types of frame, the cross frame and the diamond frame seem to be in equal demand, while another distinct type seems to be coming into favour as a combination of the strong points of the two others." The general problem of frame design is to build a structure in which applied forces are taken up as tension and stress, not as torsion or bending. Such is the principle of the "space frame," used in bridges, tower cranes, and motorcars. In bicycle frames this is not feasible, and both the diamond frame and the cross frame offer partial solutions to the problem. In the diamond frame the main forces are taken as direct stress, even though there are bending forces in the front fork and torsion forces in the entire frame as the rider exerts pressure on the pedals. (On the high-wheeled Ordinary, the rider felt these forces through the handlebar.) The stiffness and strength of the cross frame relies almost entirely on the strength of the main tube running between front fork and rear wheel (see figure 2.25). One possible improvement within this scheme is to enlarge the cross section of the main tube. This

Figure 2.25
This bicycle of 1886 has a cross frame, which was later strengthened by adding two stays as shown in this photo. The wheels were mounted with cushion tires. Photograph courtesy of the Trustees of the Science Museum, London.

idea guided the design of some recent motorized bicycles, where the gasoline tank was incorporated in the main tube. The other possible improvement is to add further frame members, thus leaving the pure crossbar principle by making a partial triangulation. When adding such further bars or stays, the obvious advantage of the cross frame—that it was equally suitable for men and skirt-wearing women—disappeared of course.

But even with stays, the cross frame came to be considered less stiff than the diamond frame: "Makers are at last grasping the advantage to be derived from staying the frames, and few machines with the cross frame are exhibited without any stays. The most popular type of frame appears to be the diamond-shaped" (Engineer, 1889: 157–158). It was even asked whether the perfect diamond frame would not be too stiff, "thereby throwing an increased amount of vibration on the joints and connections" (Engineer, 1889: 158). By 1890, the diamond frame was evidently most popular (Engineer, 1890a: 107), but there was still discussion about what constituted the best diamond form. On the Humber of 1890, with straight tubes, a 15,000-mile journey across Europe, Asia, and the Americas was accomplished between 1890 and 1903 (Caunter, 1955: 37). This certainly provided an argument in favor of the frame with straight tubes. By 1895 this diamond frame, then known as the Humber pattern, was "the favourite the world over" (Engineer, 1896: 54), and by 1897, "finality appears to have been nearly reached in the design of frames" (Engineer, 1897c: 569).

Many more details were developed in the late 1890s, such as improved rim and hub brakes, the three-speed gear, integral butt-ended tubes, and "all-steel" frames. Nevertheless, the modern bicycle can be said to have existed since about 1897. It consisted of a diamond frame with vertical bar between saddle support and bracket, equal-sized wheels, and a chain drive on the rear wheel. Its stabilization among the relevant social group of bicycle engineers had taken eighteen years. At the beginning of this period one did not see the safety bicycle, but a wide range of bi- and tricycles and among those a rather ugly crocodile-like bicycle with a relatively low front wheel and rear chain drive (Lawson's Bicyclette). In an exhibition report in *The Engineer* in 1888, the names "ordinary bicycle" and "rear-driven safety bicycle" are used to described the various models on display. Of the "rear-driven safety bicycles" it is specified that two different types of frame, the cross frame and the diamond frame, are in equal demand (Engineer, 1888a: 118). Soon after that, the label for the low-wheeled machine is condensed to "R.D. safety bicycle"

(Engineer, 1888b: 131) and another year later it is sufficiently unambiguous to call the machine a "safety bicycle" (Engineer, 1889: 158). By 1895 the stabilization seems to have been so complete as to enable *The Engineer*'s correspondent to merely use the label "bicycle" when describing the low-wheeled machine with rear-driving chains and diamond frame (Engineer, 1895: 54). This development is nicely summed up by one of *The Engineer*'s correspondents in 1897:

> The processes of natural selection, and the survival of the fittest, have given the world the diamond frame, the rear driver, the socket steering head, and so on, and the differences between the machines of any half-dozen makers are so small that only the expert can detect them. . . . [T]he modern bicycle has been in a way crystallized out of a solution of inventors' patent devices and manufacturing processes, and little room is left . . . for change or improvement. (Engineer, 1897b: 492)

By this time, the word "safety bicycle" denoted unambiguously a low-wheeled bicycle with rear chain drive, diamond frame, and air tires. As a result of the stabilization of the artifact after 1898, one did not need to specify these details: they were part of the taken-for-granted reality of the safety bicycle.

Groups of Users

As the design of the safety bicycle stabilized, so did its use among various social groups. The status of bicyclists in Britain rose abruptly when an act of Parliament conceded that the bicycle was a carriage and therefore entitled to a place on the roads, provided a bell was rung continuously while the machine was in motion (Woodforde, 1970: 3–4). The number of cyclists grew steadily. Cycle clubs proliferated: in 1880 there were more than two hundred cycle clubs, of which seventy were based in London (Marshman, 1971). The Cyclists' Touring Club, founded in 1883, had over 20,000 members in 1886. Based on the estimate that only 2.5 percent of cyclists were members of this club (Woodforde 1970), there were 800,000 cyclists in Britain in 1886, and in 1895 more than 1.5 million. The increasing number of bicyclists formed a pressure group arguing for better road maintenance, and between 1890 and 1902 the expenditure on the main roads of England and Wales increased by 85 percent, perhaps partly due to this pressure (Woodforde, 1970: 3).

With the low-wheeled safety, more social groups started to use the bicycle. Now, not only the young and athletic rode bicycles, but "bicycling became the smart thing in Society, and the lords and ladies had their pictures in the papers, riding along in the park, in straw boater hats" (Raverat, 1952: 238). The bicycle became the accepted conveyance for getting to social and business engagements, in addition to its use for sport, racing, touring, and circulating through the parks. In fashionable circles, the bicycle became so much an object to be cherished that it was not kept in the stables or outbuildings, but housed prominently in the halls of Chelsea House, Grosvenor House, and the like. Many of these bicycles were hand-painted in bright colors.[89] As Gwen Raverat remembered:

> How my father did adore those bicycles! Such beautiful machines! They were as carefully tended as if they had been alive; every speck of dust or wet was wiped from them as soon as we came back from a ride; and at night they were all brought into the house. (Raverat, 1952: 240)

The low-wheeled Safety at last paved the way for women bicycling as well. As previously mentioned, riding a high-wheeler was considered utterly improper for a woman. But with the low-wheeled safety, the two main problems presented to women by the Ordinary bicycle, indecency and lack of safety, were solved. Also in this part of cycle history, the aristocracy took the lead.[90] Gwen Raverat remembers how her mother was probably the first woman in Cambridge to have a bicycle. Then

> bicycles gradually became the chief vehicles for ladies paying calls. They would even tuck up their trains and ride out to dinner on them. One summer evening my parents rode ten miles to dine at Six Mile Bottom; their evening clothes were arranged in cases on the handlebars; for of course you couldn't possibly dine without dressing. (Raverat, 1952: 86)

In the 1890 cycle shows an increased number of ladies' bicycles were exhibited, and it was concluded that bicycling "was becoming popular with the weaker sex" (Engineer, 1890a: 108). This trend continued, and in a report on the 1896 cycle shows, a very large number of machines for female riders was again observed (Engineer, 1896: 54). The main difference between these models and the machines for men was the absence of the top bar of the diamond frame, and a great variety of designs were tried to solve the resulting need to stiffening the frame.

Appropriate bicycle clothing became even more of an issue than it had been before, when women were riding only tricycles. Gwen Raverat remembered:

> We were then promoted to wearing baggy knickerbockers under our frocks, and over our white frilly drawers. We thought this horridly improper, but rather grand; and when a lady (whom I didn't like anyhow) asked me, privately, to lift

up my frock so that she might see the strange garments underneath, I thought what a dirty mind she had. I only once saw a woman (not, of course, a lady) in real bloomers. (Raverat, 1952: 238)

The "bloomers" referred to by Raverat were an American invention of the 1850s, consisting of Turkish pantaloons with a knee-length skirt. They were named after Amelia Bloomer, who advocated them in her magazine *Lily* (Palmer, 1958: 98). As Raverat's comments suggest, bloomers were not accepted in Britain as appropriate cycling dress. The only practical outfit that could meet with some degree of approval was the "rational dress," which had originated on the continent around 1893. It consisted of knickerbockers, long leggings, and a coat that was long enough to look feminine but would not interfere with the bicyclist's movements. But for the Mr. Hoopdrivers of the world, even this dress presented a frightful sight. When for the first time his bicycle route intersected that of the Lady in Grey: "Strange doubts possessed him as to the nature of her nether costume.... And the things were—yes!—rationals! Suddenly an impulse to bolt from the situation became clamorous" (Wells, 1896: 20–21). This sight was such an attack on his emotional and physical equilibrium that, through a rapid and complex series of maneuvers, he ended up sitting on the gravel with feet entangled in his machine.

Many were the discussions in newspaper and cycling journal columns, and the case for rational dress was even fought in court. In 1898 Lady Harberton, founder and president of the Rational Dress Society, was refused service in a coffee room. The proprietor showed her into a dirty public bar where men usually drank alone. Harberton claimed the legal right to be served and took the proprietor to court. The defense argued that there had been no discrimination on the grounds of Lady Harberton's being improperly dressed. The judge upheld that defense, but nevertheless the case played a symbolic role in the fight for women's independence and emancipation.[91] Before the end of the nineteenth century, women bicyclists and their rational dress were to be seen in the remotest parts of Britain.

Other groups became interested in the bicycle as well. The postal service, for example, made extensive use of bicycles and tricycles in the collection and delivery of letters and parcels. The military also became increasingly interested, especially stimulated by the Boer War in South Africa (1899) in which the British army used bicycles on a large scale. Several bicycles were specifically designed for use in active service. Some designers offered lightweight folding bicycles, and others incorporated a gun almost as an integral part of the frame. Substantially similar bicycles were used in the First World War.

Bicycle Industry

Since the beginning of the cycle industry, local blacksmiths and mechanics have participated in constructing small numbers of bicycles to order. The dominance of the safety bicycle did not change this situation. Paradoxically, the standardization of the diamond frame had two opposite effects on industry: it further enhanced mass production, and it strengthened the position of those small workshops.

A considerable number of local bicycle makers could offer a "homemade" product to the residents of their small village at a price somewhat lower than that of factory-produced bicycles because of their lower overhead costs. Some large companies had specialized in the manufacture of standardized components, delivering them to both bicycle factories and local workshops. Thus three classes of machines could be distinguished. First, there were the mass-produced bicycles made by bicycle factories. Only the largest of these factories manufactured all components themselves; most of them had contracted out the manufacture of saddles, tires, and the like. Second, there were bicycles made by local workshops, constructed from proprietary components made by specialized firms. And the third class of bicycles, made by special departments of factories as well as by small workshops, was known as "de luxe" machines, produced without much regard for costs (Caunter, 1955: 43–44).

The dominance of the safety bicycle further stimulated the mass production of bicycles, which had begun in the days of the Ordinary. In the U.S. bicycle industry the manufacturing of interchangeable parts received more attention than in Britain (Hounshell, 1984: 193). The U.S. industry boomed,[92] and it engaged the British and industry in a struggle for the cycle business in all the markets of the world (Engineer, 1897a: 403). The British cycle industry was rather surprised at the growing international competition, for example in the case of a Dutch bicycle exhibited at the cycle show: "This, by the way, is the first time a firm of foreign manufacturers has exhibited at this annual exhibition, and Holland is certainly the last country in the world from which we expected competition" (Engineer, 1890b: 138). Apart from the United States, the rest of the world was considered a kind of industrial backwater.[93]

Chiefly because of the Safety bicycle, a cycling boom occurred in Britain 1895–1896. The increased demand could not be met by the existing manufacturers, and financiers and entrepreneurs were attracted to the industry. Some of the larger and better-known firms were purchased by these financiers and refloated for enormous sums, far above their previous value.[94] Even a shipbuilding company was reported to enter the cycle trade (Engineer, 1897a: 403). The unprecedented demand for bicycles proved to be a little unwarranted, as most dealers ordered more bicycles than they needed because they thought that only in that way they would obtain the required number in time. When the cycle season passed, many orders were cancelled. As a result of the floatations, many companies suffered overcapitalization. Large amounts had been paid for good will, patents, and other nontangibles. In 1896–1897, a great cycle slump swept over Britain (Grew, 1921: 71–72). Two or three lean years followed, ending only when the industry started to manufacture motorcycles and cars on a larger scale.

2.11 Conclusion

The leading historical question in this chapter is: How can we understand the role of the high-wheeled Ordinary bicycle in relation to its low-wheeled ancestors and successors? In setting out the problem, I suggested that this was a technical detour in bicycle history. The technologies needed to turn the 1860 low-wheelers into 1880 low-wheelers, such as chain and gear drives, were already available in the 1860s. This term "detour," however, turned out to be a misnomer. To use that term, one should assume that the bicycles of the 1880s were unambiguously better than previous artifacts and that every development that did not lead directly to this final result must be an aberration, a detour from the right path. I have shown, however, that the Ordinary bicycle can be interpreted as having been two things at the same time: a comfortable, classy, well-working artifact, and a dangerous, accident-prone, and thus nonworking machine.

The central conceptual program for this chapter necessitated the development of a descriptive model; such a model, I argued in chapter 1, should allow us to undertake case studies whose descriptions are "thick" enough to allow us to grasp the complexity of technical development, while still allowing for intercase comparisons. Such comparisons would enable us to make generalizations on the basis of several case studies, thus working toward a theory of sociotechnical change. For such a theory I formulated four requirements (see table 1.1).

To address these two issue—one historical and particular and the other theoretical—I used the same structure as for the book as a whole

Figure 2.26
The chapter follows a staircase-like argument.

(see figure 2.26). We moved along the chapter as if climbing a staircase, each step taking us a little higher and serving as a bridge to the next. On each step, different elements were introduced that were carried along and put to use on later steps. Thus I started on the ground floor with a sketch of the prehistory of the bicycle, merely setting the stage. The first step then provided a detailed history of the early stages of the Ordinary's history. Looking around in Coventry, we saw how the first bicycle producers established themselves, building on existing engineering practice and helped by the particular economic circumstances of the time. We saw Hillman and Starley make their Ariel and launch it with the first record-setting bicycle ride. Then we followed the development of the high-wheeled bicycle by describing its users, the "men of means and nerve," as well as those who could or would not use the Ordinary. I tried to do this in such a way, by providing enough empirical detail on the social groups involved, that we could reach the next step. In this step the concept of "relevant social group" was introduced. The central argument here was that groups relevant for the actors are also relevant for the analyst. This suggested the next step: to describe the "technical content" of artifacts through the eyes of the relevant social groups. To jump to that step, however, one small half-step was taken: I suggested that a focus on problems and solutions would be helpful in making such descriptions of artifacts. Armed with this conceptual apparatus, on the fifth step of the staircase we explored various solutions that were developed to solve the Ordinary's problems: tricycles, safety Ordinaries, and othcr radically different machines. This was a fairly straightforward step, once we had come as far as having relevant social groups and the focus on their problems and solutions as a vantage point. To move further proved difficult, however—how could we understand the coming of the safety bicycle? We climbed onto another step, where I introduced the concept of "interpretative flexibility." This was done by capitalizing on previous steps, especially on the decision to describe artifacts as constituted by relevant social groups. Armed with this concept, I deconstructed the air tire on the seventh step into an antivibration device and a speed-enhancing device, which resulted in a specific picture of the competition between safety bicycle and Ordinary. To reach the last level, the concepts of "closure" and "stabilization" were developed on the eighth step. Finally, having climbed all the way, we were able to give an account of the social construction of the safety bicycle.

Would it have been possible to tell this historical story without the conceptual intermezzi? Would it have been possible to develop the conceptual framework without the detailed empirical studies? I think not. Each step, empirical or conceptual, builds on the previous ones. Of course, each step also has its own small story to tell or little argument to develop; thus I have included some historical details just because they are interesting in terms of the history of the bicycle, and I have included some methodological discussions just because they link this framework to debates about a theoretical basis for technology studies. To stretch the metaphor to its limits: The staircase was not meant to be a narrow one that compelled you to move on, haunted by claustrophobia. I hope it was more of a flight of broad steps, on each of which you could move around and dwell for some time, depending on your specific interest in the views offered from there.

Where have we landed, on top of these stairs? How does the result compare to what was formulated as the aim of this book—to work toward a theory of sociotechnical change? The social constructivist descriptive model developed in this chapter meets at this moment two of the four requirements for a theory of sociotechnical change. The focus on relevant social groups and the concept of interpretative flexibility will ensure that the model meets the requirement of symmetry, while the concepts of closure and stabilization allow it to meet the change/continuity requirement. Thus the agenda is clear: We now want to use this descriptive model to generalize beyond the confines of one case. To do that, I shall now turn to the case of Bakelite. In the course of this case study I will introduce two further concepts that will help us meet the remaining requirements for a theory of technological development.

3

The Fourth Kingdom: The Social Construction of Bakelite

3.1 Introduction

"God said 'Let Baekeland be' and all was plastic." It is hardly possible to conceive of a technology that seems, at first blush, more contradictory to the social constructivist claims of chapter 2 than the invention of the first truly synthetic plastic by Leo Henricus Arthur Baekeland in 1907.[1] In standard accounts the search for a synthetic plastic material is described as lasting some forty years, a period during which numerous chemists failed, and that search climaxed with a creative flash in Baekeland's private laboratory.[2] Neither relevant social groups nor social construction processes figure in these accounts; it was, rather, the solitary individual Baekeland who created this invention, Bakelite. And what an individual! Baekeland's personalality and history have every element that would tempt the student of plastics to essay a biography of this "grand duke, wizard, and bohemian"[3] rather than a sociological analysis of the case.[4]

This will be the first aim of the chapter: to show that even in the case of "an individual inventor," a social constructivist analysis yields fruitful results. How can we understand the social construction of Bakelite despite the obvious individual achievements of its inventor? Or more specifically, how did Baekeland succeed in making a synthetic plastic after numerous chemists with access to the same resources had failed? To answer this question adequately, I will need to go beyond the descriptive model introduced in chapter 2. There, only the loose components of a social constructivist analysis of technology were provided; components that need to be combined into a coherent conceptual machinery before they can do real work. As one needs a diamond frame to make a bicycle out of the wheels, chain, steering mechanism, and saddle, so we need additional concepts to build a theory with the relevant social groups, interpretative flexibility, closure, and stabilization. It will prove crucial in making this jump from the SCOT descriptive model to an explanatory scheme that can account for Bakelite's social construction to forge a link between the individual actor's thinking and acting and the social dimensions of the SCOT account. This is the second aim of this chapter: to introduce the concepts of "technological frame" and "inclusion," which relate the interactions of individual actors to the social processes that form relevant social groups. To develop this theoretical framework, it will be necessary to give a detailed account of the relevant social groups in which Baekeland was involved. And this is the third aim of the chapter: to present a comprehensive historical study of the first commercial plastic, its inventor, and the relevant technical, scientific, and cultural context.

Bakelite merits this attention even aside from the methodological argument of this book. As I hope to show, it represents a significant aspect of the foundation of one of the most important science-based industries in the United States and the world, and the man Baekeland is an equally intriguing combination of engineer-scientist, immigrant-American, manager-financier, and sailor-automobilist. This study is, therefore, also inspired by an orthodox "inventor's biography" interest.

Leo Hendrik Baekeland was born on 14 November 1863, in the old Flemish city of Ghent. His "poor but honest"[5] parents made great sacrifices so that their son could receive a splendid education, and Leo did not disappoint them. He graduated with honors from the Ghent Municipal Technical School, was awarded a scholarship by the City of Ghent to study chemistry, supported himself by serving as bottle-washer for the chemistry professor, acquired a Ph.D. summa cum laude at the age of twenty-one, was appointed associate professor of chemistry in 1889, and made a good marriage with Céline Swarts, the daughter of the head of his department. And this is only the beginning of what was to become a truly American-dream career. Baekeland won a traveling scholarship, visited the United States, invented a photographic paper that he sold for $1 million to George Eastman, built a private laboratory, and invented the first synthetic plastic, which brought him fame as "the father of plastics" and a great number of academic awards. In the meantime, he had been naturalized as an American and "tried to live up to all that this involves."[6] He served, for example, as a member of the U.S. Naval Consulting Board and on numerous other committees. Nevertheless, "Baekeland liked simplicity. He rose early and retired early ... and was usually at work before other members of his staff" (Kettering, 1946: 292).

There are good reasons for attributing to Bakelite the glorious role of guidepost by which mankind entered the "Fourth Kingdom" of synthetic materials.[7] Bakelite was the trademark for a molding material that Baekeland patented in 1907. The features that made Bakelite famous and that pointed out the broader possibilities of synthetic materials were its being almost completely insoluble, infusible, and unaffected by other chemicals. It was, moreover, an excellent insulator for electricity and heat, and it was also very hard. It provided a new, versatile, and relatively cheap molding material. Its main drawback was that it was neither elastic nor flexible, and was very brittle. In modern chemical terms, it would be characterized as a thermo-hardening polymeric. A condensation product of phenol and formaldehyde, Bakelite was truly synthetic, whereas the cellulose for a semisynthetic molding material such as Celluloid was produced from natural raw materials. Although it has now been superseded in most applications by newer synthetic materials, Bakelite is still used today—for example, in my clarinet mouthpiece and the black handles of my high-tech stainless steel cooking pans.[8]

This chapter has a staircase structure similar to that of chapter 2 (figure 3.1). We depart from the "platform" we had reached by the end of the previous chapter, in the sense that I shall assume that the elements of the SCOT descriptional model are readily available. The first three empirical steps describe the cultural, scientific, and industrial background of the case. The earliest molding materials, which were all of natural or seminatural origin, determined to an important degree the cultural context of chemical work in this field. Celluloid set the industrial stage for synthetic plastics research and development; and early chemical work connected to the reaction of phenol and formaldehyde provided the scientific context. Against this three-colored background, the agenda for work on plastics was set by the fire hazard of Celluloid. Many chemists sought to produce a substitute for Celluloid by controlling the phenol-formaldehyde reaction, but all failed. At the first conceptual step the concept of "technological frame" is introduced to link the thinking and actions of individual actors to the social processes constituting the relevant social groups. To make this concept work for the Bakelite case, we then need to sketch Baekeland's biographical background. Thus the sixth step presents his earlier research on photographic paper and on the electrochemical Hooker cell. Baekeland, and any other actor, can only be understood as belonging to several different relevant social groups at the same time. To capture this aspect, which I will argue causes much of the dynamics of technological change, the concept of "inclusion" is introduced in the seventh section. In the empirical steps that follow, these concepts of "technological frame" and "inclusion" are used to describe

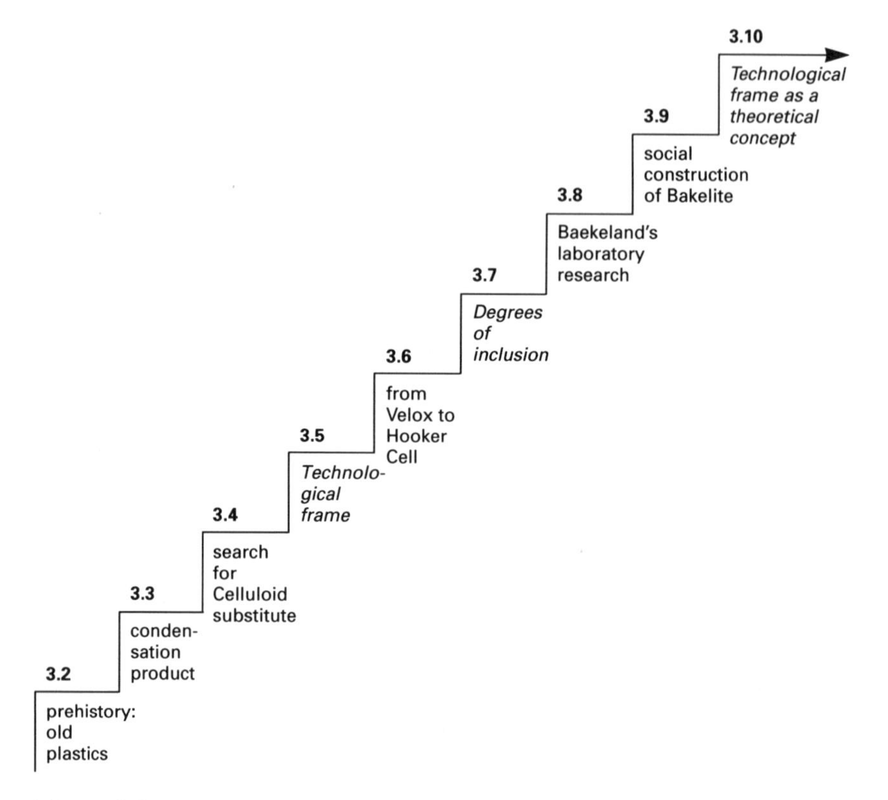

Figure 3.1
The structure of the argument in chapter 3.

Baekeland's laboratory research and the further social construction of Bakelite. The last, again conceptual, step is devoted to a more general discussion of the theoretical framework introduced along the way.

In addition to published primary sources such as patents, articles, and published historical studies, of which Robert Friedel's (1983) account of "The Making and Selling of Celluloid" is clearly the most outstanding scholarly contribution, I have made use of the manuscripts collected in the Baekeland papers at the National Museum of American History.[9]

3.2 Prehistory of Synthetic Plastics

Old Plastics and New Relevant Social Groups

The history of human use of plastic materials is as long as the history of humanity itself.[10] The Egyptians used resin, a plastic of vegetable origin, to varnish their sarcophagi, and the Greeks made jewelry out of amber, a yellowish fossil resin. Other applications were the making of buttons and combs out of horn, cow hooves, and tortoiseshells. Hooves, for example, were ground and then colored with a water-based dye and compression-molded in hot dies. Tortoiseshell was first softened by boiling it in water, then it was stretched out and often laminated together to gain the desired thickness. Final shaping was done in iron dies.[11] Although these applications were more mundane than varnishing sarcophagi, they were still quite expensive and were produced for a small elite. These buyers of luxury and fancy consumer goods form the first relevant social group we encounter in this case study. The two areas of application I have mentioned—a basis for varnishes and a molding material for small solid objects—have provided and continue to provide important markets for the plastics industry. Some of the natural plastics, such as shellac,[12] could be used for both purposes, while others, like rubber, are used for only one.

Until the mid-nineteenth century, the use of plastics had been confined to luxury and fancy goods, ranging from shellac-lacquered scent boxes to ivory jewelry. However, the vulcanization of rubber created new markets. Natural or "india" rubber, as it is retrieved from the bark of many tropical trees (most notably the *Hevea brasiliensis*), cannot withstand temperatures above 60°C (when it becomes soft and sticky) or below 10°C (when it becomes hard and brittle). In the vulcanization process rubber is heated with sulphur, which renders it more flexible and durable over a wide range of temperatures.[13] By varying the amount of sulphur, the flexibility and hardness of the material can be controlled—more sulphur makes it harder and less flexible. Vulcanization made rubber into a versatile material, suitable for many new applications.

This brought new relevant social groups into the picture. In the second half of the nineteenth century rubber was, for example, increasingly used for electrical insulation. The range of applications especially widened with the advent of "hard rubber," also known as "vulcanite" or "ebonite," which was patented by Nelson Goodyear in 1851. It contained a radically larger amount of sulphur, some 30 percent, and opened up possibilities for several new industrial purposes for which none of the older natural plastics had been employed. Apart from its use as an electrical insulating material, "hard rubber" was used for the internal coating of chemical apparatus and car battery storage compartments, and for the manufacture of surgical instruments and artificial teeth. The Crystal Palace Exhibition in 1851 in London featured a room full of rubber furniture, including ebonite chairs and bureaus.

In the slipstream of new technical applications for vulcanized rubber, shellac was also developed into a material that could be used for more than merely making luxury goods; the manufacture of phonograph records is probably its best-known use, but electrical insulating parts was another application. The new relevant social groups on the plastics market were primarily interested in molding materials for mass production of technical parts and equipment. The black color of the aptly named "ebonite" made it less suitable for producing the luxury goods for which tortoiseshell and ivory had been used. However, the partition between luxury goods and industrial applications and the associated materials was not complete: the comb industry did find "hard rubber" an attractive material because it allowed combs to be produced in a less wasteful manner than the older materials (Friedel, 1983: 26). Nevertheless, the conclusion can be drawn that although plastic materials had previously been restricted to the jewelry-wearing upper classes, they now found favor among new relevant social groups. This, however, created a problem.

New groups showed a great interest in employing these novel molding materials, which did not have the limitations of the traditional materials such as steel, glass, wood, and ceramics. Their interest pointed to the need for materials that had not the coldness, weight, and chemical activity of metals; nor the fragility and costs of ceramics and glass; nor the transience of natural materials; but that had artistic appeal and offered the possibility for coloring. Although none of the industrial natural plastics then available could meet all these requirements, there clearly was a rapidly growing demand for such molding materials. And consequently, the exotic location of the sources of shellac and rubber led several chemists and industrialists to perceive an imminent scarcity of natural plastics. "We [are] exhausting the supplies of india rubber and gutta percha, the demand of which [is] unlimited, but the supply not so," remarked the chairman of a meeting of the Royal Society of Arts in 1865.[14] At this meeting, Alexander Parkes had given a lecture on his new plastic material Parkesine, which was the first of a series of variants produced in an attempt to solve this scarcity problem by trying to modify nitrocellulose.

Nitrocellulose could be produced rather cheaply from paper, wood fibers, or rags. It is soluble in ether-alcohol and it can be very combustible, depending on the extent to which it is nitrated. The lesser nitrated forms (known as pyroxylin) are quite soluble and less flammable, whereas the highly nitrated forms (known as guncotton) are very explosive and hardly soluble. Its importance as an explosive substance immediately caught the world's attention when the Swiss chemist Christian Friedrich Schönbein found a commercially viable production process in 1846. In a letter to Faraday, Schönbein reports: "I am enabled to prepare in any quantity a matter, which next to gunpowder, must be regarded as the most combustible substance known.... I think it might advantageously be used as a powerful means of defence or attack. Shall I offer it to your government?"[15] The explosive character of nitrocellulose was immediately investigated by special military research commissions in most countries (Friedel, 1983). Somewhat later, several chemists and inventors also started to explore the possibilities of dissolving nitrated cellulose in a mixture of alcohol and ether. The "collodion," as this solution was called, was a clear fluid with the consistency of syrup that, when poured out and allowed to dry, resulted in a transparent film. Several applications—such as wound plaster, a means to render fabrics waterproof, and a basis layer for photosensitive materials—were successfully developed (Kaufman, 1963; Friedel, 1983).

Parkesine

Alexander Parkes is revered by the British plastics industry as one of its founding fathers.[16] Parkes was the first to attempt to manufacture "various articles having properties analogous to those made from india-rubber or gutta percha."[17] His material, dubbed Parkesine after its inventor, was made from pyroxylin, oil, and sometimes additional components. Solvents play a prominent role as an additional patent shows: "In manufacturing parkesine some difficulties are experienced in consequence of the high volatility of the solvents heretofore employed. According to my present Invention I employ as solvents of the pyroxyline in this manufacture nitro benzole, aniline, and glacial acid...." (Parkes, 1865a: 1–2). After mixing the various components, the mass was spread out to cover an object or pressed into a mold and then left to dry. No heat was applied, and Parkes explicitly drew an analogy with the "cold vulcanization" process for rubber that he had patented in 1846. Additionally, in his 1865 patent and in his 1865 paper for the Royal Society of Arts, Parkes (1865a: 2) mentioned that "I also according to my Invention render the ordinary volatile solvents more suitable for use by the addition of camphor; by this means I obtain to some extent the same advantage as by the use of a less volatile solvent." This employment of camphor, which later would play an important role in a patent litigation, "exercises an advantageous influence on the dissolved pyroxyline, and renders it possible to make sheets, &c., with greater facility and more uniform texture, as it controls the contractile properties of the dissolved pyroxyline" (Parkes, 1865b: 83). Parkes also addressed the problem of its combustibility, and claimed "that it could be made almost uninflammable" by adding specific substances (Parkes, 1865b: 85).

Parkes's aims for the new material were not clear. In his 1855 patent he said that "my object is to employ collodion or its compounds for manufacturing purposes generally." He envisaged his Parkesine not only as a substitute for rubberlike materials, but also as "suitable for bookbinding, button manufacture, and other applications where a hard, strong, brilliant material is required" (Parkes, 1855: 4). Thus Parkes did not distinguish clearly between the different uses to which Parkesine could be put, and in his paper for the Royal Society of Arts (1865b: 81) he claimed that he had "succeeded in producing a substance partaking in a large degree of the properties of ivory, tortoise-shell, horn, hard wood, india rubber, gutta percha, &c., and which will, . . . to a considerable extent, replace such materials, being capable of being worked with the same facility as metals and wood." The mentioning of ivory and tortoiseshell besides rubber and gutta-percha among the materials to be replaced supports the impression that he was aiming at the manufacture of luxury goods as well as the mass production of industrial components.

In the beginning, Parkes seemed to stress Parkesine's use for the production of fancy articles. For example, at the 1862 World Exhibition, he presented "medallions, buttons, combs, knife & handles, pierced and fret work, inlaid work, card cases, boxes, pens and penholders."[18] However, by 1866, when Parkes tried to persuade investors to put capital into a newly incorporated Parkesine Company, the prospectus hardly mentioned Parkesine as a beautiful material for making "works of art." Instead, it stressed applications such as its use for making carding, roving, and spinning rollers; for insulating telegraphic wires; for the manufacture of tubing; and as varnish and coating for iron ships (Friedel, 1983). Parallel to this shift in focus away from the fancy applications, Parkes tried to make his material as cheap as possible (Friedel, 1983; Dubois, 1972). We may now, retrospectively, consider this as a straightforward and obvious move to serve a larger market. For contemporaries this was not so clear, however. Indeed, we will see that the invention of celluloid a few years later was inspired by the demand for an ivory substitute for nonindustrial purposes. This shift proved inadequate to secure the involvement of the new social groups of users. His eagerness to show the applicability of Parkesine to a wide variety of purposes meant that he placed less emphasis on finding a dependable chemical production process for at least one specific form of Parkesine. Thus the plastic was not produced with a constant quality, and a great number of the items sold by the new company were returned as unacceptable because of shrinkage, twisting, and distortion (Worden, 1911). In 1868 the Parkesine Company was liquidated.

Ivoride

A second variant of a nitrocellulose plastic was closely linked to Parkesine. The manager of the Parkesine Company, Daniel Spill, attributed the failure of Parkesine to the fact that they had not made their material white enough. If it could be made whiter, Parkesine would appear as a more credible substitute for ivory. In 1869 he founded another company, and with only minor changes in the manufacturing process he continued the production of what by now was called Xylonite. This venture fared no better than the previous one and was abandoned in December 1874. Spill had an unshakable faith in his material, however, and established yet another company in 1875. This time he did succeed in finding a small but rather stable market for what he now called Ivoride (Kaufman, 1963).

The emphasis placed by Spill on the need to make a credible substitute for ivory deserves more attention, because it is an idée fixe common to all of the early plastics researchers. Parkes said that "it was an important feature with me to make an artificial ivory."[19] About John Wesley Hyatt, the inventor of Celluloid, Baekeland observed that "celluloid and the celluloid industry with all that it implies, after all meant merely a big parenthesis in the life of Hyatt in his quest of a perfect billiard ball which would replace the expensive ivory article" (Baekeland, 1914a: 90). And is was Baekeland himself who at last succeeded in making an acceptable substitute for the ivory billiard ball. Robert Friedel (1983) made a detailed analysis of the scarcity of ivory in the second half of the nineteenth century. He concluded that, although the widely perceived "ivory problem" in the nineteenth century was largely a myth, there might well have been a "billiard ball problem." Billiards was one of the few important indoor sports of the time, and billiard balls made up the largest source of demand for ivory. The extremely high quality of ivory needed to make billiard balls (sometimes no more than one tusk of fifty was suitable) could well have added to the potential scarcity. Indeed, the New York billiard ball producers Phelan and Collander offered a $10,000 award to "any inventive genius [who] would discover a substitute for ivory, possessing those qualities which make it valuable to the billiard player."[20]

Celluloid

The Parkesine and Ivoride stories primarily serve the purpose of helping me to sketch the cultural background of plastics development; the story of the third seminatural plastic needs to be recounted because it shaped most of the relevant social groups that were to be the main actors on the plastics stage.

Celluloid, as the third variant of solution to the scarcity problem of natural plastics, was developed by John Wesley Hyatt in Albany, New York. As the popular story goes, Hyatt's research was triggered by Phelan and Collander's offer of a $10,000 prize for a substitute for ivory. Hyatt first tried several well-known plastic compositions, such as wood fiber with shellac. Then he also started to experiment with collodion, using this to coat billiard balls made of another substance. Apart from problems of unequal shrinkage and brittleness, this also revealed another problem of using guncotton, as Hyatt remembered in 1914:

> In order to secure strength and beauty, only coloring pigments were added, and in the least quantity; consequently a lighted cigar applied would at once result in a serious flame, and occasionally the violent contact of the balls would produce a mild explosion like a percussion guncap. We had a letter from a billiard saloon proprietor in Colorado, mentioning this fact and saying he did not care so much about it, but that instantly every man in the room pulled his gun. (Hyatt, 1914: 159)

Although this experimentation did not result in a satisfactory substitute for ivory, an important consequence was Hyatt's acquisition of familiarity with processes for molding plastics under heat and pressure. This experience made Hyatt aware of the problems of liquid collodion solutions such as the ones Parkes and Spill had used: the drying process inevitably caused shrinkage, which cause difficulties when molding solid objects. Looking back in 1914, Hyatt recounted:

> From my earliest experiments in nitrocellulose, incited by accidentally finding a dried bit of collodion the size and thickness of my thumb nail, and by my very earnest efforts to find a substitute for ivory billiard balls, it was apparent that a semiliquid solution of nitrocellulose, three-fourths of the bulk of which was a volatile liquid and the final solid from which was less than one-fourth the mass of the original mixture, was far from being adapted to the manufacture of solid articles.... (Hyatt, 1914: 158)

After Hyatt had taken out several patents describing such processes, in 1870 a patent was issued that referred to "the use of finely committed camphor-gum mixed with pyroxyline-pulp..., [and] rendered a solvent thereof by the application of heat" (Hyatt, 1870). Together with his brother, Isaiah S. Hyatt, Hyatt founded the Albany Dental Plate Company in 1870. They advertised "a newly-invented and patented material for Dental Plates or bases for artificial teeth, that cannot fail to delight every dentist who desires a better material for the purpose than hard rubber" (*The Dental Cosmos* 13 (1871); cited in Friedel, 1979: 53). The dental plates did have various imperfections: some had a strong camphor taste; some became soft in the mouth (sufficiently so for the teeth to become loose); and plates were found to warp after having been adjusted to the patient's mouth (Friedel, 1983). Although these dental plates were far from satisfactory, the concerted effort to produce a material with specific, consistent qualities resulted in the Hyatt brothers forming the Celluloid Manufacturing Company to produce Celluloid in semi-finished form (rods, sheets, tubes, etc.). During the period 1872–1880, the Hyatts granted licenses to different companies for the production of Celluloid consumer goods, with each company devoting itself to a narrowly defined market (Friedel, 1983).

The stabilization of (one of' the two rival plastics—Ivoride/Xylonite and Celluloid—in large part took place in a patent controversy. This controversy, between Spill and Hyatt in the United States, concerned who had priority in the "invention" of the use of camphor in the production of a plastic from nitrocellulose. An analysis of the debate nicely illustrates how actors, in the heat of a controversy, deconstruct and subsequently construct facts and artifacts that for later generations come to have such solidity and unambiguity that it is hard to think of them as not being "caused by nature." Throughout this book patents play an important role. From a historical-sociological point of view, the recent work by Bowker is a central resource, but earlier historical studies have also contributed to an understanding of patents' roles in technology development and industrial science.[21] Here I will be primarily concerned with patents as a means for tracing important aspects of the way in which plastics were viewed and handled; later the role of patents in building industrial companies will be more prominent.

The interpretative flexibility of the artifact Celluloid can be easily demonstrated by following the litigation process between Spill and Hyatt. For Spill, Celluloid meant a mixture of nitrocellulose and camphor that, although prepared in a slightly different way, was essentially identical to his Xylonite or Ivoride. For Hyatt, Celluloid and other nitrocellulose plastics were crucially different, and this difference was to be found in the fabrication process.

Hyatt used, he said, a "solid solution" instead of a liquid solution of nitrocellulose and camphor. This has become the accepted truth about Hyatt's Celluloid invention: he used camphor to make a "solid solution." But it was only many years after the initial patent application, when he was involved in patent litigation trials, that Hyatt used the term "solid solution" to describe the material produced in one of the first stages of the production process. He used this term to create a crucial difference between earlier nitrocellulose plastics and his Celluloid. Hyatt argued that to avoid the shrinkage caused by the drying of Parkesine, "I must initially produce a solid solution by mechanical means" (Hyatt, 1914: 158). However, this "solid solution" appeared at the time to be more of a moist mixture, and the Hyatt brothers seemed to view the making of celluloid as a drying process: "The product is a solid, about the consistency of sole-leather, but which substantially becomes as hard as horn or bone by the evaporation of the camphor" (Hyatt, 1870). The difference with Parkes and Spill is emphasized exactly on this point by, for example, Chandler (1914a, 1914b).

The patent dispute between Spill and Hyatt was resolved by Samuel Blatchford, at that time a justice of the U.S. Supreme Court, and "the most highly regarded patent judge of his time" (Friedel, 1983: 132). He decided on 21 August 1884 that neither Spill nor Hyatt should be named as the inventor of a camphor-nitrocellulose plastic, as Parkes had already covered that combination of substances in his patents. In effect, this meant a victory for Hyatt, because the judge's decision denied to Spill the novelty of using camphor and thus nullified his grounds for litigation against Hyatt. The Celluloid Manufacturing Company succeeded in putting itself on a firm financial base. The increasing stabilization of Celluloid can be traced by following its use as an intermediate material positioned between cheap but unattractive plastics such as rubber and luxurious materials such as ivory. For example, the advent of Celluloid brought combs, cuffs, and collars within reach of social groups that had been unable to afford such luxurious articles. (Luxurious, because the original cotton cuffs and collars had to be washed every day, and this was such a laborious job that servants, it seems, were needed to do it; see figure 3.2.)

This patent litigation trial has had another important effect: It further shaped the terms in which chemists thought about celluloid, its production process, and possible strategies for improving the process. Much of this contextualization revolved around the crucial role of the "solvent" camphor, which made the pyroxyline manageable. In his patents, How-

Figure 3.2
Advertisement for Celluloid. The advertisers often used anti-Chinese sentiments in the promotion of Celluloid cuffs and collars. Photograph courtesy of the Warshaw Collection of Business Americana, National Museum of American History, Smithsonian Institution, Washington, D.C.

ever, Hyatt himself did not mention camphor as a solvent—only as an additive. His later use of the term "solution" during the litigation trials must have added to the perceived importance of the role of solvents in celluloid production. I will return to this point.

Linked to the differences in how these two industrial chemists conceptualized the production processes of their plastics were differences in goals and in the resulting lines of development. Spill valued his plastic primarily as a substitute for expensive natural plastics, as is indicated by the name Ivoride and his emphasis on the need for the material to be white. Consequently, mass production by molding was not his first priority. For Hyatt, the goal of constructing a material that could be used to produce large numbers of narrowly defined products of consistent quality led inevitably to a focus on the production process and especially on the molding characteristics of the material.

Having described some of the processes that led to the eventual stabilization of Celluloid, the next step in the descriptive model is to identify problems associated with this artifact. One problem with Celluloid, in the view of certain important social groups, was never solved. This was its flammability. In 1875 a *New York Times* editorial entitled "Explosive Teeth" summarized the widespread concern about the safety of Celluloid (Friedel, 1983: 96). Most accounts of the dangers of Celluloid addressed its flammability rather than the risk of explosion. But "humorous" stories kept popping up, like the one entitled "Presents" in the *New York Times*, which described a lover who gave his sweetheart Celluloid clothing, only to see the items blown to pieces when he lit his cigar.[22] The Newark Celluloid factory had thirty-nine fires in thirty-six years, resulting in at least nine deaths and thirty-nine injuries.[23] Several reports were published in *Scientific American* on Celluloid's flammability with, however, ambiguous conclusions (*Scientific American*, 1892a,b).[24]

As I showed in the case of the development of the bicycle, problems seldom have equal pertinence for all social groups. Thus data about fires and other accidents caused by explosions where Celluloid was said to be involved were interpreted quite differently by different people (Kaufman, 1963). For example, any trained chemist would have thought it the height of folly to heat nitrocellulose under pressure, knowing its "dynamite" character. A chemistry professor, visiting Hyatt's factory, warned that if a little too much heat were applied, the substance would inevitably destroy them—together with the building and the adjacent property! Although Hyatt was skeptical, he was worried enough to put the proposition to the test:

> The following day between 12 and 1, when all were out, I rigged up for a four inch plank used as a vice-bench, braced it between the floor and ceiling, between the hydraulic press and the hand pump, intending it to shield me from possible harm. I then prepared the mold, heating it to about 500°F, knowing it would certainly ignite the nitrocellulose and camphor, and thinking I would abide by the result. The gases hissed sharply out through the joints of the mold, filling the room with the pungent smoke. The mold, press, building and contents were there, including myself, very glad that I did not know as much as the Professor. (Hyatt, 1914: 159)

However, not many users were convinced by this experiment, and national and local authorities made special safety regulations for Celluloid processing industries (Worden, 1911). Nevertheless, celluloid continued to be used, even for children's toys and cigar holders.

Throughout this forty-year period of searching for materials to replace natural plastics, there was a focus on solvents. First came Parkes's emphasis on the role of solvents in inventing Parkesine; then Parkes's efforts to tame the flame by adding specific chemicals to his solvents; then the emphasis laid by the Hyatt brothers on camphor as a key element in the invention of Celluloid; and finally, the crucial role of the "solid solution" during the patent litigation between Hyatt and Spill. Ongoing research on Celluloid, as a continuous problem-solving process—aimed at, for example, modifying the composition of celluloid and thus optimalizing the production process—overlapped with trying to find substitutes for camphor and with studying the employment of these other solvents. I shall return to this focus on solvents, as it will prove to be an important element in the explanation of the developments described in the next section.

3.3 The Phenol-Formaldehyde Condensation Product and Its Interpretative Flexibility

Evidently, Celluloid's flammability posed a problem for practically all relevant social groups. From the earliest efforts by Parkes in 1865, many "Celluloid chemists" devoted their research to solving this problem by trying to modify the Celluloid production process or by inventing substitute materials that were less flammable; in the next section I will follow these efforts in detail. In this section, our attention will shift to another strand of chemical research: experiments on the condensation reaction that occurs when phenol and formaldehyde are combined. This temporary shift in focus is necessary for understanding why Celluloid chemists were going to consider this reaction a potential candidate for producing a replacement for Celluloid.

During his studies of phenolic synthetic dyes, A. Baeyer observed that aldehydes and phenolics react with one another, and he investigated the possibilities of making dyes on the basis of these reactions. In three articles, a large number of reactions were described.[25] The purpose of this research is evident from the detailed account he gave of the coloring that occurred at various points of the reactions: "a red oxidation product is formed," "colors cotton blue," "a white precipitation," "blue flocks," "colorless." However, none of the reactions described was considered a success, and no synthetic dye was made. Consequently, there was not much reason for mentioning these details on colors, only if this reporting was simply part of the accepted culture in this relevant social group—color seemed to be one of the most important meanings of any chemical substance.

In reporting how various phenolics reacted with formaldehyde, Baeyer devoted only half a sentence to the reaction between phenol and formaldehyde: "Thus, for example, phenol gives a colorless resin"[26] (Baeyer, 1872c: 1095), and he discarded this substance, and others, because it was very difficult to crystallize. Crystallization formed part of the standard procedure for studying potential dyes because it made it possible to establish more accurately the chemical composition of the product, and that was used in designing production processes. As Baekeland summarized this research,

> the earlier investigators, like Ad. Baeyer and others, were especially on the lookout for substances of definite chemical constitution, which could be easily isolated, crystallized and purified for the development of their purely scientific work; if they obtained noncrystalline bodies, of resinous appearance, this was merely considered as a drawback, and constituted an unpleasant obstacle in their theoretical research work. (Baekeland, 1911b: 933)

It would take another sixty years, until well after the successful commercial stabilization of synthetic plastics, before Staudinger's (1926, 1961) macromolecular theory would reveal the exact nature of the chemical structure of these resinous substances.

Apart from the problem that the resinous condensation products were not analyzable, there was something in the economic context that made Baeyer's last article on the formaldehyde-phenolics reaction only of academic interest. At the time, formaldehyde could not be produced on an industrial scale; it was merely a laboratory curiosity. More than one page of his third article is used by Baeyer to describe how he produced methylene representatives of formaldehyde to carry out his research. Chemically this was considered similar to using a watery solution of "formaldehyde."[27]

Arthur Michael, an American chemist, did not mind the limited availability of formaldehyde, for he had primarily academic questions in mind. After having briefly reviewed the work of Baeyer and his pupils, Michael concluded that "absolutely nothing is known" about the constitution of these resinous compounds (Michael, 1883: 340). He wanted to study questions related to vegetable resins: "Has this resin-formation from aldehydes and phenols a connection with the formation of resins in the vegetable world? In other words, have the contents of the cells the property of forming these resins? And do they resemble in their properties those found in nature?" Thus for Michael the sticky reaction products were not "an unpleasant obstacle in [his] theoretical research work," but the very object of his research, because he hoped that the study of these synthetic resins would lead him to a better understanding of natural resins.

Michael found that mere traces of acid were enough to start the condensation reaction, and that the acid acted as catalyst: "In one experiment twenty grams of these substances were completely converted into the resin by a single drop of acid" (Michael, 1883: 341). Besides acids, alkalis could also be used as condensation agents. Finally, he observed that the intensity of the reaction depended largely on the condensation agents used. Michael's article concluded that: "the above results make it extremely probable that the formation of at least some of the resins in the vegetable world is due to aldehydes and phenols coming in contact with the contents of the cells, as both of these classes of compounds are undoubtedly among the products formed in plant-life" (Michael, 1883: 349). This statement makes clear that he was not concerned with any form of embryonic plastic, nor with a potential synthetic dyestuff, but only with an analogon to vegetable resins.

In 1886 substantive improvements were made in the production of formaldehyde, and it became possible to buy several liters of 30–40 percent formaldehyde solution. By 1888 the German firm of Mercklin & Lösekann had begun to produce formaldehyde on a commercial scale.[28] Thus, as the chemist W. Kleeberg noted at the beginning of his paper, "since this formaldehyde has become readily available, I have ... carried out some experiments on its reaction with phenolics" (Kleeberg, 1891: 284). His description of what he saw to be the condensation process when formaldehyde reacted on phenol came fairly close to what Baekeland later would report: "Under a quite strong development of heat, a rose-red gummy mass is separated off, which solidifies when cooling down. The ground product loses all of its color when boiled repeatedly in water or alkali. It is completely insoluble in alkali and is hardly soluble in all other usual solvents. When heated it carbonizes, without melting" (Kleeberg, 1891: 284).

He found it impossible to analyze the condensation product because it was not soluble and he could not purify it. Hence he turned away from the formaldehyde-phenol reaction and devoted the rest of his article to reactions between formaldehyde and other phenolics that proved easier to analyze.

It is not clear what meaning the reaction and the condensation product had for Kleeberg. He did refer to Baeyer's work, and his only explicit reason for conducting these experiments was the increased commercial availability of formaldehyde. This suggests that he was not primarily interested in academic questions of, for example, chemical structure or of the origin of vegetable resins, but that he had industrial applications in mind. Considering his quick denunciation of the condensation product when it proved impossible to analyze, these applications were probably in the realm of synthetic dyes, like Baeyer's.

Two other chemists, O. Manasse and L. Lederer, approached the "rose-red gummy mass" from a completely different perspective. Manasse and Lederer ran into the condensation reaction when they, independent of one another, tried to develop a production process for phenolalcohols. These phenolalcohols were of general interest "because of their existence in nature" (Lederer, 1894: 223), but they were also of great commercial interest "because of their close relationship to aromatic oxycarbon acids and oxyaldehydes, which play an important role in the dyestuff industry and in medical practice" (Lederer, 1894: 223–224). Until then, the production of phenolalcohols had been carried out by reduction of the respective aldehydes. This process was rather problematic because of its raw materials and its low efficiency (Manasse, 1894). Lederer also places Kleeberg's work in the context of this search for a phenolalcohol production process when he described how several efforts were made to apply the now readily available and highly active formaldehyde. However, "these efforts failed throughout: almost always 'awkward resins' were produced" (Lederer, 1894: 224). Whereas Kleeberg had used acids as condensation agents, Manasse and Lederer employed bases, and they attributed their success in controlling the condensation reaction to this difference. Manasse's speculations about the exact nature of the reaction would form Baekeland's starting point for his theoretical work on the formula of Bakelite.

The research described in this section resulted in a rather discouraging body of knowledge concerning the condensation reaction of phenol with formaldehyde. The scattered bits of knowledge on the reaction could not be integrated in Kekulé's structure theory, because that required an exact description of the chemical constitution. This did not particularly bother most chemists, however, because they assumed that the resinous substance was impure and would, once purified, provide nice, neat crystals. Thus there seemed not much point in pursuing the matter.

The previous paragraphs show the interpretative flexibility of the phenol-formaldehyde condensation product. For Baeyer and his collaborators, and probably also for Kleeberg, it was a potential synthetic dye, albeit a failed one. For Michael it presented an analogon to vegetable resins and thus provided an instrument for biochemical research. For Manasse and Lederer the condensation product was just an accidental by-product of the reaction to produce phenolalcohols—an "awkward resin" to be avoided, without any interest of its own. The fourth artifact, and the one that has figured most prominently in histories of plastics, is phenol-formaldehyde resin as an embryonic plastic, a form of "proto-Bakelite." This artifact, however, did not yet exist in the 1880s and 1890s; it would be another decade before it was constructed by chemists searching for a celluloid substitute.

We will now take up the plastics story where we left it—in the search for a substitute for Celluloid. In the course of this search, attention was drawn to the phenol-formaldehyde reaction. But first, all efforts stayed within the tradition of Celluloid research and production.

3.4 In Search of a Celluloid Substitute

There were several reasons to search for a substitute for Celluloid. As mentioned, first and foremost was its flammability, as in the case of "a lady, standing near a bright fire, [who] had one of the [celluloid] buttons of her dress ignited by the heat, whereby her dress was scorched."[29] But Celluloid was in other ways susceptible to heat, even when not catching fire. At a temperature of 100°, Celluloid softened considerably and showed slight swelling. When the temperature rose further, it decomposed, "swelled out, emitted dense fumes of camphor, and charred" (*Scientific American*, 1892b). This chemical instability, which could sometimes even occur at room temperature, added to the necessity for finding a substitute.

During research aimed at developing new plastics, chemists concentrated on trying to find other solvents that would temper the flammability of the nitrocellulose. This focus on solvents extended the tradition of thinking about nitrocellulose plastics. This preoccupation with solvents was further enhanced by the high price of natural camphor. Bonwitt (1933: 163–172) listed some 100 different chemicals with associated derivatives that had been patented as camphor substitutes.[30]

However, it proved impossible to find an adequate camphor substitute. Thus the search process for a Celluloid substitute was broadened to include the search for materials to replace the other component, nitrocellulose. For example, one class of plastics developed in this period used vegetable and animal proteins such as casein, a solid milk component, or blood proteins as substitutes for nitrocellulose. The casein plastics, first patented in Germany in 1879, were very popular and became known as *Kunsthorn* (artificial horn).[31] It is important to see that even in following this line of research, the dominant problem-solving strategy still focused on finding the most suitable solvent. Thus after having identified the possibility of using these proteins, most further research was aimed at developing different modifications of plastics by using specific solvents. In accord with Hyatt's own nonchemical background and modus operandi, this work was not chemical in any sophisticated sense of the word; rather it was a combination of mechanical material processing and molding technologies.

Such was the situation at the turn of the century. Two lines of development are important for us here. First, the stabilization of Celluloid had created a new market for plastic materials. But the features that these plastics were required to have became more and more sophisticated, and Celluloid could not fulfill all demands. The second line was the development of chemical synthesis in the nineteenth century. At this intersection of developments, several chemists started to investigate the phenol-formaldehyde reaction again, but now for the first time with the explicit objective of synthesizing a plastic material.

The English chemist Arthur Smith (1899a,b) patented a process to produce "an adequate substitute for materials such as ebonite, wood, etc., which is, because of its resistance to most chemical agents, especially suited for electrical insulation and other purposes such as cells, partition walls, boxes or other parts of electrical equipment" (Smith, 1899b: 1). He started his patent application by referring to Baeyer's research and the colorless resins he had produced by the phenol-formaldehyde reaction. Following the standard problem-solving strategy, he tried to moderate the violent reaction by using solvents like methylalcohol or amylalcohol, in which he dissolved the reacting bodies and the acid condensation agent. But the reaction was still too violent, and Smith had to use acetic paraldehyde instead of formaldehyde, although the former was much more expensive, as Baekeland (1909a) would remark ten years later. Smith poured his mixture of reacting chemicals into a mold, in which it hardened. After the initial hardening, the product "was taken from the mold and dried at 100°C" (Smith, 1899b: 1). He claimed that, by varying the amount of solvent, he could retain the material in plastic form so as to press it in different shapes. His material could be absorbed by asbestos fibers or produced as plate material. It was insoluble by cooking in acetone, chloroform, ether, or alcohol, and it did not soften under heating. Smith recommended it for applications in electrical insulation.

I want to briefly mention patents by Lederer (1902a,b) that were also aimed at producing "horn-like substances," but did not employ the combination of phenol and formaldehyde. He pressed a mixture of acetylcellulose with phenol or chloralhydrate into molds, where its initial hardening took place under moderate heat (50° to 60°C) and high pressure. Its final hardening occurred outside the mold. It is not surprising that Lederer, for whom the phenol-formaldehyde condensation product was an accidental by-product of the reaction designed to produce phenolalcohols and thus merely something to be avoided, did not share Smith's interest in modifying "the awkward resin" into something useful, even though he was trying to develop new elastic, hornlike substances. Lederer's case, however, further illustrates the urgent need to develop a synthetic plastic.

The Austrian chemist Adolf Luft followed a path very similar to Smith's efforts to soften "the awkward resin" (Luft 1902a,b). Luft also treated this phenol-formaldehyde condensation product with solvents such as "acetone, alcohol, glycerin, camphor and organic acids," but, in contrast with Smith, only after the condensation reaction had occurred. He claimed in his patent of "a process to produce plastic materials" that his material was especially suitable as a Celluloid substitute (Luft, 1902b). He paid considerable attention to retrieving the costly camphor from the process. After the treatment with weakening solvents, the "transparent and more or less plastic" mass was put into molds and dried at temperatures of about 50°C. By employing other materials as fillers, Luft could produce materials with different properties—some of them potential substitutes for ebonite.

Although I have not found any indications that Luft's phenolic resin was a commercial success, it is reported to have formed the starting point of one strand of the British plastics industry. Some of Luft's material found its way to England and into the hands of James Swinburne. This electrical engineer, who was much concerned with the quality of insulating materials, recognized its potential features and in 1904 founded the Fireproof Celluloid Syndicate to carry out research in this field. This research did not yield results in the domain of plastics, but he developed a very hard lacquer, and in 1910 the research syndicate was transformed into the Damard Lacquer Company, its name invoking the quality of the lacquer. This enterprise would merge in 1927 with two British licensees of Baekeland to form the company Bakelite Ltd. (Kaufman, 1963).

Several more chemical patents, dated between 1902 and 1905, described processes to produce phenol-formaldehyde plastics.[32] All of these patents treated solvents as the crucial element in harnessing the violent condensation reaction. Also, all of them considered the hardening phase of the reaction as a drying process—that is, they stipulated that temperatures of around 100°C were to be employed.

Swinburne's making a lacquer in the process of research primarily aimed at developing a solid insulating material is not atypical for this field of chemical engineering. These two main areas of application—varnishes and molding materials—have been associated with plastics from prehistoric times. Another phenol-formaldehyde varnish was patented by C. H. Meyer of the German chemical firm Louis Blumer as a "shellac substitute" (Blumer, 1902a,b). He boiled phenol and formaldehyde with a relatively large amount of acid (one molecule of acid, two molecules each of phenol and formaldehyde) and obtained a product that was soluble in alcohol, ether, and other solvents normally used in varnish production. The Blumer firm marketed this lacquer base successfully. Similar processes were patented by De Laire (1905) and Bayer (1907).

Such was the situation shortly after the turn of the century. Many chemists were patenting phenol-formaldehyde substitutes for celluloid and similar plastic substances. None of them, however, succeeded in building a successful commercial enterprise around the new materials. And then Baekeland took the stage.

3.5 Technological Frame

Now it is time once again to stop the flow of the historic tale. In this section I will sketch the concept of "technological frame," including its empirical operationalization. Later, the closely linked concept of "inclusion" is introduced and the theoretical background of technological frame is discussed.

What does it mean to be member of a relevant social group? How does the SCOT description of a case in terms of relevant social groups, problems, and solutions relate to interactions among individual actors? This is the issue that the concept of "technological frame" should help us solve. In so doing, it will draw the creative process of invention into the analysis, though not primarily in psychologistic terms.[33]

Reviewing the plastics story so far, we saw the relevant social group of celluloid chemists: strongly motivated by the goal of producing a plastic material; initially aiming at the consumer market but gradually also at the industrial market; trying to modify the production process of celluloid to harness its flammability and to develop new applications; searching for cheaper raw materials; seeking to exploit the condensation reaction between phenol and formaldehyde, apparently promising a fully synthetic plastic, etc. Such were the diverse activities and interactions of the celluloid engineers. It is this diversity of interactions that the concept of technological frame is meant to capture.

A technological frame structures the interactions among the actors of a relevant social group. Thus it is not an individual's characteristic, nor a characteristic of systems or institutions; technological frames are located between actors, not in actors or above actors. A technological frame is built up when interaction "around" an artifact begins. Existing practice does guide future practice, though without logical determination. If existing interactions move members of an emerging relevant social group in the same direction, a technological frame will build up; if not, there will be no frame, no relevant social group, no future interaction. Thus the artifact Parkesine did not give rise to a specific technological frame, because the interactions "around" it came to an end before really taking off. The opposite happened to Celluloid: its stabilization was accompanied by the establishment of a relevant social group of "Celluloid chemists." The continuing interactions of these chemists gave rise to and were structured by a new technological frame.

A technological frame comprises all elements that influence the interactions within relevant social groups and lead to the attribution of meanings to technical artifacts—and thus to constituting technology. Following the example of the celluloid chemists, these elements include (to begin with, at least): goals, key problems, problem-solving strategies (heuristics), requirements to be met by problem solutions, current theories, tacit knowledge, testing procedures, and design methods and criteria. The analogy with Kuhn's "paradigm," among other concepts, is obvious, and I will return to such analogies.

The elements mentioned in the previous paragraph will obviously be part of the technological frame of a relevant social group of engineers. In developing the descriptional model, however, I have argued that all social groups should a priori be treated as equally relevant. This implies that the concept of technological frame needs to be applicable to groups of nontechnologists as well—for example, consumers, managers, journalists, politicians. One way to secure this universal applicability is to incorporate elements such as users' practice and perceived substitution function. The use of the high-wheeled Ordinary as a machine with which to impress women was an important element of the technological frame of the relevant social group of sport cyclists and thus helped to form the Macho Bicycle. Similarly, the perspective of Celluloid as a substitute for ivory, tortoiseshell, and shellac was an important element in the technological frame of users of Parkesine, Ivoride, and Celluloid.

A third category of elements needs to be included: the artifacts themselves. The interactions in relevant social groups are never governed by cognitive and social factors alone. Artifacts, as stabilized in previous construction processes, play a crucial role too. Celluloid formed, as I will show, the cornerstone of the Celluloid chemists' technological frame. With the stabilization of an artifact, criteria of what defines this artifact as a working machine will also emerge. Such criteria form crucial elements of the technological frame being built up at the same time. Thus the technological frame comprises the actors' criteria for "working" and "nonworking," rather than our own hindsight knowledge.

How is this concept of technological frame to be operationalized for empirical research? The prescription that it includes such a wide variety of elements, from material and technical to social and cognitive, seems to make it vulnerable to the criticism of being a catchall concept: everything is included, so that applying the concept does not make a difference, does not add perspective, does not provide insight. This question will be addressed in two stages. First, in the remainder of this section I will argue that it is possible to give a quite unambiguous characterization of the technological frame of a relevant social group. Then I will show that the concept does indeed do some "real work": that it makes a difference whether a technological frame is one way or another, and that describing the technological frame of a relevant social group does indeed explain a particular course of events.

Technological frame is a theoretical concept: it is used by the analyst to order data and to facilitate the interpretation of the interactions within a relevant social group. Like other concepts such as "culture" or "form-of-life," technological frame will be most effectively used when the analyst focuses on situations of instability, controversy, and change. This builds on the same argument—why artifacts in the SCOT descriptive model are described in terms of problems and solutions. These problems and solutions then can be interpreted as generated by a tech-

Table 3.1
Tentative list of elements of a technological frame.

Elements
Goals
Key problems
Problem-solving strategies
Requirements to be met by problem solutions
Current theories
Tacit knowledge
Testing procedures
Design methods and criteria
Users' practice
Perceived substitution function
Exemplary artifacts

nological frame. To describe a technological frame, it will help the analyst to start with the tentative list of elements that comprise a technological frame (see table 3.1).

Thus, for example, it was the Celluloid chemists' explicit goal to produce a substitute for natural plastics, aimed at the production of fancy articles and dress items. Once the exemplary artifact Celluloid had stabilized, key problems were the price of the solvent camphor, the flammability of the product, and its molding characteristics. Many of the solution criteria were set by the standards of the natural plastics—color, lack of shrinkage and distortion, price. In the course of the stabilization of celluloid, and the accompanying technological frame, an additional criterion for any new material became its aptness for being molded. As I described, there was no chemical theory involved in the interactions of the Celluloid engineers. Tacit knowledge of Celluoid chemists involved the application of heat and pressure, but to specific maxima. Everybody knew, for example, that the molding press had to be heated up to about 100°C, and no higher, because that would damage the Celluloid. See table 3.2 for a summary of the Celluloid chemists' technological frame.

This list of technological frame elements can only be tentative. In each new case, in each new relevant social group, additional elements may need to be incorporated to give an adequate interpretation of the interactions. Also, not all elements listed in the table may be relevant for a specific group; "current theories," for example, was an empty category when describing the Celluloid chemists' technological frame.

Table 3.2
Elements of the technological frame of Celluloid chemists

Elements of the technological frame	Technological frame of Celluloid engineers
Goals	Production of fancy articles
Key problems	Price of the solvent camphor, the flammability and molding characteristics of Celluloid
Problem-solving strategies	Modification of the solvent in the reaction
Requirements to be met by problem solutions	Set by the standards of the natural plastics: color, lack of shrinkage and distortion, price, aptness for being molded
Current theories	No chemical theory
Tacit knowledge	Application of heat and pressure without specific maxima
Perceived substitution function	Natural plastics
Exemplary artifacts	Celluloid; production machinery such as presses, preheaters

The analogies and differences with other concepts can be seen quite clearly now.[34] Technological frame is evidently one of the many children of Kuhn's (1970) disciplinary matrix and probably most similar to Constant's (1980) tradition of practice. However, technological frame differs from these theories in two important respects. First, technological frames are more heterogeneous than disciplinary matrices and related concepts. Although disciplinary matrices contain symbolic generalizations and metaphysical assumptions as well as values, their character is primarily cognitive. Technological frames are not purely cognitive, but also comprise social and material elements.[35] Second and most important, the concept of technological frame is meant to apply to all relevant social groups, not only to engineers.[36] To bring out this feature more clearly, it might have been better to employ the phrase "frame with respect to technology" rather than "technological frame," but the burden of linguistic clumsiness seemed too great.

Now we are prepared to continue tracing the history of plastics. Here Baekeland enters the story. When he eventually turned toward the phenol-formaldehyde reaction, he first followed the same path as the Celluloid engineers Smith, Luft, De Laire, Fayolle and others. But he was able to break away from that tradition and find new ways. These "new ways" actually were old ways for Baekeland—they are explained by the technological frame in which he had worked until then. That is why I will relate part of Baekeland's biography, to describe the relevant social groups and associated technological frames that structured his early interactions.

3.6 Leo Hendrik Baekeland: From Velox to Hooker Cell

This brief biographical prelude will lead us from research on photographic paper in Belgium and the United States to electrochemical research at Niagara Falls. To give some color to the picture, and because there is generally more to a person's history than relevant social groups and frames, I also sketch some of Backeland's personal background and development.

Baekeland's Belgian Years

Baekeland was an active amateur photographer. In these early days of photography it was very helpful to have some chemical knowledge and skills through which to control the processes of image making; photography was only for the professionals and the most ardent amateurs. By the end of the 1870s, the wet collodion process began to be replaced by the dry gelatin plate process. The important advantages of the latter were that the plates were factory-sensitized. This made the photosensitive materials no longer highly perishable, and it also made the work of the photographer decidedly less complex. (During the wet collodion period, the photographer had to carry a dark tent in which he coated the plates with photosensitive collodion immediately before making a photograph; see figure 3.3.) But developing the negative image on these new dry plates was still subject to a host of variables, many of which were not adequately understood and led to poor images or no image at all.[37] Baekeland set out to find a process that would be simpler and more foolproof.

Baekeland thought of a dry plate that carried not only the photosensitive emulsion, but also, on its other side, two different chemicals that, when dissolved in water, would combine to form a developer. The two developer components were covered by a protective layer to prevent oxidation. After making the exposure, the plate needed only to be immersed in water to dissolve the protective layer and the chemicals. Baekeland received a Belgian (1887a) and a German (1887b) patent on this process. Together with a wealthy friend he started an enterprise for manufacturing these plates.[38] The episode that followed is worth

Figure 3.3
A field photographer and his equipment, comprising a camera, dark tent, glass plates, and chemicals. Photo courtesy the Johns Hopkins University Press.

recounting in some detail, because it highlights the interconnectedness in Baekeland's life of business, academia, and family.

The firm "Baeckelandt en Cie., Scheikundige Produkten" was established on 1 January 1888 in Ghent.[39] The young Baekeland would soon find out that often there is a gap between a feasible idea and its successful implementation into a production process. The company was constantly short of money. Deliveries were missed. The emulsion layer sometimes separated from the glass plates. Customers complained that the quality of the images was unreliable and that sometimes there were no images at all.[40] Many years later, Baekeland would recollect that

There was a time in my life when, as a young teacher of chemistry, I was just as cock-sure as some of my older colleagues that everything was as simple as it appears in the textbooks. This lasted until I tried to make bromide of silver for commercial purposes in the preparation of photographic emulsions. Stas had already published the fact that there were three or four different varieties of bromide of silver, all of exactly the same chemical composition, yet differing in their physical properties. By the time I got fully engaged in the manufacture of silver-salt products for photographic purposes, I had come to the conclusion that instead of three or four varieties of bromide of silver, there existed, perhaps, a hundred varieties, but that of those hundred varieties, there was only one which might probably keep me from the poor-house, and assure me a success as a manufacturer. (Baekeland, 1916: 184)

Baekeland devoted all his free time to the company. But this created new problems. Professor Théodore Swarts objected to Baekeland's entrepreneurial activities, probably in his two capacities as head of the department and as future father-in-law. Baekeland then sent a letter to the ministry of education (responsible in Belgium for appointing university professors), offering his resignation from his job as a university assistant.[41] This must have been followed by intense negotiations, involving not only Swarts and Baekeland but also civil servants of the ministry. Three weeks later Baekeland withdrew his resignation, conceding to the ministry that "it is better if I do my best to live on good terms with my future father-in-law."[42] Now the planning for an academic career and a wedding could proceed.

Baekeland had been quite serious about his plan to become managing engineer of the firm "Baeckelandt en Cie., Scheikundige Produkten." His appointment as assistant professor was to end by September 1889, and he was prepared to use that opportunity to leave academia. However, in July the university had already suggested to the ministry that Baekeland be appointed associate professor of chemistry.[43] Part of the agreement that led to the withdrawal of Baekeland's resignation was that he accept this position. Indeed, this appointment and the wedding were so connected during the negotiations that the university's rector wrote an additional letter to the ministry urging that the official nomination be received before the wedding.[44]

But a third plan was interfering. In 1887 Baekeland had participated in a government-sponsored competition among alumni of the four Belgian universities of the prior three years, and he had won first prize in chemistry. Part of the award was a travel fellowship, and after visiting the United Kingdom in the spring of 1889, he planned to visit the United States in the fall. In the same letter to the ministry in which Baekeland withdrew his resignation, he confirmed the final arrangements for this trip. Letters were crossing one another in a rather confused bustle, and Everyone seemed keen to interfere.[45] The wedding between Leo

Baekeland and Céline Swarts was held on 8 August 1889,[46] and two days later the young couple sailed on the *Westernland* of the Red Star Line from Antwerp to New York. On 25 September 1889 Baekeland was appointed associate professor,[47] but it was too late to stop him from leaving the university and reentering business.

Baekeland's Early Years in the United States: Velox Photographic Paper

Soon after their arrival in the United States, Baekeland visited a camera club in New York, where he met Richard Anthony and Charles F. Chandler. The firm E. and H. T. Anthony & Company produced gelatin dry plates and bromide printing papers. Chandler, a well-known chemist and professor at Columbia University, was a consultant to Anthony's firm.[48] Anthony offered Baekeland a post in his factory. He accepted and wrote to the education ministry in Belgium: "I have found an opportunity to put to practice an industrial process I have invented. Thus I foresee now that I will not be able to specify exactly my date of return and that these activities will withhold me for some time from fulfilling my duties as associate professor."[49] He then asked for honorable discharge, but with the possibility of returning to Ghent University with his professor's title. After the ministry asked him, in vain, to reconsider, this honorary title was granted on 30 November 1889.[50]

At the Anthony firm he worked on bromide emulsions, designed processes for making photographic chemicals, and tried to improve machine-coated printing papers.[51] Most of this work was rather routine chemical engineering, but Baekeland was also involved in research designed to develop celluloid films as a base for photographic emulsions.[52] Besides this development of celluloid film as a substitute for dry plates, another technical development shook up the photographic industry of the day: the process of printing a positive image on paper. The factory-sensitized gelatin dry plates had been generally accepted, but factory-sensitized printing papers were not introduced into the United States before the mid-1880s and did not become widely adopted until the early 1890s. The first kind of factory-produced photosensitive paper was coated with a silver bromide gelatin. This paper was called "developing-out paper": the glass negative was placed above the gelatin bromide paper and exposed to sunlight for only a very short time; then the paper was taken into a darkroom where the latent image was developed by chemicals, and fixed. Anthony's was one of the first U.S. firms to market this printing paper, which signaled a revolutionary change in the printing stage of photography. Soon the "paper war" would start.[53] Baekeland stayed two years at Anthony's and left in early 1892 to become an independent consultant.

The next period was very hard for Baekeland. Céline had returned to Belgium in 1890 for the birth of their first child. The separation resulted in an emotional rift, and only in 1892 did Céline and her daughter return to the United States; Céline had been deeply hurt by what she considered Leo's lack of affection and feeble efforts to encourage her return.[54] Baekeland was also unsuccessful professionally. There is no account of any customer seeking his consultancy, and he divided his attention over a number of "half-baked inventions," ranging from a tin extraction process to a safety explosive. When he suffered a serious illness and had, according to his own retrospective account, time for reflection, he changed his tactics: "Instead of keeping too many irons in the fire, I should concentrate my attention upon one single thing which would give me the best chance for the quickest possible results" (Baekeland, 1916: 184). He turned to his old love, photography, and decided to work on the development of new kinds of printing paper.

Baekeland garnered the financial support of Leonard Jacobi, a scrap metal dealer from San Francisco. Together in 1893 they founded the Nepara Chemical Company in Yonkers, New York. Their company was one of several that began photographic paper manufacture at the beginning of the "paper war" in the U.S. photographic industry. The Nepara Company started the production of silver bromide printing paper of the "developing-out" type. The "paper war" was fought with various means—patent suits, patent pooling, patent licensing agreements, price cutting, and company mergers (Jenkins, 1975). Baekeland started research on other types of printing paper. Several bromide and chloride emulsions were studied in great detail.[55] The usual production process of bromide and chloride emulsions involved precipitation and ripening (by heating) of the gelatin solution, followed by washing to remove the soluble salts. This last step was necessary to remove salts that would otherwise crystallize on the film, thus impairing transparency. Baekeland found, however, that in chloride emulsions the heating and washing steps, although they increased the sensitivity of the emulsion, had a disastrous effect on the tone and general gradation of the image. He then succeeded in finding a formula that resulted in the salts not crystallizing. Thus he was able to omit the washing step entirely—an act of photographic heresy. He also refrained from ripening the gelatin solution and thus produced an emulsion with high image quality but an unusually low light sensitivity. This low sensitivity was completely contrary to the prevailing trend of pushing the sensitivity of films and papers as high as possible. But Baekeland realized the advantages: his paper was so slow that it could be handled in the subdued artificial light of a darkroom. But its sensitivity was still high enough that it could be exposed effectively when such an artificial light source was brought close. Thus Baekeland developed the first artificial light "developing-out printing paper": the paper was exposed briefly in artificial light and then easily developed at a safe distance from the same light. The whole process of printing could be carried out in the darkroom, and hazardous exposure in varying sunlight was circumvented.

This episode allows us a first glance at Baekeland's research style, as it had been formed by his Belgian academic training. Kaufmann (1968) analyzed his laboratory notebooks with respect to this work and showed how Baekeland was willing to try virtually all possible variations, but that he was careful to commit his heresies only one at a time. He thus obtained a fairly complete picture of the relevant variables and their influence on the final result. Materials purchased from vendors were analyzed to ensure that they met the claimed specifications, and if they did not, Baekeland prepared his own basic ingredients. Changes in formulas were introduced only on the basis of test results from earlier batches, and when the new results did not seem to follow logically from the modification, the experiment was repeated.

By April 1894 Baekeland recorded in his last notebook entry that he had found "an excellent coating." But then, as Baekeland later noted, "There is an enormous difference between preparing a few sheets of paper in the laboratory for one's own use, and a successful manufacturing process which can be carried on day by day in wholesale, steady commercial production."[56] While simplifying the emulsion and making it more reliable, he ran into scaling-up difficulties. One chief problem was controlling the effects of the U.S. climate. Although he produced excellent paper in temperate weather, this proved to be a practically hopeless task in the hot summer. Using artificial cooling, as others had done, did not have the required effect. After more research, Baekeland found that the trouble was caused by the high humidity. As he remarked later, "In winter the air is so dry as to cause electric sparks and abundant static electricity by friction, . . . in the summer months the air is often so saturated with moisture that many objects and machinery condense humidity on their surfaces at temperatures as high as 76°F" (Baekeland, 1916:

185). Baekeland installed a refrigerating system over which the air for the coating room could be drawn, first to extract its moisture by precipitation as ice, after which the temperature was raised again by leading the air over heated pipes. Silver chains were trailed over the paper on the coating machines to carry off the static charge that accumulated, especially in winter, because of the friction between paper and frame (Kettering, 1946).

What is important to note at this point is that Baekeland's involvement with photographic chemistry implied a specific way of approaching problems. First it meant a systematic investigation of all variables. Second, it showed a careful strategy for upscaling processes from laboratory experiment to full-scale production. Especially on the first point, the technological frame that Baekeland was now being socialized into differed substantially from that of the celluloid engineers.

The new paper was marketed under the name "Velox." Baekeland was convinced of Velox's qualities, but professional photographers were not. For this relevant social group the use of sunlight was an essential element of photography. All photographic techniques—from the first daguerreotype to the wet and dry collodion processes to the gelatin plates and papers—had assumed the use of sunlight, and professional photographers refused to use the new printing paper by gaslight. Baekeland tried to convince them and had an intensive correspondence. One professional photographer wrote back that "I have tried every photographic paper in existence and I have been more or less successful with all of them until I tried yours, but it is hopelessly no good. You cannot blame my insuccess on faulty manipulation, because I am professor of chemistry at——College."[57] As Baekeland himself realized later, "most of these people knew too much about photography and, on this account, never gave themselves the trouble of even glancing at the printed directions which were sent to them" (Baekeland, 1916: 185). Another letter received by Baekeland illustrated this neglect of crucial requirements in using the Velox paper: "I am a professional photographer of twenty-five years' experience. Your paper is the greatest photographic swindle of the age. You claim your method of printing is several hundred times faster than albumen paper, and here I have kept a print in the printing frame for several hours in the sun and I can hardly see a faint image."[58] But then, at last, success came from quite unexpected quarters, as a new relevant social group entered the game. Amateur photographers started to use Velox paper, and they did go to the trouble of reading the directions.

They studied the detailed comments that Baekeland provided in additional publications and distributed from his Nepara Chemical Company laboratory.[59] These extra instructions are fine examples of an effort to fight the tacit knowledge of the usual printing paper and to replace it with new rules for to making Velox prints. Baekeland mentioned, for example, the amazing rapidity with which images appear in the developer:

> Being accustomed to the slower development of dry plates, the beginner at once proceeds to dilute or to restrain his developer accordingly, and by doing so falls into the error of producing greenish-blacks. The image should appear quickly—that is to say, it should acquire the necessary strength in a few seconds, which allow ample time for examining the print and transferring it to the hypo-bath. It is useless and even harmful to wash the prints after development, because by doing so development goes on, and as the developer which clings to the paper is now more diluted, and consequently will give another tone, unpleasant results will be produced. (Baekeland, 1897: 2)

Increasingly, these amateurs produced excellent pictures—often better than those the professionals turned out using their sunlight-printing process. It took another five years, but then the business really began to prosper. By 1899 the Nepara Chemical Company needed new factory buildings, and Baekeland decided to extend the old building and to build a completely new two-story building in addition.[60] Obviously this commercial success indicating the emergence of a new market of amateur photographers made quite an impact on the photographic industry in the midst of the paper war.

To counter the success of the Nepara Chemical Company, in 1898 Eastman Kodak introduced its own artificial light paper ("Dekko"), and the Photo Materials Company marketed its "Azo" paper (Jenkins, 1975). In July 1898 Eastman Kodak purchased the Photo Materials Company. Despite these actions, Baekeland's Velox paper continued to threaten the market position of the established photographic manufacturers. Then other strategies were tried by the two leading photographic paper producers, Eastman Kodak and American Aristotype. First they tried to control the foreign supply of raw paper, which came almost entirely from the General Paper Company in Belgium, and were moderately successful. Then they started to create a large holding company by merging and purchasing smaller manufacturers.

Several offers were made to Baekeland.[61] There is a much-repeated story that when Baekeland was invited to Eastman's office in Rochester, he was prepared to ask for $50,000 and to settle for no less than $25,000;

he almost fainted when Eastman offered him a million dollars.[62] This anecdote, however, gives a much too naive picture of Baekeland. In a letter to his friend Remouchamps, Baekeland wrote on 18 June 1899 that Eastman "has already four times made proposals but since we have such a prosperous business, we set rather high conditions and even last week there was still a gap of one million."[63] Two weeks later, he wrote that the negotiations had proceeded and that he would probably become "one of those disgusting capitalists who live off their interests. To get our millions I have to sign an agreement that during the next twenty years I will never ever found a photographic manufacturing business, nor give consultations on this subject. I will probably sign this within a few days."[64] Baekeland seemed fairly indifferent about the outcome of the final negotiations. If Eastman did not purchase his company, Baekeland planned to capitalize it, keeping a controlling number of shares and arranging his life so as to be just as free as he would have been had Eastman bought him out.

On 8 July 1899, Baekeland and Eastman signed the agreement: the Nepara Chemical Company was purchased for $750,000, and Baekeland would refrain from any future involvement in the photographic business.[65] Baekeland had good reasons to sell "this business which I have raised myself and which has caused me so much sweat and sorrow."[66] Eastman surely was nervous about Baekeland as a competitor, but Baekeland had just as many reasons to be nervous about Eastman. Velox was a registered trademark, but it was not patented because Baekeland reasoned that he did not have the money to defend his patents anyway. Also, the strategies employed by Eastman and American Aristotype to control the raw paper market seemed to be working, and an impulsive journey by Baekeland to the General Paper Company in Brussels, around Christmas 1897, had not succeeded in completely averting this threat. Third, Eastman and other companies had marketed their own gaslight papers, at lower prices than Velox. But most important, as is obvious from his personal letters, Baekeland was exhausted by almost a decade of intensive research and even more so by the increasing burden of managing a business. Eastman's offer was his ticket to return to the laboratory, as he wrote to an old friend: "Whether I will ever go into business again, I really do not know; but I will certainly establish myself a research laboratory, where I shall be able to work on subjects which interest me most."[67]

Although $750,000 was a large sum for Baekeland, it was a bargain for Eastman, for he bought not merely a product and a trademark, but a whole technology. He acquired a range of different chloride and printing papers of varying sensitivity, tonal quality, and surface texture, plus all the necessary manufacturing specifications. Velox opened the way to a completely new system for exploiting the amateur photography market. Until that time, customers sent back their exposed films for developing and printing by Kodak. Now the printing process had become so easy that film and camera dealers could set up darkrooms in their shops and process film themselves: the "photofinishing" trade was emerging.

Electrochemical consultant

Before finalizing his plans for building a laboratory, Baekeland and family embarked on board the SS *Friesland* on 8 November 1899 to sail to Antwerp. What was initially planned as a two-year family holiday, bringing the family (including their five-year-old son, four-year-old daughter, and Leo's elderly mother) through Europe from Norway to Italy, finally became more of a working sabbatical.[68] Baekeland visited several laboratories in Switzerland, Italy, and Germany. He spent most of his time with Georg Karl von Knorre at the Electrochemical Laboratory of the Technische Hochschule in Charlottenburg near Berlin, refreshing his knowledge of electrochemistry. Thus he returned to one of those numerous "half-baked inventions" (the electrolytic process for tin extraction) that he had been exploring before he decided to concentrate on photographic paper. This European family trip ended with a cycling holiday in northern Italy and a visit to the Paris World Exhibition in May 1900.

After returning to the United States, Baekeland bought the estate "Snug Rock" in North Yonkers, situated on the Hudson. In a separate building on this estate he built his own new laboratory, "modest but conveniently equipped" for all kinds of chemical research (see figure 3.4). There is no detailed evidence about the research he conducted during the first years after he established his laboratory. In the fall of 1901 he mentioned "a very important discovery, already seven patents,"[69] but no patents were awarded to him around that time, and I have not found concrete information about this discovery. Baekeland did devote some of his time to finishing his previous scientific research; having not published anything since 1898, in 1904 he published eight articles on photographical chemistry and one on an electrochemical topic, which probably built on his work in the German laboratories.

Electrochemistry in the United States was developing rapidly, especially since the opening of the giant hydroelectric station at Niagara

Figure 3.4
The Yonkers laboratory with, after March 1908, a library at the main floor (note Kékulé's portrait on the wall) and the main lab on the first floor. This photo was probably taken after the fire on 2 March 1909. Photo in Baekeland Papers, courtesy National Museum of American History, Smithsonian Institution, Washington, D.C.

Falls in 1895. In Baekeland's student days electrochemistry was limited to the electro-disposition of a few metals such as copper, nickel, silver, and gold from aqueous solutions. New developments included a process for extracting aluminum from alumina (by C. M. Hall and P. Héroult), a process for making sodium on a commercial scale (by H. Y. Castner), and the invention and production of carborundum (by E. G. Acheson). The related industries were all located near Niagara Falls.[70] The American Electrochemical Society was founded on 3 April 1902.

In February 1904 Baekeland was hired by Elon H. Hooker as an independent consultant, immediately after Hooker had decided to join forces with Elmer A. Sperry and Clinton P. Townsend to develop their electrolytic process for producing sodium hydroxide.[71] The electrolysis of salt was one of the most promising problems in early electrochemistry: the pioneers hoped to produce sodium hydroxide (or caustic soda), for which a large market existed in the manufacture of soap, wood alcohol, and other products. An electrolysis of salt, or rather brine, would also yield chlorine, which could be used for bleaching. Four years earlier Townsend, who worked at the patent office, had patented a relatively simple electrolytic cell for this purpose. However, he needed money and engineering assistance to refine his idea in order to turn it into a commercial process. Sperry had considerable experience both in raising funds and as an electrical and mechanical engineer; hence Townsend decided to share his patent rights with Sperry in 1901. After two years of research and engineering they enrolled Hooker, an engineer from Cornell with considerable influence and wealth.

Baekeland was employed by Hooker to supervise the construction and operation of a pilot plant. This plant was to be built in an unused wing of a Brooklyn boiler house of the Edison Electric Illuminating Company, which could provide the necessary electricity. In April 1904 two scale 1 : 1 models of the Hooker cell were installed. Within a year Baekeland's research had convinced Hooker of the commercial viability of the process, and in December 1904 he bought the relevant patents from Townsend and Sperry.[72] There was still a long way to go, however, before a full-scale production plant could be built.[73] Research continued under the direction of Baekeland. The greatest attention was devoted to making the process more robust. As Baekeland described later, in a presentation before the Society of Chemical Industry in 1907, "Even if a cell shows a maximum efficiency it may yet prove a commercial impossibility: if the cell is too delicate to be operated by men of average skill, or if the cost of maintenance and repairs is too high, or if the initial cost of the cell plant becomes exorbitant."[74] In his way of operating, the influence of the photochemistry technological frame can be recognized: as in the crucial years of Velox research, Baekeland spent most of his time studying the different variables, testing the components of the cell, and optimizing the process (see figure 3.5).[75] One of the most persistent troublemakers was the chlorine, which kept leaking out of the apparatus. At times the fumes became so noxious that the building had to be evacuated. The brine carrying a large amount of chlorine gas was very corrosive. When the test cells were dismantled, many of the pipes were found to be paper thin.

The two models were tested under varying conditions, day and night, for months. His first two U.S. patents were based on this research (Baekeland, 1906a,b). One patent specified a method for circulating the corrosive brine to extract the chlorine gas, while keeping the salt content at the necessary high level. The other described a more robust diaphragm. The next step was to construct during 1905 near Niagara Falls a small plant, designed to produce five tons of caustic soda per day. The Townsend cells now incorporated the two improvements patented by Baekeland. In 1906 the plant was scaled up further, providing more opportunities to test the cells under less favorable factory conditions. Baekeland would later summarize this strategy as "Commit your blunders on a small scale and make your profits on a large scale" (Baekeland, 1916: 186). In 1907 it was decided to quadruple the plant size, and by early spring of 1910 the enlarged plant was up to capacity and the company was making a profit for the first time. Hooker was to become one of the major American chemical manufacturers. Baekeland had stopped his day-to-day involvement before that time, however, when most of the basic problems seemed to have been solved. He remained a consultant to Hooker for several more years.

During this period Baekeland worked increasingly within the technological frame of elecrochemical engineers; his involvement is summarized in table 3.3.

3.7 Different Degrees of Inclusion in a Technological Frame

Evidently, Baekeland was a member of different relevant social groups. Typically everybody will be a member of several relevant social groups, successively and at the same time. Baekeland was, for example, successively a member of the relevant social group of photo chemists and of electrochemists. And at the same time he was member of the relevant

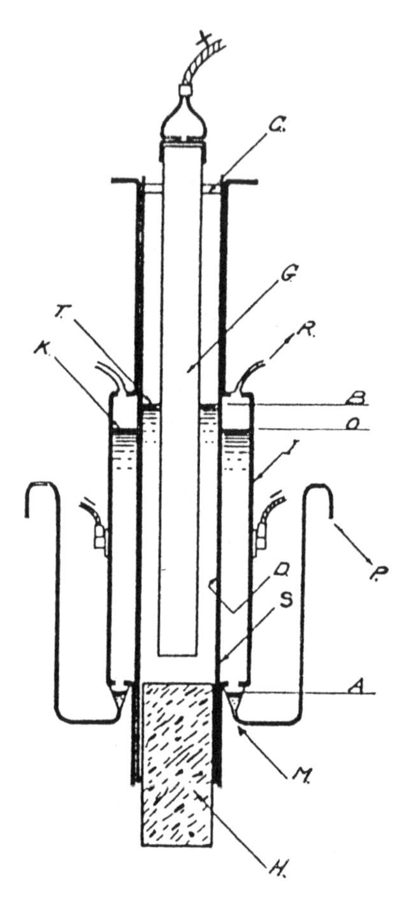

Figure 3.5
Vertical section of the Townsend cell. The anode space, containing saturated brine (T), is enclosed by a lid (C), two vertical diaphragms (D), and a concrete base (H); the diaphragms are in close contact with perforated iron cathode plates (S); the cathodes are fastened to iron sides (I), bulging outward in the middle, thus forming the cathode compartment, containing kerosene oil (K). When electric current is sent through the cell, chlorine gas is formed in the anode compartment and hydrogen and sodium hydroxide in the cathode compartment. The clever trick of the cell is the effective separation of the various products of the electrolysis. The brine, due to differences in hydrostatic pressure, will move through the cathodes and the diaphragms. In the cathode compartment the aqueous caustic solution is accumulated below (in A) and the hydrogen gas moves upward (through R), while the kerosene oil and the diaphragm prevent chemical reaction between the three products.

Table 3.3
Elements of the technological frame of electrochemical engineers

Elements of a technological frame	Technological frame of electrochemical engineers
Goals	Flow production of chemicals
Key problems	Corrosion, reaction efficiency, volume of production output
Problem-solving strategies	Specific design of diaphragms; careful examination of all variables; stepwise scaling up from laboratory to pilot to production scale
Requirements to be met by problem solutions	
Current theories	Basic inorganic chemistry; electrochemistry; fluid dynamics
Tacit knowledge	Industrial flow processing
Perceived substitution function	
Exemplary artifacts	

social group of electrochemists and of pioneering motorists (see figure 3.6). If a technological frame structures the interactions between the actors of a relevant social group, how does this work out for an individual actor who is a member of different relevant social groups? This is the issue to be addressed in this intermezzo.

Let us first focus on the case of successive membership in different relevant social groups, which can best be described as a successive socialization into different cultures. As mentioned, any photographer had to be half a chemist in the early days, and on the production side the emphasis was even more on the chemistry of developing and printing, rather than, for example, optical physics Being an amateur photographer, having professional chemistry training, doing chemical research in a photographic production environment—all these activities contributed to Baekeland's socialization into the culture of the relevant social group of photo chemists. Baekeland thus acquired the technological frame of photo chemists. A similar process occurred when he started his electrochemical studies at Charlottenburg, Berlin, and as a full-time researcher and consultant at Niagara Falls. Without a very close analysis of his work it is not possible to trace in detail how he slowly was enculturated into the new technological frame, while elements of the older technological frame moved to the background. Where elements of the two

Figure 3.6
Baekeland with wife and children in his first motor car, around 1897. Photo in Baekeland Papers, courtesy National Museum of American History, Smithsonian Institution, Washington, D.C.

frames are not contradictory, the different repertoires will continue to exist together. In the case of Baekeland it seems safe to assume that there was not much friction between the two technological frames and that the electrochemists' frame was more or less added to his previous enculturation.

The case of contemporaneous membership in different relevant social groups, and thus involvement in different technological frames at the same time, is more interesting. It is to describe this situation that I propose the concept of "inclusion in a technological frame." The degree of inclusion of an actor in a technological frame indicates to what extent the actor's interactions are structured by that technological frame. If an actor has a high degree of inclusion, this means that she thinks, acts, and interacts to a large extent in terms of that technological frame. It is expected that actors who are contemporaneously member of different relevant social groups will have different degrees of inclusion in the associated technological frames.

Inclusion is not a monodimensional concept. Because a technological frame has a variety of elements, two actors both with rather high degrees of inclusion may still have different "scores" on different elements of that frame. So there can be no easy measure of the degree of inclusion.[76] The concept will be employed to explicate the analyst's interpretation of the pattern of interactions of an actor. Later it will do some work when I address the problem of the obduracy and malleability of technical artifacts. At this moment the introduction of the inclusion concept is merely a consequence of the observation that actors are typically members of different relevant social groups, often at the same time.

3.8 Baekeland's Laboratory Research into a Phenol-Formaldehyde Plastic

The success of his Velox enterprise had not only given Baekeland the material means for a comfortable life and a well-equipped laboratory, it had also returned his gusto for pursuing vastly differing projects at the same time. Besides his electrochemical consulting work for Hooker, Baekeland pursued many other interests between 1900 and 1907. Though he undoubtedly became an electrochemist of some sort, his inclusion in that technological frame stayed relatively low.

He spent much of his time writing—letters to friends and colleagues, laymen's articles on photography, technical papers for formal presentation (mostly on photography), and general commentaries on science and education.[77] He also devoted much time to tinkering with his motorcar. He led a family motor tour through Europe, from Edinburgh to Naples, which was a true odyssey in those days.[78] And finally, he carried out a variety of chemical research projects—on the effects of X-rays and ultraviolet radiation on organic compounds,[79] on the chemistry of preparing food from soybeans,[80] on an improved nitrocellulose base for movie film, on the impregnation of wood with "sulphite liquor,"[81]—and he started the research that led to the now famous Bakelite patents of 1907.

I shall analyze this period of laboratory research in detail, using Baekeland's laboratory notebooks, diaries, and correspondence. One reason to do this is its importance as a key step in building the modern plastics industry. Another reason is the insight it provides into the practice of one highly skilled engineer, which is in itself worth our attention. The third reason, most central to the project of this book, is to study the way in which Baekeland successfully constructed not only a new plastic *but also a specific historical account of that invention.* The case will show how the construction of Bakelite not only involved chemical experiments, financial expertise, and legal patent maneuvers, but also the writing of history in a specific way.

Starting Research on Synthetic Resins

Baekeland had become interested in the reaction between phenol and formaldehyde in about 1902, two years after establishing his private laboratory at Snug Rock.[82] This was probably one of his reasons for hiring Nathaniel Thurlow as a research assistant in December 1904, for previously Thurlow had been involved in studying phenolic bodies.[83] First they familiarized themselves, by reading publications and through experimentation, with the state of the art. Theoretically, the chemistry of phenol-formaldehydes was in disarray. Chemists in Europe were mixing various aldehydes, phenol, solvents, acids, and alkalies, under different pressures and with or without applying extra heat to dry the product. The results ranged from sticky syrups to unmanageable solids that defied chemical analysis.

Baekeland tried to find patterns in the chaos of different results. As could be expected from someone with a high inclusion in the technological frame of photo chemists, he set out to map the role of all possible variables in the reaction instead of (as Celluloid chemists did) just substituting one solvent for another, or (as dye chemists had done) aiming at chemical analysis. Condensation reactions happened with a wide range of inputs: phenol or related materials such as cresol, and formaldehyde or relatives such as methylal, methylene acetate, or hexamethylentetramine. The products could be soluble or insoluble, could soften or harden when heated, could be crystalline or not. The type of condensation agent, acid or alkaline, obviously made some difference, but what kind? One researcher had used neither one! The addition of solvents made the reaction more controllable, but the drying afterward became so tedious as to render the result commercially useless. More heat of course would speed up the reaction, but this led to Kleeberg's dead end. The abandonment of the use of any solvents seemed a possible way out; but this had recently been tried by the Englishman Henry Story (1905a,b), who surprisingly found that the drying and hardening process still took several weeks to months.

There was only one line of research that seemed, to Thurlow and Baekeland, to offer a potentially positive result. This concerned the search for a shellac substitute. Thus they followed the lead of recent work done by Swinburne, Blumer, and De Laire aimed primarily at developing soluble compounds to be used as lacquers. Thurlow, who seems to have been doing most of the work, found that by avoiding large amounts of formaldehyde (as Blumer and De Laire had done) and acid, he could better control the reaction. Using very small amounts of acid as a condensation agent, he produced a soluble, fusible "shellac substitute" in the fall of 1906. An agreement was signed in which Thurlow transferred all rights to Baekeland,[84] and a patent was filed on 18 February 1907 (Baekeland, 1907a). It is interesting to note that the patent did not describe a lacquer but "a method of indurating fibrous and cellular material"—an impregnating substance.

To study the possibility of commercially producing this indurating condensation product, which he named "Novolak," Baekeland established relations with the chemical firm Berry Brothers in Detroit. Hooker was involved in some way as well, probably as a potential financier. Thurlow did most of the research at this stage. Baekeland's diary contains some entries in which Thurlow is said to have reported certain progress in his research, and when problems arose, Baekeland offered advice.[85] Although Baekeland once visited Berry Brothers, he was hardly involved in the Novolak research.[86] Baekeland later would give a specific interpretation to this lack of active involvement in the Novolak research that resulted in an overly rational account of his search for modifying the unmanageable "red-rose mass" of Kleeberg. I shall first give my reconstruction of these first days in the laboratory, and then return to a discussion of Baekeland's interpretation, whereby I will make use of the different degrees of inclusion he had in two technological frames.

Exciting Days in June 1907: Invention of the First Bakalite

I do not know what made him act on that very day, but after Thurlow had left for another trip to Berry Brothers on 18 June 1907, Baekeland turned to phenol-formaldehyde himself. I doubt he had planned his turn to phenol-formaldehyde beforehand. It is fairly certain that he had not told Thurlow of his plans.[87] During the previous days he had been experimenting with the impregnation of wood with sulphite liquor.[88] He started a fresh laboratory notebook on 18 June, which suggests that he experienced a dramatic, unplanned change of events.

For his ongoing impregnation research, Baekeland had prepared wooden blocks, measuring 1″ × 1″ × 10″. Each was divided into two pieces of 5″, only one of which would be treated with the impregnating material. Thus he was sure that the planned strength tests were conducted on comparable pieces of wood. He carried out seven experiments with two types of wood during the first day: he soaked one block of a pair with a mixture of equal volumes of commercial phenol and formaldehyde; heated it subsequently to 140–150°C; measured the absorbed quantity and measured the relative strength.

After four days, Baekeland made an entry on this period in his diary, nicely capturing in his not impeccable English the thrill of this work: "spent all these days in my laboratory and found many interesting things. . . . I consider these days very successful work which has put me on the track of several new and interesting products which may have a wide application as plastics and varnishes. Have applied for a patent for a substance which I shall call Bakalite."[89] Thus one could say that it was in these four days that the "discovery" of Bakelite took place. However, although Baekeland obviously recognized what he had found as a potential plastic (he had even given it a name already), this substance was, as will become clear, quite different from the Bakelite that was eventually constructed.

What had happened? On the second day, before continuing his experiments, Baekeland had taken another look at the pieces of wood he had impregnated the previous day. Suddenly his attention was caught by something strange that he described in his lab notes: "the surface of the blocks of wood does not feel hard although a small part of gum that has oozed out is very hard."[90] He reasoned that this might be caused by the formaldehyde evaporating before it could react with the phenol. Remembering previous experiments, done when reviewing and checking the literature, he conjectured that "the proper way would be to impregnate [the wood] with the viscous liquid which is obtained by boiling $CH_2O + C_6H_5OH$ together without a catalytic agent."[91]

To see to what extent this would be possible, he immediately set up a series of experiments aimed at identifying the true nature of the viscous liquid and at describing the drying process that would take place when heating it:

> I have heated in sealed tubes a portion of this liquid so as to determine whether there is a further separation of H_2O or whether this is simply a phenomenon of drying, and if the liquid is simply a solution of the hard gum in excess of phenol, then by simple open air evaporation I shall be able to accomplish hardening while I shall not succeed in closed sealed tubes.
>
> I have also heated an open tube rammed with a mixture of asbestos fiber and liquid (see further).
>
> Also a sealed tube rammed with mixture of asbestos fiber and liquid (see further).
>
> Everything heated 4 hours at 140°C–150°C.[92]

The results must have been startling.

The notes on the third day were made in a decidedly different tone, as if he were making a public statement: "In order to facilitate matters I shall designate in the future the different products by a special name."[93] He then described triumphantly four different products. Product A was the liquid condensation product first obtained. It was soluble in alcohol, phenol, and acetone. It could be heated somewhat, but would then become pasty, although remaining soluble in alcohol. Product B was an elastic rubberlike product, insoluble in alcohol but soluble in other mixtures of solvents. It could be softened by heating. Product C was an infusible, insoluble hard gum "which seems to be the last condensation product." It could be softened slowly in boiling phenol. Finally product D "is insoluble in all solvents, does not soften. I call it Bakalite and is obtained by heating A or B or C in closed vessels."[94] From this moment on, he must have worked almost continuously for two days. The previous sentence, under the heading of "June 20," is followed by some thirty-three pages of notes, before the heading "June 23" appears. On Saturday 22 June he "reluctantly . . . had to participate at the Shore dinner of the Verein Deutscher Chemiker."[95] But on 23 June he "spent all day busily in my laboratory research work on Bakalite and similar problems' again."[96] On 24 June he jotted down "poor sleep since several days."[97]

The lab notes of June 20 and 21 show that Baekeland envisaged many commercial possibilities for the products he was studying. In the very beginning he once described product D as "a nice smooth ivorylike mass,"[98] but clearly he was more concerned with industrial applications than with finding a substitute for luxury plastics. We can see here the effect of his being also included in the electrochemist technological frame. Producing raw materials for industry was the main goal in that frame, while celluloid chemists focused primarily on consumer goods. "Insulating masses," "molding materials," "a substitute for celluloid and for hard rubber," a "new Linoleum," and "tiles that would keep warm in winter time" are but a few of the applications mentioned, scattered through the thirty-three pages of these two days.

One of the elements of Bakalite that stabilized to some degree relatively early (although it was later briefly to be questioned again) was the need to heat the products A, B, and C well above 100°C to get product D. It took him much longer, however, to decide on the need for pressure. At first he only acknowledged the need for closed vessels: "I am satisfied now that open air heating does not give the same hard mass as enclosed tube heating. Open air drying may give a more transparent mass but it lacks that hardness and strength which is so characteristic for the mass baked in closed vessels and which is real Bakalite."[99] To reach this conclusion, he carried out several experiments in which he heated product A in open and sealed vessels. Very detailed observations of one of the open-tube experiments pointed toward the necessity of a closed vessel and some pressure to prevent the evaporation of phenol.[100]

Baekeland's research exemplifies his inclusion in the photographic chemists' and electrochemists' technological frames. His uncompromised focus on reaction variables, geared toward bulk production, is radically different from the type of research that plastic engineers typically were involved in. Also the identification of his products was done in terms of reaction variables, rather than by chemical analysis, as the dye chemists would have done.

A large amount of time was spent by Baekeland in sorting out the place of Novolak in his new system of four products. Already on 20 June he observed that the aqueous liquid that separated when product A was formed, "leaves on condensation a residue of fusible and soluble gum similar to soft Novolak."[101] This pointed to a similarity between A and Novolak, which he did not investigate further at that moment. Later that day, he conjectured that his product B, when mixed with normal Novolak, might improve its lacquer qualities. On 23 June he did set up an experiment to check this and subsequently concluded that "I am now definitely convinced that a mixture of 1 vol. normal Novolak solution in wood alcohol + 1 vol. acetone solution of B gives a varnish which dries to a tough adherent film much superior to Novolak alone."[102]

After he had also further studied the relation between Novolak and product A, he summarized triumphantly: "This settles the whole varnish matter in a most simple way."[103] Accordingly, he started to draft the patent applications describing his "Bakalite invention." This first relatively stable meaning of Bakalite, as embodied in these patents, was however not to be the final one.

Before continuing the historic tale, it is illuminating to contrast the previous analysis of the first four days with Baekeland's own account, given nine years later.[104] Baekeland's retrospective account differs on two important points. First, he presented his research as being aimed, from the beginning, at developing a plastic molding material. In my analysis, he had not made that clear-cut choice against Novolak. Second, he presented his research strategy as a set of rational decisions phrased in terms of studying the role of the solvent in the reaction. In my analysis, the path was not only less rational and linear than suggested, but also much less centered around the solvent.

In his 1916 description of the discovery of Bakelite, Baekeland described his negative evaluation of Novolak. He "took the point of view that their qualities were not better than those of the natural resins. In fact, in practical experiments on a large scale, I had found that in many respects, their qualities were inferior to those of shellac, which, furthermore, could be purchased at a more advantageous price."[105] In April 1907, however, he mentioned in his diary Thurlow's "successful attempt to make Novolak on a commercial scale."[106] Further entries show that he supported this work and believed in it, although he was only marginally involved himself. We have seen that much of his early research on Bakalite was in effect devoted to improving Novolak and to developing other impregnating materials. And much of the research that followed was similarly directed.

But when Baekeland was so convinced of the futility of pursuing the study of soluble resins like Novolak, he had something to explain: how did he get around to studying the impregnation of wood? His rational reconstruction draws heavily on a review of previous research on the phenol-formaldehyde reaction. Accordingly, he places much emphasis on the solvent, as could be expected from someone with some inclusion in the celluloid chemists' technological frame. Thus his 1916 account starts as follows: "I was particularly attracted by the failure of Kleeberg and hoped to find a suitable solvent for his worthless product.... So I concluded to purchase or prepare further solvents of every kind so as to determine whether I could not do something with the substance of Kleeberg, but after many attempts finally had to give it up as a hopeless task."[107] In my analysis, Baekeland and Thurlow had studied solvents only when they were familiarizing themselves with the problem more than two years earlier. Baekeland continued his retrospective account by explaining how he got to impregnating wood—his now famous idea of going to the mountain because the mountain was not coming to him:

> Then I changed my tactics. I reasoned that if nothing could be done with the substance when it was once produced in a flask or any other vessel, I should attempt to carry out the reaction so as to produce the substance right on the spot where I wanted it. For example, I reasoned that if I could make the synthesis inside of the fibres of wood by injecting first the two reacting raw materials, and then start the reaction, I might be able to incrustate the fibres of the wood with an unusually hard and inert substance, and give to the wood new properties.[108]

He thus switched the historic sequence: first research on phenol-formaldehyde plastics and then impregnation of wood; my analysis suggests that it was just the other way around.

Why did he give this historically inadequate account? Rather than trying to give a psychologistic explanation, I will suggest an explanation in terms of Baekeland's increasing inclusion in the technological frame of the celluloid chemists. When he started his research on phenol-formaldehyde, his work was primarily structured by the technological frames of photo chemists and electrochemists. In those frames there was no obvious reason to focus on the solvent in the condensation reaction. Rather, a detailed analysis of all reaction variables was to be expected. In the construction process of Bakelite that followed, however, Baekeland interacted with the relevant social group of celluloid chemists (partly by reviewing the literature, partly, as we will see, in person during patent litigation) and gradually got more and more involved. His degree of inclusion in the celluloid technological frame increased, and the history of the invention of Bakelite was cast in terms that were in accordance with that frame: building on Kleeberg's work, aiming at a molding plastic, focus on the solvent. Thus this difference between Baekeland's actual work and his account of it nine years later can be understood to mirror the increasing degree of inclusion in a technological frame.

The Second Bakalite

It was several years before a form of Bakalite stabilized for a longer period. We thus have here a slightly different case as compared to the bicycle, where I demonstrated the interpretative flexibility by showing how the various relevant social groups constituted different artifacts. In this case the process is less one of closure—limiting the interpretative flexibility among artifacts that exist in different relevant social groups. We will see a series of quite different Bakalites, the first four of which did not stabilize longer than several weeks or months. Only the fifth, by that time called Bakelite, did stabilize for a long period of time, although of course further modified as more relevant social groups got involved. It seems more appropriate to view this as a stabilization process, because it did involve the gradual "condensation" of one specific meaning of the artifact Bakelite, though more changes of meaning were involved than only the dropping of modalities.

Baekeland could not devote all of his time to the Bakalite research. Besides a two-week trip to Canada, he also spent much of early July trouble shooting the electrolytic cell in Niagara. But after coming home in the early morning of 11 July, he immediately became engaged in the Bakalite work again. First he informed Thurlow about his "discovery of Bakalite." The second thing he did was to prepare two samples of B for Berry Brothers.[109] This clearly shows his interest in continuing the work on Novolak by trying to improve it by adding condensation product B. He then corrected the first patent specifications on Bakalite, and drove with his son to the notary public to sign them and to the post office to send them by special delivery to his patent attorney.[110] His diary entry of this day ends with: "Unless I am very much mistaken this invention will prove important in the future. I intend to concentrate all my efforts on this subject during this summer."[111] Four days later he received a telegram that the Bakalite patents had been filed at the patent office on 13 July 1907 (Baekeland, 1907b,c).

These patents, often called the "heat-and-pressure patents," formed the basis of Baekeland's later control of the phenol-formaldehyde plastic market. This is not to say, however, that these patents described the actual Bakalite process in any precise sense. That process was, as we shall see, still to be constructed in the following years by Baekeland and his collaborators, but also by other relevant social groups. The first patent (1907a) describes a method of making "insoluble products," the second (1907b) focuses on an "indurating product." Both patents mention the separation of water as a crucial step in the reaction process and both cite the subsequent simultaneous application of heat and pressure. In addition, the first patent describes the material as "a hard, compact, insoluble and infusible condensation product," and it specifies that articles can be formed from the intermediate reaction product before the final hardening by the combined action of heat and pressure. It also mentions the possibility of adding filling materials. The only reference to a condensation agent is the last claim in the first patent, where it is said that to cause "the water to separate from the mixture of a phenolic body and an aqueous solution of formaldehyde [one should add] to said mixture a metallic salt soluble in water."

Baekeland now continued to study many process variables systematically. The *time* needed to heat the intermediate products was found to be one to three hours, depending on the specific qualities needed.[112] In studying the role of *condensation agents*, he had started on 23 June with four experiments, using respectively a plain mixture, HCl, $ZnCl_2$, and NH_3.[113] Without any explicit reasoning he continued to study only the acids HCl and $ZnCl_2$, and he seems to have forgotten completely about the possibility of an alkalic agent until 31 August 1907. The *temperature* at which the intermediate products were to be heated was studied extensively although Baekeland had already realized that here was a major difference between his process and the processes patented by all other chemists. Their 100°C resulted in a mere drying process, while his 140° to 160°C stimulated a chemical reaction that might otherwise not occur. By 21 June, Baekeland had made a three-page list of possible *filler materials* with which he could produce "compound Bakalite."[114] On 18 July he started further experiments on "compounding Bakalite."[115] These were very successful, as he noted in his diary: "great excitement for me this morning when opening 14 tubes with different Bakalite compounds."[116] Of course, another major subject for study were the optimum *proportions of phenol and formaldehyde.* A recurrent theme in Baekeland's research during the first months was the question of whether an excess of phenol (as used by Story and Thurlow) was needed, and whether something other than the optimum mixtures could explain the results obtained by other chemists.[117] On 14 July, Baekeland wrote in his notebook: "I now can conclude that Story proportions are not right for Bakalite and that mine equal parts are the real thing."[118]

It is fascinating to see in Baekeland's laboratory notebooks how great the interpretative flexibilty of the various "facts" still was, long after the "moment of invention" as symbolized by the filing of the patents. Baekeland continued to be unsure of crucial elements in the process, and he studied virtually all parameters.

By the end of July, he was reaching tentative conclusions again. Bakalite and "the method of making the same" were, for the second time, acquiring stable meanings. Now the "Bakalite invention" comprised the following:

• Product A, the liquid soluble condensation product, was made by boiling equal proportions of phenol and formaldehyde with a few drops of HCl. A typical batch was made of 250 cc phenol, 250 cc formaldehyde, and 4 cc HCl.

• A few drops of HCl gave no reaction, whereas a lot resulted in a wild reaction.

• Product A was distinctly different from Novolak, although their production processes were very similar; Novolak could not be made to "bakalize" into product D. The difference between Novolak and A was probably caused, Baekeland conjectured, by the continuous stirring during the Novolak production that made all phenol take part in the reaction, while in the production of A some phenol separated into the supernatant liquid.

• Improved Novolak could be produced by adding product B to ordinary Novolak.

• Heating A or B for about three hours in a closed vessel at a temperature of 140°C–150°C gave products C and D (the difference between these two had vanished by this time). HCl favored this hardening process.

• Filler materials seemed promising to produce compound Bakalite, but further research was necessary.

This new meaning of Bakalite was stable enough for Baekeland to "have some misgivings as to my Bakalite patent specifications," filed on 13 July.[119] He continued his research, however, and this led again to further changes in the meaning of Bakalite.

The Third Bakalite

A major shift in the meaning of Bakalite occurred around 31 August 1907. "In order to determine whether alcalinity will favor the production of good A,"[120] Baekeland added some ammonium carbonate instead of acid to a 1 : 1 mixture of phenol-formaldehyde. The results were promising. He then tried KOH and NH_3, which confirmed the conclusion that alkalic condensation agents activated the formation of product

A. He also found that adding an acid to the thus produced product A transformed it into Novolak. Continuing these experiments the next day, a Sunday, he observed that the A obtained by boiling neutral, acid, or alkalic phenol-formaldehyde would yield Bakalite in all cases, but that there were advantages to using alkalic agents: HCl produced a more violent reaction because the mixture was self-heating, thus creating CH_2O gas bubbles, which made the final Bakalite spongy. The Bakalite produced from "alkalic" A was generally of better quality: "magnificent transparent and yellow, very hard." "This opens a new field" he concluded in his diary.[121] Baekeland subsequently devoted much time to developing new ways of making Novolak, mainly by using the alkaline agent NH_4Cl (which, Baekeland supposed, did set HCl free during the reaction). Almost all of his experiments in this month were concentrated on making "Bakalite A" and Novolak, which shows again how interested he was in producing varnishes. After a month of hard work in his laboratory, albeit frequently interrupted by his usual other affairs, he started to draft two new patent specifications for the use of alkaline condensation agents on 25 September 1907. These patents were filed on 15 October (Baekeland, 1907d,e).

Another important modification was also developed in the course of these studies: the addition of pressure to the "bakalizing" process. Until now, Baekeland had merely used sealed tubes and no extra pressure. During the series of experiments aimed at clarifying the role of alkalic condensation agents, he often had made Bakalite that was "near perfect, but for some air bubbles." On 3 September, while testing various alkalics, he had heated a sealed tube for three hours at 190°C. The result was "a magnificent & hard mass" and he attested to "the great influence of heating under pressure."[122] Two weeks later, he returned to this observation and repeated the experiment to confirm his initial conclusions. On 17 September 1907 Baekeland concluded, with several exclamation marks in the margin of his laboratory notebook, that pressure was needed to prevent the material from becoming spongy because of air bubbles and to make the hardest material.

Thus we can identify here the third stabilized meaning of Bakalite, now prominently featuring the role of base condensation agents and the importance of combined application of heat and pressure in the final hardening phase.

During his research Baekeland showed examples of Bakalite to visitors, discussing the possibility of selling raw Bakalite A to firms that could then manufacture the final products. It is to this phase of the social construction of Bakalite, when new social groups started to interact, that we now turn.

3.9 The Social Construction of Bakelite

Immediately after filing his first patents, Baekeland began to show examples of Bakalite to visitors. He obviously was not very secretive at this stage. Almost any chemist visiting Snug Rock was guided into the laboratory and shown pieces of Bakalite and the patent specifications. He even carried some examples around—when traveling to Niagara, for instance, "in the evening on train [he] showed Mandel samples of Bakalite."[123]

Enrolling Other Manufacturers

The first serious business contacts were related to various compound Bakalites: "Carborundum Bakalite," "Graphite Bakalite," and "Cork Bakalite." A typical example of a connection with other firms is a visitor from the Cork Company of Pittsburgh:

> Went with motor to fetch him and brought him up to my house. Showed him my samples of Cork Bakalite. He seemed much interested and desired to go further in the matter. He told me they are using now a German patent for compressed cork consistence in the use of nitrocellulose solution[;] drawbacks are slow drying, evaporation in vacuo, limited thickness of slabs on account of irregular evaporation and great cost of raw materials ... I showed him my samples and he seemed to recognize advantages as to [?] and cheapness and simplicity of process. He promised to furnish me with samples of different grades of pulverized cork and I promised him to make Bakalite Cork samples with it.[124]

Similar arrangements were made for the testing of Carborundum Bakalite with the Norton Emeri Company and the Carborundum Company.[125] These two companies were so eager to use Bakalite in their production of grinding wheels that Baekeland (1907f) first drafted a patent to protect this specific application before continuing negotiations. In the letters Baekeland subsequently sent to both companies, together with the samples of Carborundum Bakalite, he gave many details of the production process and the relative proportions of Bakalite and carborundum (and possible alternative choices). He also suggested specific construction methods for the main intended product, grinding wheels.[126]

Also in November, Baekeland and Thurlow began to devote more time to experimenting with molding Bakalite. One of the results was a more clearly defined sense of the value of differentiating between the three different types of Bakalite (A, B, and C; D had by now conflated with C). On 4 December a patent was filed that Baekeland called "the intermediary product patent"[127] (Baekeland, 1907i). Here he described the possibility of "arresting the reaction when the initial condensation product has been transformed into a mass which is solid at all temperatures and hard when cold, soft and elastic when heated but infusible, insoluble in alcohol, glycerin and formaldehyde" (Baekeland, 1907i), and then subsequently "shaping said product and hardening the same by adequate application of heat" (Baekeland, 1907i). This is one of the elements of the "Bakelite invention" that is generally considered crucial, and was indeed presented as such by Baekeland himself in his formal presentation to the New York section of the American Chemical Society in February 1909. And although some of the basic vocabulary was available with the differentiation between A, B, and C Bakalite, it was only half a year later that Bakalite acquired this fourth meaning of a sophisticated molding material. This focus on molding can be decribed as a further increase in the inclusion of Baekeland in the technological frame of the celluloid chemists.

The contacts with industry now also became more focused on molding. In the Carborundum Bakalite patent (Baekeland, 1907f), for example, there was no explicit mentioning of molding grinding wheels. Rather, the emphasis was on the basic abrasive material that might simply be applied to the surface of the grinding tool.[128] When three executives of Norton Emeri Co. visited Baekeland in December 1907, they brought along two small iron molds for molding Bakalite grinding wheels.[129] Thurlow and Baekeland worked intensively with them: "I showed them liquid A Bakalite, all my samples of B and C Bakalite and explained to them how A gets transformed in C and B and how we proceed. We then showed them how to make a wheel."[130] Then the stage of detailed business negotiations had apparently been reached. Baekeland

> told Mr. Higgins that provisionally I would not propose the matter to their competitors so as to give them a chance to determine whether it was worth while for them to obtain the exclusive sale of Bakalite for grinding wheel purposes. I told them that I intended to sell them liquid Bakalite at 25¢ a lb. in large lots and if they were willing to contract for regular large annual quantities we might make an arrangement on that basis.[131]

Baekeland also told them that he would insist on the word "Bakalized" or "Bakalite" being used in conjunction with products manufactured by other companies.

Figure 3.7
When the initial problems were solved, Bakelite would indeed redeem the old promise of synthetic plastics: to replace ivory in billiard balls.

In January 1908 the first problems arose. Baekeland received letters informing him that some billiard balls made from Bakalite were too heavy and not elastic enough (see figure 3.7). The Norton Emeri Company wrote him that they did not succeed in making the grinding wheels hard enough.[132] One day later Baekeland took an early train to Worcester, Massachusetts, to the Norton Emeri Company to see for himself. Upon arrival, the general manager showed him the test results and commented that Bakalite was not as hard as shellac. Baekeland, however, thought "their mixing and heating defective. Concluded it would be best to carry an experiment out in our lab."[133] He returned with the sleeper train. Further experiments in his own laboratory and again in Worcester seem to have solved the problems.

The case of the grinding wheels was typical. In the year following his initial findings, Baekeland explored possibilities for applying Bakalite in more than forty industries. Besides the applications already mentioned, he also made bearing liners, coating for pumps and containers to protect them against chemicals, fireproof coating, floor tiles, phonograph records, electrical insulators, valve seats, knobs, buttons, and knife handles. But although these applications seemed effective in Baekeland's laboratory, they often gave rise to problems under manufacturing conditions in other companies. Highly included in the technological frame of celluloid chemistry, these engineers were not able to handle a material that lost its malleability so completely. And techniques used in manufacturing hard rubber articles either fractured the Bakalite or broke the mold.

Although the classic applications for plastics such as knobs, buttons, and knife handles do appear on the list of sample articles, Baekeland was primarily interested in developing his Bakalite into a versatile and precise molding material for mass production under industrial conditions. At this point a new technological frame started to emerge, together with a new relevant social group of Bakelite engineers. As Baekeland stated in his triumphant presentation before the Chemist Club, Bakelite's "use for such fancy articles [as knobs, buttons, knife handles] has not much appealed to my efforts as long as there are so many more important applications for engineering purposes" (Baekeland, 1916: 157). The first precision-molded articles were made in collaboration with the Loando Hard Rubber Company of Boonton, New Jercey (called the Boonton Rubber Company in Baekeland's notebooks). This firm's trade consisted of recycling the rubber of bicycle tires; they produced rubber for other companies and manufactured electrical parts themselves. After R. W. Seabury of the Loando Company visited Baekeland for the first time in January 1908, they quickly got down to business. The first articles molded by Seabury were made with asbestos and wood flour as fillers. These articles were precision bobbin ends, which could not be made satisfactorily from rubber asbestos compounds. From 12 February 1908 onward, increasing amounts of liquid and solid Bakalite A were sold to Loando—the first firm to which Bakalite was sold on a commercial basis; the order of 12 February formed the opening entry in the account book kept by Céline Baekeland.[134]

These circumstances were promising, but Baekeland was still in doubt about how to set up his business. On 18 March 1908 he wrote in his diary: "Almost every day since I invented Bakalite I have been thinking about the best method for developing this into a substantial business that would not involve me too much neither financially nor otherwise."[135] Most nagging was the uncertainty about the patents:

> If I only knew that my patents will be granted and that there will be no interference cases which may drag me into long and costly litigation, the whole matter would be more simple and I would not hesitate making the necessary expenses to forge ahead and engage one or two more assistants and equip my laboratory so as to be able to work on a larger scale and investigate all possible industrial applications.[136]

He had to wait until 16 November 1909 for his first Bakelite patent to be allowed (Baekeland, 1909f). A larger set of Bakelite patents, including the most central ones, were granted two weeks later (Baekeland, 1907a,c,e,f,i).

Bakelite's Public Presentation

The account book shows a slow increase in commercial sales of Bakalite in the course of 1908, listing two more firms as customers. In 1909, however, sales rose dramatically; deliveries went to many new customers, and total production increased accordingly. This drastic development was undoubtedly caused by Baekeland's first public presentation of what he now called Bakelite. On 5 February 1909 he was to present his paper entitled "The Synthesis, Constitution, and Uses of Bakelite" before the New York section of the American Chemical Society, at the Chemists Club. From notes in his diary, it is clear how important Baekeland considered this event. The day before, he wondered: "I hope I am not making a mistake by thus sending my work boldly into publicity. I trust on the strength of my patents."[137] He spent the next morning polishing his samples, reading newspapers and a book borrowed from his son, and then taking a brief nap. No normal work day—he "dictated no letters"—but a deliberate concentration on the big event. In the afternoon he drove to New York with two of his lab assistants. They first unpacked the samples at the Chemists Club, then had dinner at the Savoy and returned before the meeting started at 8:30 P.M. Baekeland's paper was the third and last to be read that evening.

Baekeland was a well-known and respected chemist and industrial engineer by that time. The announcement of his paper had drawn a large audience to the Chemists Club that Friday night. His presentation was a didactic tour de force, combining the reading of an ambitious scientific paper with the demonstration of Bakelite samples and their various properties. In his own words: "I did not read my paper which covers

7000 words but spoke outright. Experiments went excellent and numerous attendance seemed to take much interest in everything and at the end I received an applause which I would call an ovation and for which I felt very thankful."[138] Most reactions were very positive. Elmer Sperry, for example, said that this whole matter, interesting as it might be from a chemistry standpoint, was much more intriguing from an engineering standpoint; during the plenary discussion he illustrated this point with examples. Baekeland returned home after midnight, but he could not sleep and wrote a ten-page report in his diary. The next day there were articles in newspapers such as the *New York Herald*, the *Sun*, and the *Tribune*, one bearing the headline "Here's to $O_7H_{38}C_{43}$. The Bakelite A to B to C."

The paper he presented on that memorable evening was published three months later in the new *Journal of Industrial and Engineering Chemistry* (Baekeland, 1909a). It was a three-part article: It reviewed the history of phenol-formaldehyde condensation research, explained the Bakelite manufacturing process, and discussed Bakelite's chemical structure. The historical part has had quite an influence on subsequent accounts of plastics history. Most historians have, for example, accepted the cast of characters as presented by Baekeland in this paper. Thus most of them missed Arthur Michael's work, which would have disrupted the scheme of gradually accumulating knowledge about the phenol-formaldehyde condensation reaction.[139] This scheme was presented by Baekeland with skillful rhetoric, leading to the apotheosis of his invention of Bakelite. In his first sentences he set the stage by emphasizing in one sweep its familiarity and its complexity:

> since many years it is known that formaldehyde may react upon phenolic bodies. That this reaction is not so very simple is shown by the fact, that according to conditions of operating or to modified quantities of reacting materials, very different results may be obtained; so that bodies very unlike in chemical and physical properties may be produced by starting from the same raw materials. (Baekeland, 1909a: 149–150)

He continued by sketching a nicely linear development from Baeyer via Kleeberg to Luft and Story. This historical setting thus placed him squarely within the relevant social group of celluloid chemists. We shall now see how this functioned as a stepping stone for the newly emerging technological frame plus relevant social group of Bakelite engineers.

The second part of the paper began: "This will close my review of the work done by others and I shall begin the description of my own work by outlining certain facts, most of which seem to be unknown to others, or if they were known their importance seems to have escaped attention. Of these facts I have made the foundation of my technical process" (Baekeland, 1909a: 153). The Bakelite process he then described constituted a further, fifth modification of the meaning of Bakelite. Building on his historical introduction, he now set up the Bakelite process in terms of previous researchers. He distinguished two classes of resinous products: "the first class includes the products of the type of Blumer, DeLaire, Thurlow, etc.,... [and] the second class includes the products of Kleeberg, Smith, Luft, Story, Knoll as well as my own product." In the subsequent argument he stressed the crucial role of heat and pressure, elaborated on the use of alkaline condensation agents as valuable for the manufacture of solid condensation products, and specified the role of the "intermediate condensation product" Bakelite B. Thus the fifth stabilized meaning of Bakelite stressed its versatile character by specifying the process rather than the product. This added to the gradual differentiation between the celluloid and the Bakelite technological frames. Baekeland described:

> The preparation of these condensation products A and B and their ultimate transformation in C for technical purposes constitute the so-called Bakelite process. This can be described easily:
>
> I take about equal amounts of phenol and formaldehyde and I add a small amount of an alkaline condensing agent to it. If necessary I heat. The mixture separates in two layers, a supernatant aqueous solution and a lower liquid which is the initial condensation product. I obtain thus at will, either a thin liquid called *Thin A* or a more viscous mass, *Viscous A* or a *Pasty A*, or even if the reaction be carried far enough, a *Solid A*.
>
> Either one of these four substances are my starting materials and I will show you now how they can be used for my purposes.
>
> If I pour some of this *A* into a receptacle and simply heat it above 100°C, without any precaution, I obtain a porous spongy mass of *C*. But bearing in mind what I said previously about dissociation, I learned to avoid this, simply by opposing an external pressure so as to counter-act the tension of dissociation. With this purpose in view, I carry out my heating under suitably raised pressure, and the result is totally different.
>
> This may be accomplished in several ways but is done ordinarily in an apparatus called a Bakelizer. Such an apparatus consists mainly of an interior chamber in which air can be pumped so as to bring its pressure to 50 or better 100 lbs per square inch. This chamber can be heated externally or internally by means of a steam jacket or steam coils to temperatures as high as 160°C or considerably higher, so that the heated object during the process of Bakelizing may remain steadily under suitable pressure which will avoid porosity or blistering of the mass. (Baekeland, 1909a: 156)

Baekeland then stressed the multitude of applications, first giving examples of impregnated wood, cardboard, or pulp board, then showing its use as chemical-resistant lining and coating material, and finally demonstrating Bakelite as molding material: "As to Bakelite itself, you will readily understand that it makes a substance far superior to amber for pipe stems and similar articles. It is not so flexible as celluloid, but it is more durable, stands heat, does not smell, does not catch fire and at the same time is less expensive" (Baekeland, 1909a: 157). This brought him to discuss at length the other element in the fifth meaning of Bakelite: its character as a molding material.

Although Baekeland had perceived its potential use for molding at an early stage, a detailed argumentation and exemplification was only realized in the course of 1908 and formulated in his 1909 paper. He first argued the general importance of molding materials:

> The great success of celluloid has mainly been due to the fact that it can easily be molded.... The addition of camphor and a small amount of solvent to cellulose nitrate was a master-strike, because it allowed quick and economic molding. In the same way white sand or silica would be an ideal substance for a good many purposes, could it be easily compressed or molded into shape and into a homogeneous mass. But it *cannot*; and therefore remains worthless. (Baekeland, 1909a: 157)

Bakelite C does not mold, and researchers before Baekeland had tried to solve this problem "by the admixture of solvents and subsequent evaporation, but we know now that these very solvents imply most serious drawbacks" (Baekeland, 1909a: 158). Baekeland then explained the important cost factor of the time that an article actually spent in its mold, and he showed how his process reduced this time to a minimum:

> As stated before, the use of bases permits me to make a variety of A that is solid although still fusible.... A mixture of the kind is introduced in a mold and put in the hydraulic press, the mold being heated at temperatures preferably about or above 160–200°C. The A melts and mixes with the filler, impregnating everything; at the same time it is rapidly transformed into B. But I have told you that B does not melt, so the molded object can be expelled out of the mold after a very short time and the mold can again be refilled.... At the end of the day's work or at any other convenient time all the molded articles are put in the Bakelizer and this of course without the use of any molds. (Baekeland, 1909a: 158)

In the Bakalizer the articles then were transformed into the hard, infusible, and insoluble Bakelite C (see figure 3.8).

In the last part of his paper Baekeland discussed the chemical structure of Bakelite, drawing mainly on his success in synthesizing Bakelite

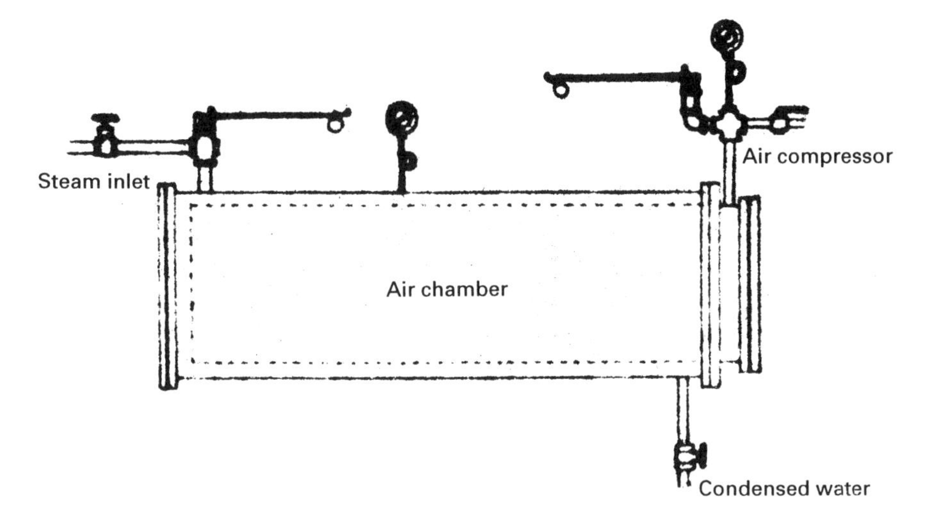

Figure 3.8
The Bakelizer: "an apparatus consisting essentially of an inner chamber where the objects are placed and in which, by means of a suitable pump, air can be compressed to 100 or 120 pounds. . . . A steam-jacket heats the chamber to a temperature of 140° to 180°C (Baekeland, 1909b: 596–597).

via different routes. He concluded that Bakelite is a polymerized oxy-benzyl-methylen-glycol-anhydride: n($C_{43}H_{38}O_7$). This final, sixth meaning of Bakelite—its chemical formula and as such the rock-bottom identity from a naive point of view—would later be called false when chemists gained more insight into the nature of macromolecular structures and polymerization processes in the 1920s.[140]

Also the very subject of his first public presentation of Bakelite contributed to Baekeland's rewriting the history of his research process, as did the 1916 paper. In choosing the molding, Kleeberg-type of Bakelite to be presented first, he further played down his own efforts directed toward developing varnishes and shellac substitutes. The second paper, delivered at the general meeting of the American Electrochemical Society on 8 May 1909, was devoted again primarily to solid Bakelite applications, now categorized in four groupings: block-working, impregnating, coating, and molding (Baekeland, 1909b). In this paper Baekeland discussed at much more length a wide variety of applications and concluded by showing the relative advantages of Bakelite over older materials such as rubber, shellac, and porcelain. The last paper of this publication campaign was presented before the Chemist Club again, on 14 May 1909

(Baekeland, 1909c). It dealt with the soluble and fusible family of condensation products—the varnishes and shellac substitutes.

Upscaling Production

The success of the first paper in February, enhanced by the coverage in the general press, caused the mail to pour in.[141] The account book shows an exponential increase in orders, mainly of five types of Bakelite ("solid A," "pasty A," "dissolved A," "liquid A," and "high-voltage insulating Bakelite").[142] This first commercial production and the "reduction to practice" was carried out with a small staff, which now consisted of Nathaniel Thurlow, Lewis Taylor (family chauffeur and general helper), August Gothelf (an analytic chemist), and Jim Taylor, John Hickey, and Lawrence Byck (assistants since early 1909). The circumstances under which this production went ahead are captured in the words of Byck, describing how the production facilities consisted of one small cast-iron jacketed still, little "Old Faithful" (see figure 3.9).

> "Old Faithful," complete with reflux and distillling condenser and receiving tank, was set up in a corner of Dr. Baekeland's garage, adjacent to the laboratory. After one serious laboratory fire,[143] the Doctor decided he would rather the garage burn down than the laboratory. "Old Faithful" had an agitator and this required motive power, but electric power lines had not yet reached Harmony Park (although electric light of the 1900 vintage was installed). So the Doctor acquired a steam engine from an old White Steam automobile....
>
> A still must be charged with raw materials. This was easy. Fortunately, the major bulk raw materials in use at that time, cresol and formaldehyde, were both liquids. A hand operated gear pump on tripod legs easily straddled a barrel of raw material standing on a portable platform scale.... In making the first varnishes, addition of the alcohol at the crucial moment had to be made much more quickly than was possible with the little pump. So the alcohol was dumped onto the hot resin, through the open manhole, by hand from buckets. This was always an interesting, not to say exhilarating, moment. Lewis Taylor did this, invariably with the entire staff (and frequently the Baekeland family) as audience—at a safe distance!... Alcohol vapor fires were commonplace.[144]

Obviously, the forty-gallon capacity of this "pilot plant" could not meet rising production demands. By the end of 1909, there were dozens of customers who had placed repeat orders of what were considered commercial quantities.

Not all efforts to apply Bakelite were successful, not even if they were carried out by people from outside the relevant social group of celluloid chemists. The attempt of the Protal Company, for example, managed by F. G. Wicchmann (who had cxperience as a chemist in a sugar refinery), proved a complete failure. The company was on the verge of bankruptcy

Figure 3.9
The first reaction vessel, "Old Faithful." Photo in Baekeland Papers, courtesy National Museum of American History, Smithsonian Institution, Washington, D.C.

when Baekeland made his first public presentation on 5 February 1909. Wiechmann read about it in the press and hoped that Bakelite might save his enterprise. He did some experimenting, collected extra funds, and bought manufacturing equipment. He was planning to produce a Bakelite material with vegetable protein filler. However, when Baekeland visited the factory, he "received a mental shock when I saw that equipment; most of the machinery was either all wrong, or was installed in an absurd way, so as to make it unsuitable for profitable manufacturing."[145] In 1910 the Protal Company was dissolved.

Bakelite was not only finding application as a substitute for celluloid, shellac, hard rubber, and similar existing materials, but it was increasingly used in new situations. The electrical industry was expanding rapidly and the automotive industry was in its fast-growing infancy, and both urgently needed new insulating materials that were electrically better, resistant to chemicals, able to withstand heat, and amenable to mass production. Electric lighting and telephone service were expanding, wireless telegraphy was on its way, amateur photography was booming, the phonograph had taken the public by storm—and all these new technologies provided a potential market for Bakelite.

Baekeland was acutely aware of these commercial possibilities. One way to meet the increased demand was to issue licenses for the production of Bakelite. Most industries, however, had problems in working with Bakelite. Indeed, Baekeland complained that he "had to go again all through the same old experience as in the introduction of Velox" (Baekeland, 1916: 186). That is, he would not be able just to issue licenses on a royalty plan for the use of his patents; it even proved impossible to sell only the partly finished product (Bakelite A); he had to go to the trouble of completely manufacturing Bakelite C himself. Thus Baekeland turned, again, from being a research chemist into an entrepreneur.[146]

In the spring of 1910, a prospectus was drafted that described two financial schemes for the establishment of a Bakelite company.[147] The prospectus listed possible sources of profits, totaling $770,000 per year, but added that "there is every reason to believe that this estimated amount is much too small, and that with proper effort this income could easily be doubled." The General Bakelite Company was founded on 29 September 1910, opening its office in New York on 5 October. Baekeland transferred his patent rights in the United States, Canada, and Mexico to this company and received a modest sum of cash and enough shares of stock to control the company. By that time the company had still no other factory but the pilot plant in Baekeland's private Yonkers laboratory. In the process of establishing the General Bakelite Company, however, Baekeland had approached the Roesler & Hasslacher Chemical Company. This firm was the prime U.S. importer of phenol and cresol. One of R & H's subsidiary companies, the Perth Amboy Chemical Works in Perth Amboy, New Jersey, had space available in one of its buildings. To make the fit even more ideal, the Perth Amboy Chemical Works was the only domestic producer of formaldehyde at the time.

It took five months to rehabilitate a ramshackle corrugated metal building at Perth Amboy and to install the production equipment. Experiences on the start-up day, 23 February 1911, illustrated some of the risks of employing new workers to upscale production. Byck remembered an action by a new employee, J. J. Frank, on that first day:

The very first large batch of the product to be made at Perth Amboy was laminating varnish. It was an historical event.... Everything went wonderfully well—the resin was reacted, dehydrated, finished, the alcohol was added. Dr. Baekeland asked whether the resin was all dissolved or was there a possibly a lump of "B" around the agitator. Quick as a flash J. J. grabbed a [kerosene] lantern, swung open the manhole door and held the lantern in for a look-see. There was a mild BOOM, the hot alcohol vapor ignited, flames spurted from the manhole.... Fortunately, J. J. was not badly hurt; he lost his hair, eyebrows, eyelashes, moustache, and his job. The batch of varnish was saved.[148]

In March 1911 sales were $2,826, and in April this amount had already doubled to $5,612. Also in April they received an order for $4,000 worth of pipe stems, but there was insufficient equipment to fulfill it; only two months after starting production, the plant proved too small. A new three-story building was erected in Perth Amboy in 1914, and this building was doubled in length and raised to five stories in 1917. Sales soared during this period (see figure 3.10).

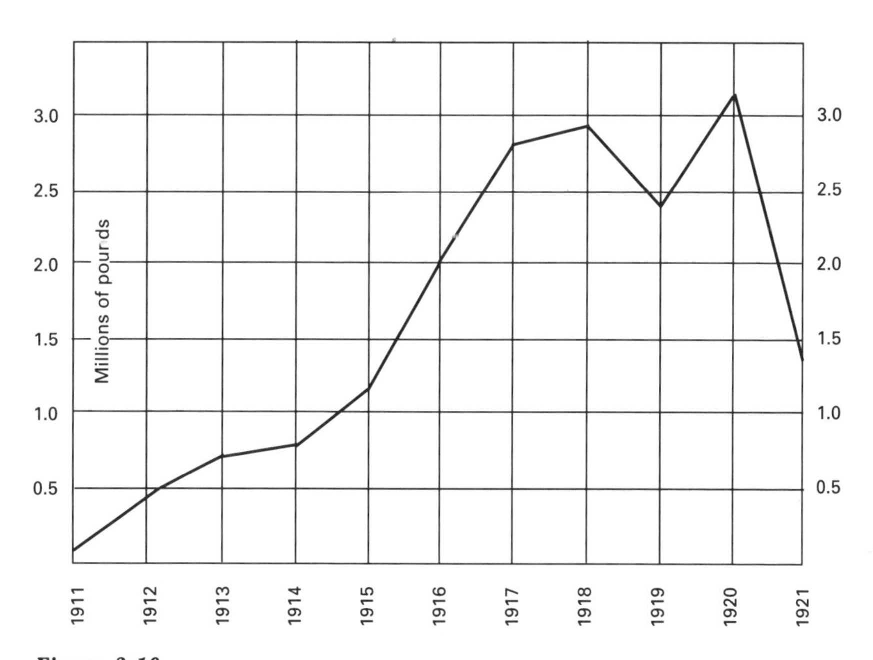

Figure 3.10
Net sales in pounds of thermosetting products (that is, without lacquers, cements, etc.), 1911–1921, by the General Bakelite Company (Byck, 1952: 5).

From 1911 to 1917, several developments contributed to the further stabilization of Bakelite. First there were a number of patent litigations, which further formed the newly emerging relevant social group of Bakelite engineers—partly because more chemists were enrolled from the celluloid chemists' group, and partly because the trials helped to explicate the Bakelite technological frame. Then there was the further collaboration with other industries. Third was the influence of World War I, and finally there was the influence of two more relevant social groups on the construction of Bakelite, industrial designers and consumers. I will briefly discuss these five strands of development that contributed to Bakelite's stabilization.

Patent Litigation

Almost at the same time that Baekeland's patents were issued, other researchers filed patents for phenol-formaldehyde condensation products. Most of these were primarily drying processes with all the related drawbacks.[149] The first really competing patents were filed by the German firm Knoll and Company (1907, 1908a,b) and their chemist Hans Lebach (1908). The similarities between this resin and Bakelite were obvious. Lebach, for example, also distinguished three types of condensation product: "Resit A," "Resit B," and "reine Resit" ("pure Resit"); the trade name was to be "Resinit" (Lebach, 1909a). For patent purposes it was of course crucial whether the two artifacts were indeed identical. The cards were ambiguously dealt: Lebach's U.S. patent was filed on 21 December 1908, more than a year *after* Baekeland's key patents, but Lebach's first patent, in Germany, was reportedly filed in February 1907 (Lebach, 1909c: 1599), five months *before* Baekeland's U.S. patents. The inventor who considered himself first had to argue for the identity of the two artifacts, while the one thinking himself second could only win by stressing the difference.

Baekeland asked whether there was "a real difference" (*ein tatsächlicher Unterschied*) (Baekeland, 1909d). In an article Lebach responded by extensively discussing Baekeland's analysis of the chemical formula for Bakelite and then suggesting some modifications. Surprisingly Lebach (1909c: 1600) then agreed with Baekeland about Resinit and Bakelite being almost identical, but his reason for doing so became clear in the same sentence: "Bakelite and Resinit are not very different at all, apart from the fact that Resinit, as is clear from the date of our German patent application and albeit Baekeland's contrary idea, was somewhat earlier conceived than Bakelit." Finally he discussed his process that provided, he argued, a more simple, inexpensive, and small-scale method for hardening in the final production phase than Baekeland's heat-and-pressure process. His was the acid-hardening method: a 20–30 percent solution of HCl applied to a Resinit A or Resinit B mass would start the hardening process, but slowly enough to allow the molding, shaping, or casting of objects. After hardening, the acid could easily be washed out by rinsing with an aqueous soda solution.

Baekeland (1909e) cast doubt on the identity of Resinit and Bakelit, suggesting that Lebach's Resinit was a mixture of different chemical bodies rather than one polymerization product such as Bakelite C. Further, he doubted the technical feasibility of the acid-hardening process. This had, according to Baekeland, implications for the question of which product had come first:

> Dr. L. claims that Resinit was conceived before Bakelite. I agree completely with him if he means with the name Resinit any insoluble, infusible condensation product of phenol and formaldehyde. Indeed, he then can go back many years and cite the work of Baeyer, Kleeberg, Luft and Story and others, as I have done in my first article.
>
> However, it is not merely the issue to produce a specific chemical body. The problem is much more complex, since the goal is to produce and manufacture a product in such a way that it can be used reliably for very specific technical purposes.[150]

Lebach replied in the same issue of the *Zeitschrift für angewandte Chemie* by sustaining his claim of the identity of pure Resinit and Bakelite C, but he stressed again that his acid-hardening process was different from the Bakelite process. Baekeland would not be able to protect anything other than his heat-and-pressure process, Lebach concluded (1909d).

He was not right. As would become clear in the next few years, the patent situation was unusually satisfactory for Baekeland, because the process did not depend on a single patent or a single set of claims. Instead, the patents were so related that it was almost impossible to carry out anything practical without infringing on at least three or four patents at the same time. Moreover, the priority patent claimed by Lebach was not granted. On 1 November 1909, Baekeland agreed with Knoll & Company (of which Lebach was an employee) and with Rütgerswerke AG to found a German firm as licensee for the production of Bakelite. Thus the Bakelit Gesellschaft mbH. was established on 25 May 1910, and Lebach soon joined its staff. After this commercial closure of the controversy, the scientific controversy was closed as well. Baekeland (1911b, 1912, 1913) publicly recognized the value of the acid-hardening process for specific purposes. In doing so, the identity between Bakelite C and acid-hardened Resinit was constructed after all.

The two other patent struggles, against the competing phenol-formaldehyde resins Condensite and Redmanol, resulted in 1922 in similar arangements when former competitors joined forces with Baekeland.[151] The first thread of this complex story starts in 1909 with J. W. Aylsworth, chemical consultant for Thomas A. Edison. Aylsworth was trying to develop a material for the manufacture of gramophone discs, which were to replace wax cylinders. He read Baekeland's papers, started research himself, and was granted several patents (Aylsworth, 1909, 1911a–h, 1912, 1913, 1915). Together with Kirk Brown, a wealthy man of considerable business experience, he formed the Condensite Corporation of America on 23 September 1910, which started production in an old Edison battery plant in Glen Ridge, New Jersey. Some six months later Brown and Aylsworth received notice from the General Bakelite Company that Condensite was infringing on the Bakelite patents and that they were accordingly warned to stop production immediately. In July, Baekeland (1911a) published a detailed and rather critical analysis of Aylsworth's Belgian patent. Brown called on Baekeland and they agreed that the Condensite Corporation would continue production under license of the General Bakelite Company, and that this license would be exclusive. Practical implementation of this agreement failed and another suit was filed, this time adjudicated in favor of Condensite. The General Bakelite Company agreed to pay royalties for use of two more recent patents by Aylsworth (1913, 1915). It was also agreed—and this would turn out to be consequential for the second development—that the General Bakelite Company was to bring suit against any party that infringed the Bakelite or Aylsworth patents (Haynes, 1945b).

The second thread of the big American patent struggle over phenol-formaldehyde resins started with a young Canadian chemist, L. V. Redman. Redman developed phenol-formaldehyde varnishes, applied for patents, and started a business in 1914 with Adolph Karpen, a Chicago manufacturer of furniture. The Redmanol Chemical Products Company began production of transparent cast resin and later expanded into the laminating and molding fields. Formica was an early Redmanol account. Inevitably the question arose of whether Redmanol was infringing on the Bakelite patents. The opening fight was conducted in academic circles. Redman et al. (1914) published a long and studious article that first reviewed all academic and patent literature, then carved out a niche for further study, and finally presented the new synthetic plastic. So much attention was devoted to the theoretical discussion of the chemical structure of the various condensation products that the last sentence almost could escape unnoticed: "Application for basic and process patents on these resins and their uses have been made" (Redman et al., 1914: 15). Baekeland (1941b: 167) hit back immediately: "The paper in question [by Redman et al.] would have had increased importance if some of the opinions and statements expressed therein were more in accordance with facts." Page by page, almost sentence by sentence, Baekeland took the Redman article apart, bringing in Aylsworth's and Lebach's patents as allies where necessary. After taking care of all the academic issues, Baekeland concluded that

> If we add that all the infusible materials made from phenol (hydroxybenzol) and formaldehyde and ammonia, or hexamethylentetramin, dry or wet,[152] have the same specific gravity, the same color, the same appearance, the same resistivity to solvents and chemicals, and that up till now, no property has been mentioned which is not common to all these products, it becomes easy to draw conclusions as to their absolute identity. (Baekeland, 1914b: 170)

Of course, there is no self-evident, unambiguous way in which these articles allow us to decide which party has "truth" on its side; a favorable legal finding, however, was granted to Baekeland by Judge Thomas I. Chatfield, who declared the Bakelite patents valid and infringed.[153]

Redmanol now seemed to be in a very weak position. Karpen is attempt to get a license was denied because of the exclusive Condensite license. He offered to sell the Redmanol company, but the General Bakelite Company did not want it. He offered to buy Bakelite, but they did not want to sell. It turned out that Karpen had actually been buying control of stock in the Condensite Company during the suit because he had anticipated this negative outcome. Thus he confronted his previous enemies Baekeland and Brown with a ménage à trois, whether they liked it or not. The only sensible way out of this deadlock was a merger of the three firms, and on 3 March 1922 the Bakelite Corporation of Delaware was named as a holding company to own the stock of all three. The main actors shared the highest executive positions in the new firm, Baekeland continuing as president. The editor of the *Journal for Chemical and Metallurgical Engineering* was jubilant:

> The chemical industry will be the first to welcome this amicable settlement of a historic controversy. We cannot help feeling that the merger opens the way to a wonderful opportunity for further progress and expansion. The constituant companies will continue on a competitive basis of manufacture; but there will be an

exchange of information obtained in the several research laboratories. Thus the conditions will be most favorable for progress in the art of manufacture and for service to the industries consuming the product.[154]

These patent struggles and commercial rearrangements did contribute to the stabilization of Bakelite, as can be seen in Baekeland's (1911b, 1912, 1913) description of Bakelite after the closure of the controversy with Lebach and his review of phenol-formaldehyde resins after the resolution of the conflict with Aylsworth and Redman (Baekeland and Bender, 1925). A last example of how Bakelite was further socially constructed in these patent suits is the "invention" of a new entity: "reactive resins." Judge Chatfield defined these as "condensation products which can be transformed by the application of heat alone, or transformed and controlled by the application of heat and pressure, or transformed by the application of heat followed by evaporation or drying, into an infusible or insoluble product, without the addition of chemical substances to carry on the reaction."[155] Chatfield identified the substance produced by the Redmanol Company as this "reactive condensation product," which implied that it was not a Story-type substance but identical to one of Baekeland's materials and indeed infringed on his patents.

The merger between the three American companies was followed by a similar one in England, where Swinburne's Fireproof Celluloid Company Ltd. combined with Mouldensite Ltd. and Redmanol Ltd. to form the Bakelite Ltd. company. Besides the German and British companies, already mentioned, by the end of the 1920s other foreign affiliations were the Japan Bakelite Company Ltd. (formed in 1931, extending work that had started under license by the Sankyo Company in 1915), the Bakelite Corporation of Canada Ltd. (formed in 1925), and the Societa Italiana Resine in Milan.[156]

Further Collaboration with Industrial Social Groups

The second way in which Bakelite was further constructed after 1911 was through work at the Perth Amboy plant and close collaboration with other industries. Much of the early efforts went into the production of "Transparent," a cast resin for smokers' articles. Other projects were postponed, and as Byck remarked: "Looking backward now, it seems as though possibly a disproportionate amount of time and effort were spent on this problem, at a time when so many other, larger, more important industries were clamoring for attention. But the early cast resin was a high-priced, profitable item; there were good customers eager to buy and none knew how large this business might grow."[157] Molding Bakelite was still causing trouble—there was poor mold release, and the various molding materials tended to entrap air and to blister. But, Byck commented, these new molding materials "were 'fast molding'—only about five minutes' cure was required for small pieces, not counting the cooling under pressure. Yet for that period and considering the properties of the molded pieces compared to other available materials, this was phenomenal. Sales continued to expand."[158] The employment of Lebach's acid-hardening and similar techniques further decreased the molding time, but introduced new problems such as bad mold staining, fouling, and mold etching.

These difficulties with molding Bakelite resulted in the relative commercial importance of laminated varnishes (see figure 3.11) during the first two decades of the business (see table 3.4). It was not until 1930 that the sales volume of molding materials surpassed the volume of laminat-

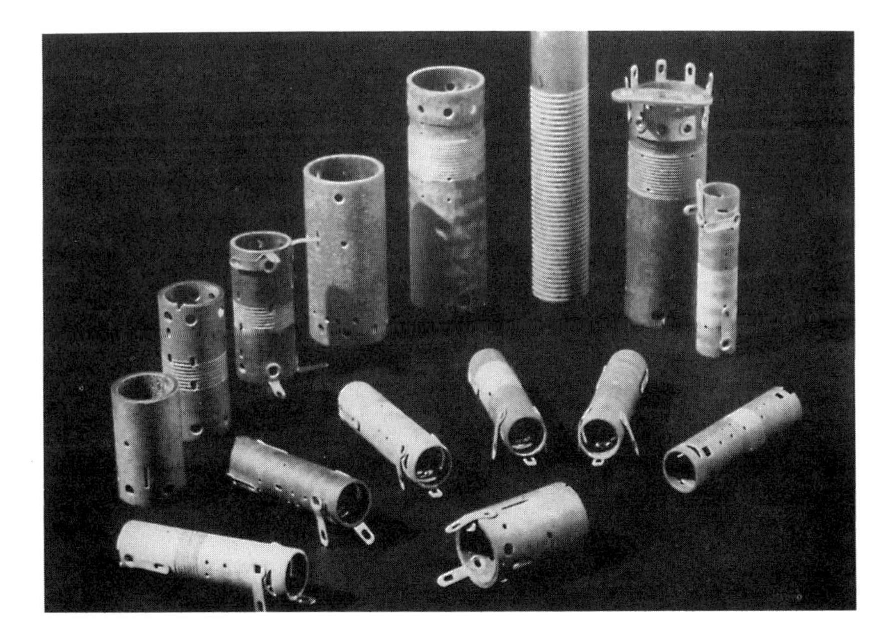

Figure 3.11
Laminating plastics were produced by bakelizing superimposed layers of paper or fabric that had been impregnated with Bakelite. The photo shows a group of electrical coil forms produced from paper-base laminated tubing. Photograph courtesy of the Warshaw Collection of Business Americana, National Museum of American History, Smithsonian Institution, Washington, D.C.

Table 3.4
Sales volume of the General Bakelite Company, classified by important product categories

Year	Laminating varnish	Molding material	Cast resin	Other	Total	Unit sales price ($)
1911	53,918	4,365	2,576	39,031	99,890	.370
1912	162,704	131,234	4,661	160,474	459,073	.407
1913	393,188	174,521	26,725	113,342	707,776	.419
1914	395,507	261,347	14,968	111,831	783,653	.390
1915	551,361	445,425	26,022	172,183	1,194,991	.405
1916	1,099,336	630,986	40,297	233,913	2,004,532	.433
1917	1,960,921	588,731	53,696	214,182	2,817,530	.445
1918	2,089,766	542,950	101,601	185,635	2,919,952	.517
1919	1,292,130	684,103	140,508	224,355	2,341,096	.607
1920	1,986,248	756,296	143,449	247,245	3,133,238	.650
1921	929,237	197,725	24,346	207,426	1,358,734	.481

Source: Byck, 1952: 9.

ing varnishes.[159] Most of the products that I will discuss in the remainder of this section were indeed made of laminating Bakelite.

Also, some intended applications failed, and the meaning of Bakelite was accordingly revised. Some fillers were less successful than was thought. Bakelite paper was a failure and was dropped; after long experimentation and customer trials, the phonograph discs were also dropped. Novolak varnishes for finishing wood and liquid Bakelite for impregnating wood were disappointing and gradually faded from the scene. This eclipse of the varnish element from the meaning of Bakelite explains why Baekeland, when giving a reconstruction of the early days of his Bakelite research (Baekeland, 1916), played down his search for varnishes and overemphasized his early recognition of the value of molding Bakelite.

Two new industrial fields were crucially important for Bakelite's prosperous development: automobile and electric technologies. The automotive ignition people took to Bakelite. Its insulation properties, chemical resistance, and qualities as a molding material soon convinced most producers. Remy Electric and Charles "Boss" Kettering were among the first customers of the Loando Hard Rubber Company, but soon established their own molding plants. Kettering used Bakelite for his "Delco" ignition and starting system that revolutionized the automobile field. It

Figure 3.12
Typical Bakelite switches. Courtesy Museum Boymans-van Beuningen, Rotterdam; Catalogue No. 288, 1981.

has even been claimed, with some exaggeration, that the modern ignition system was made possible by Bakelite. Acquainted with Bakelite in this way, automotive engineers soon found other applications: steering wheels, door handles, instrument panels, magneto couplings, timing gears, heaters, gearshift knobs, and radiator caps.

The same qualities (with less emphasis on Bakelite's chemical resistance) made it possible to capture the electric industry at large (see figure 3.12). Switches used to be made of rubber, hard rubber, porcelain, pressed paper, and even brass. These were all quite dangerous. Rubber products could not stand temperatures above 50°C. Under the influence of sunlight, sulphur evaporated from all vulcanized rubber products, which was harmful for metal parts. Porcelain could stand very high temperatures but was brittle and difficult to mold, especially in combination with metal components. Pressed paper, like wood, would absorb moisture and thus become electrically conducting. Since the first molded articles—tiny bobbins for the support of movable field coils produced by the Loando Company for the Weston Electrical Instruments Corporation (1909)—a plethora of products was developed, from industrial applications such as third-rail insulators for subway tracks (1910), molded meter covers (1914), and circuit breaker insulation and dash pots for elevators (1921) to numerous parts for home appliances such as toasters, washing machines, electric irons, vacuum sweepers, and ventilators. The telephone industry did not only use Bakelite for molding the casing of telephones, but also in a multitude of special parts in the exchange: grasshopper fuses, relaying insulators, pulse machine drums, dividers, sender finders, selectors, and sequence switches.

The booming telephone and radio industries, with their need for small precision-molded components, gave further stimulus to the application of Bakelite for electrical purposes. Already in 1915 a Bakelite advertisement featured a large commercial radio set equipped with a Bakelite laminated panel. For radio amateurs "Bakelite" quickly became a household word: lamp sockets, headphones, assembly boards, molded dials, coil forms, and numerous other parts were made of Bakelite materials. Headsets gave way to horns, some of which were molded, and a host of new uses developed for Bakelite: static eliminators, lightning arresters, inside aerial frames. Then the "furniture period" began and consoles, highboys, and lowboys invaded the family room. New Bakelite materials were developed and surfed the crest of the radio wave.

Although Baekeland had studied grinding wheels as early as 1908, it was not until 1921, when the cold molding process was perfected, that abrasives bonded with Bakelite became commercially feasible. The new wheels, first advertised in 1922, could operate at 50 percent higher speeds than usual, which resulted in important savings of time and labor in all sorts of metalworking industries.

Packaging and closure was another field of application, now routinely associated with plastics but originally developed by Bakelite engineers. Lightness in weight and resistance to chemicals were important, but perhaps even more attractive were the possiblities for molding unusual shapes, using a variety of colors, and incorporating trademarks in the closure (see figure 3.13).

The enrollment of these new relevant social groups—most prominently the automotive industry and the radio and electricity industries (see table 3.5)—suggests a rather uncomplicated growth of Bakelite production and spread all over the world. This was, however, not the case: World War I disrupted this scenario and played a complicated role in partly hampering and partly stimulating the further development of Bakelite.

World War I

The third important strand in the social construction of Bakelite after 1911 was World War I. It is as clear that the war was highly influential on Bakelite's development as it is difficult to assess this influence unambiguously. First there were some consequences at the institutional level. The Alien Property Custodian Act affected the General Bakelite Company because of its intimate relationship with the Roesler & Hasslacher Chemical Company, who had large German stockholders. It became necessary to divorce the two firms. Perth Amboy, located on the waterfront, was in a military area and under guard day and night. The influ-

Figure 3.13
"De vergulde hand" soapbox. Form and decoration (with the trademark) of refillable boxes were used to enhance the recognition of the product. Photograph courtesy of Collection Becht, Naarden, The Netherlands.

ence of the war was evident in nonmarket terms as well. And especially when the great "Spanish flu" broke out in 1918, which was to exact a death toll of about 24,000 in the U.S. army (compared to 34,000 deaths on the battlefield)— business virtually came to a standstill.

Then there were the effects on the plastics market. On the one hand, there was an increase in demand, for example, for ignition systems for military trucks and airplanes (Redman and Mory, 1931). On the other, there was the restraining influence of scarcities of raw materials. Most manufacturers had large supplies of phenol, but nevertheless its price rose rapidly from $0.08 to $1.88 a pound. When the prewar supplies were exhausted, cresol was used, despite its slower reaction characteristics. Formaldehyde also became scarce, resulting in an eightfold price rise.

There is a remarkable difference between the developments in the United States and in Europe. The general scarcity of raw materials, including shellac and rubber, caused a rather wild search for substitutes in Europe. These substitutes were mostly of inferior quality, and their composition and technical properties generally changed every month. At

Table 3.5
Sales volume of the General Bakelite Company, classified by important use categories

Year	Automotive	Electric and radio	"Smokers articles" cast resin	"Bedstead" lacquer	Other	Total
1911	8,524	77,516	2,576	5,420;	5,854	99,890
1912	103,145	231,679	4,661	104,386	15,202	459,073
1913	181,163	420,616	26,725	56,381	22,892	707,776
1914	203,521	481,336	14,968	43,902	39,926	783,653
1915	276,683	726,370	26,022	78,633	87,283	1,194,991
1916	533,952	1,201,770	40,297	41,673	186,840	2,004,532
1917	698,314	1,849,398	53,696	12,778	203,344	2,817,530
1918	467,374	2,154,159	101,601	16,072	180,746	2,919,952
1919	663,060	1,336,432	140,508	12,907	188,189	2,341,096
1920	762,224	2,005,597	143,449	9,128	218,480	3,133,238
1921	379,515	772,615	24,346	4,132	178,126	1,358,734

Source: Byck, 1952: 10.

the end of World War I the market was flooded with these wartime substitute plastics, damaging the regulatory function of the market as well as the public image of synthetic plastics. In their names these new products sought to stress the similarity with established products such as Bakelite (see figure 3.14). Hence many people thought that Bakelite was merely a mass of glued and compressed wood and paper pulp; the high price went toward paying for patents and the industry's high profits (Micksch, 1918; Brandenburger, 1934). The image of synthetic plastics was profoundly corrupted for more than a decade. The German word *Ersatz* ("substitute") had become, in both Dutch and English, synonymous with "inferior quality," and Bakelite was considered by many to be "ersatz." Accordingly the sales of Bakelite and other truly synthetic plastics in Europe (especially in Germany) lagged behind sales in the United States.

In the United States the public image of plastics did not suffer such damage. Chemistry had come out of the classroom and laboratory and was enjoying the positive connotation of the "machine age."[160] In addition to this difference between Europe and America, another circumstance helped the prolific development of especially Bakelite. During the war the scarcity and high price of phenol prompted various companies,

Der Name Bakelite

ist uns durch Warenzeichen geschützt. Dieser Schutz erstreckt sich auch auf Zusammensetzung des Wortes „Bakelite“ mit anderen Worten, wie „Bakelite-Lack, Bakelite-Ersatz“ und dergl.
Wir werden unsere Warenzeichenrechte entschieden schützen und warnen vor unberechtigtem Gebrauch.

Bakelite Gesellschaft m. b. H., Berlin W. 35
Lützowstraße 32.

Figure 3.14
Many "Ersatz materials" had names similar to Bakelite. This prompted the German Bakelit Gesellschaft m.b.H. in turn to publish advertisements like this (from Kunststoffe 14 (1924): 14).

including the General Bakelite Company, to build new plants for the manufacture of phenol. Given the lengthy planning, building, and production preparation cycle, many of these plants started production only by the end of or even after World War I. This then resulted in large surplus stocks of phenol.[161] The resulting low phenol prices were very advantageous for the marketing of Bakelite products.

The Relevant Social Group of Industrial Designers

The fourth strand of development in the stabilization of Bakelite is the role played by industrial designers.[162] In the 1930s industrial design emerged as a profession. What we now associate with "industrial design" was until then typically restricted to designing luxury products that were produced in relatively small quantities. The new profession of industrial designers was born out of a lucky conjunction of an economic imperative for manufacturers to distinguish their products and the new streamlined (see figure 3.15) or machine-age style, "which provided motifs easily applied by designers and recognized by a sensitized public as 'modern'" (Meikle, 1979: 39; see figure 3.16). Specifically, the plastics industry sought to enroll the social group of designers to improve the image of their new materials and to distance it from possible "ersatz" connotations. The General Bakelite Company, for example, organized in 1932 a series of symposia, conferences, and courses for industrial designers (Dubois, 1972: 184–185; Meikle, 1979: 80–82). One of the main goals of

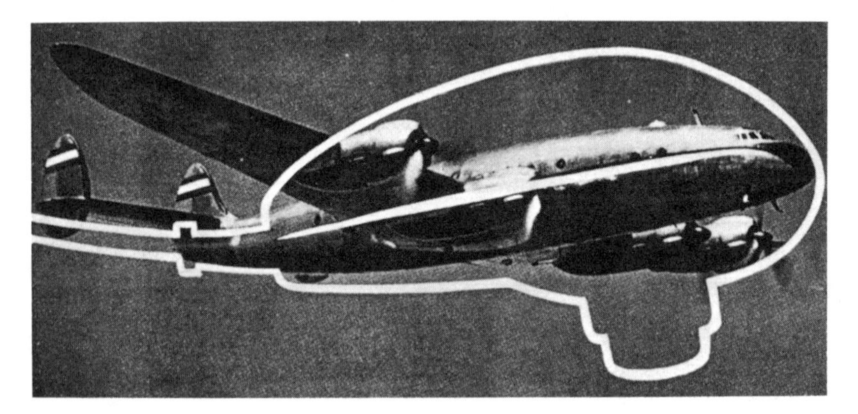

Figure 3.15
Top: The Philishave (type 7735), known as "the egg" (1948–1951), was made of Bakelite. *Bottom*: Philips advertisement saying, "The aerodynamical world of the man with his Philishave." Photograph courtesy of Collection Becht, Naarden, The Netherlands.

Figure 3.16
The use of the same design elements in the "Silvertone" radio in plastic (1938, Clarence Karstadt) and the Electrolux vacuum cleaner in enameled steel and cast aluminum (1937, Lurelle Guild) is obvious. The vacuum cleaner photograph is courtesy The Brooklyn Museum, 86.15, Gift of Fifty.

these symposia was to teach the new industrial designers, often artists by training, the technical constraints of molding plastics and thus help them get established. Accordingly, this development was a two-sided coin: not only did the industrial designers help to shape plastics; the technology of molding plastics also shaped industrial design, in terms of both professionalization and content—the specific machine-age or streamlined style.

The plastics industry's desire to upgrade the image of plastics coincided with the more general aim of distinguishing products by giving them a modern look. The General Bakelite Company directed advertisements at manufacturers, stating explicitly that "You can profit from the vogue for simple designs."[163] During the Great Depression, manufacturers had another goal: to produce better products for less money. Industrial design thus had to contribute to a reduction of the manufacturing costs of mass production while maintaining products' exclusive and luxurious appearance. And of course, using the new synthetic plastics was one straightforward way of doing just that, as the General Bakelite Company argued in their advertisement: "attractive modern designs" could be combined with "less costly molds." Another way to add a gloss of exclusiveity to a product was to use a variety of cheap metal inlays, which recalled of the expensive hand engravings in products made of wood, metal, or natural plastics (see figure 3.17). However, this deliberate promotion by plastic producers of the modern machine-age style (see figure 3.18) did not preclude the use of other design styles.[164] "The Material of a Thousand Uses," as Bakelite was called in the advertisements, was also used in more imitative forms (see figure 3.19).

As I mentioned, plastic products were not only molded into the modern shape, but the plastic molding material also shaped the modern style. One problem with molding products was the exact matching of parts, because the smallest difference in size would be obvious. The "old" solution of filing away the protruding rim was not acceptable because it would make the mass-production process too expensive. The designers' solution was to make small facet rims that camouflaged the difference in size (see figure 3.20). Another of Bakelite's technical features directly influenced the streamlined aspect of the modern style. The molding mass was very viscous, and the designer could guarantee an easy flow of the material through the mold. This implied the use of rounded rather than sharp edges (see figure 3.21) and of streamlined forms generally.

But the advent of Bakelite shaped modern industrial design in other ways as well.[165] The relevant social group of industrial designers directly

Figure 3.17
Three cigar boxes, made with one mold but with different inlays so as to give them a more exclusive character. Courtesy Museum Boymans-van Beuningen, Rotterdam; Catalogue No. 288, 1981.

(a)

(b)

(c)

(d)

Figure 3.18
Four Bakelite examples of machine-age style:
(a) Functionalist Ekco radio (1934, Wells Coates).
(b) Streamline Sonora radio (after 1945).
(c) Moderne (since 1966 called "Art Deco") Philips radio (1933).
(d) Biomorphic "Radio Nurse," a shortwave transmitter for home or hospital use (1938, Isamu Noguchi, The Brooklyn Museum Collection, 88.67).
Photographs a–c courtesy of Collection Becht, Naarden, The Netherlands.

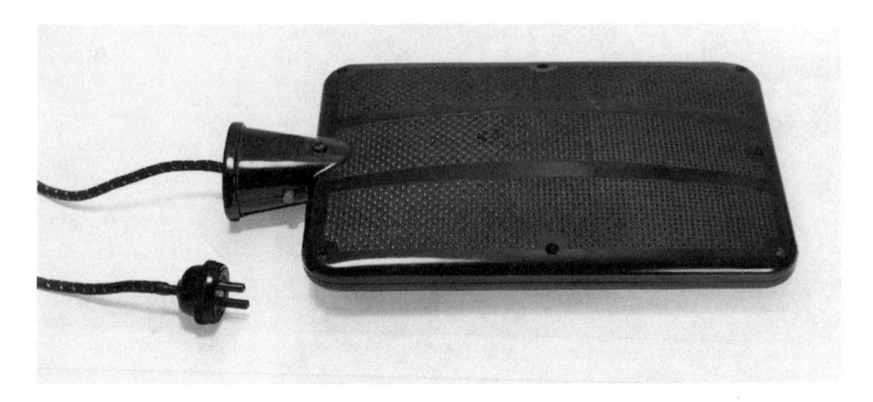

Figure 3.19
Designs were also based on the ability of Bakelite to imitate (semi-)natural materials: Electric hot water bottle made by R. A. Rothemel, Ltd. Photograph courtesy of Collection Becht, Naarden, The Netherlands.

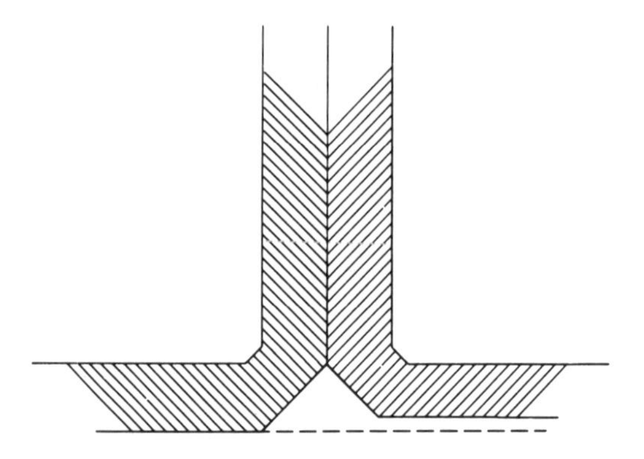

Figure 3.20
A facet rim will camouflage the small difference in size between two molded parts of one product. Another solution was to sandblast products, so as to give the surface a texture that would serve as camouflage.

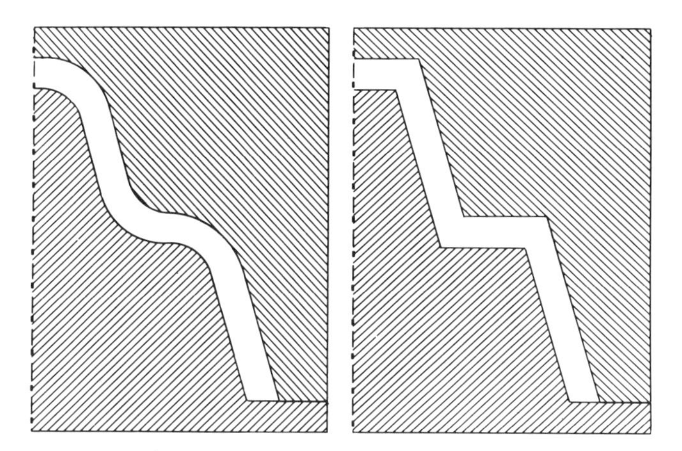

Figure 3.21
To allow a good flow of the viscous plastic through the mold, sharp edges and narrow passages were to be avoided.

benefited from the Bakelite campaign. Bakelite advertisements, for example, used the headline "Raymond Loewy says" and included a small photograph of the designer with a brief sketch of his career. "The 'New Deal' in industrial design is establishing the triumph of beauty through simplicity" was another typical line in Bakelite advertisements. Thus the campaign sought primarily to encourage manufacturers to use Bakelite as a stylish material, but at the same time boosted the reputation of the industrial design profession. This liaison between plastics producer and industrial design was not restricted to the General Bakelite Company. Many companies followed suit, and even the industry's trade journal was redesigned from *Product Engineering* via *Plastic Products* into *Modern Plastics*, containing after 1934 virtually no technical articles but focusing almost exclusively on design.

Thus was Bakelite's meaning still further modified through interactions with the relevant social group of designers. Not only was it a versatile molding material for mass production, but it was also a stylish advocate of the new machine age. It represented first the technological optimism of the pre-Depression years and a shrewd way for the business to get back onto its feet afterward. This meaning of stylish material culminated in Bakelite products being bought by museums of modern art and recently featured in large exhibitions.[166] After a Dutch Bakelite exhibition in 1981, the prices of Bakelite products at my local flea market rose two- to tenfold. At a 1989 auction, the recommended price for a

Figure 3.22
Jumo Desk Lamp (1945), now a collectors' item. Courtesy H.U. und U. Kölsch, *Collection Historische Kunststoffobjekte.*

Bakelite Jumo Desk Lamp (see figure 3.22) was *f*2000–*f*3000 ($1,000–1,500), and the lamp was sold for *f*5980 ($3,000) (Sotheby's, 1989).

The Relevant Social Group of Customers

The fifth important route in tracing Bakelite's stabilization is its construction by the relevant social group of consumers. It is often difficult to trace social groups of consumers and the meanings they attribute to artifacts. Seldom do consumers have their own direct voice, and in most cases we have to rely on spokespersons who are, strictly speaking, not members of the social group of consumers. Who are the consumers? This problem is notorious—not only for constructivist students of technology, but also, for instance, for sales professionals and advertisers. Marchand (1985) describes how the elite of the new advertising profession tried to tackle this problem in the 1920s and 1930s. To follow the "typical consumers" developed by the advertisers would be one way for a historical sociologist to start. Another possibility is the method I used in the bicycle case: the employment of biographical materials and novelists' accounts. A third possibility is to use spokespersons such as consumers' organizations, as established in the United States in the 1920s (Marchand, 1985). Ruth Schwartz Cowan (1987) explicitly addressed this problem for the sociologist and historian of technology and showed how various sources can be used to construct an adequate description of groups of consumers.

In the case of Bakelite, there is a fourth possibility. We are lucky to have a German survey, made under the commission of the Bakelit Gesellschaft m.b.H., of public opinion about Bakelite products. This survey was based on a sample selection of the German population (Institut für Wirtschaftsbeobachtung, 1938). It seems to answer quite adequately many of the questions we would like to ask. A second special source is a study in which Dutch "consumers" aged sixty to eighty were shown various Bakelite products and interviewed about how they made use of them in the 1920s and 1930s.[167] The two studies yielded very similar results, suggesting that most of these findings would be broadly applicable to consumers in other European countries and the United States.

The first part of the German market study addressed how well known Bakelite was. "Bakelite" was, as a word, known to 76 percent of the German adult population. This word could mean quite different things, however, to different respondents. For some, "Bakelite" was identical to "synthetic resin" (*Kunstharz*) or "molding material" (*Press-stoff*). For others, these terms were taken to mean completely different things: "Bakelite" had a relatively positive connotation, but "synthetic resin" recalled the "ersatz" meaning. (And a substantial number of interviewees had indeed heard of Bakelite, but had no knowledge of synthetic resins.[168]) For some the name "Bakelite" was connected to specific objects, while for others it was primarily a general-purpose raw material. For most users Bakelite's dark color was the primary characteristic by which they could recognize it. This color also suggested its being composed of worthless materials, and that again was associated with an ersatz character. Significantly, Bakelite was better known in industrial regions than in primarily rural parts of Germany.

The second part of the market study was directed at consumers' evaluation of Bakelite. The results of this part are congruent with the results of the Intomart study, carried out some forty years later with a panel of Dutch consumers. For users who were generally positive about

Figure 3.23
Egg cups on a plate with a saltcellar in the center. "That was standing in my mother's best room" an interviewee recalled when the artifact was shown to him (Kras et al., 1981: 43). Apparently, Bakelite artifacts were not only bought for day-to-day use. Photograph courtesy of Collection Becht, Naarden, The Netherlands.

Bakelite, its noteworthy features were the elegance of the designs, its durability and hardness, and its practical usefulness. Hardly mentioned were its nonflammability, chemical resistance, and cheapness. For consumers with a primarily negative evaluation, brittleness was the main drawback. This analysis is rather in line with the meanings attributed to Bakelite by other relevant social groups. The role of Bakelite as a typical "design material" was confirmed in this study too: a beautiful item might be displayed in the house's best room rather than actually used (see figure 3.23). On the other hand, the ambivalent assessment of the value plastics is always there: in the same panel it was remarked that one would never give someone a Bakelite object as a present; rather, one would choose crystal or delftware.

3.10 Technological Frame as a Theoretical Concept

In the previous chapter I have described the history of the bicycle and shown that in so doing, the concept of "relevant social group" could play a valuable role. The primary conceptual gain was the notion of "interpretative flexibility," which allows a sociological analysis of technology. Using the various relevant social groups, the interpretative flexibility of the high-wheeled Ordinary and of the air tire could be demonstrated.

It was thus possible to give a better explanation of some aspects of the bicycle's history. We could, for example, understand why the high-wheeled Ordinary could stabilize, despite its "obvious" flaw of being an unsafe machine. The interpretative flexibility of the air tire made it possible to identify an important mechanism for the stabilization of the safety bicycle.

These descriptions were, however, quite static. Although the process of stabilization was traced by checking the decreasing number of modalities in writing about the low-wheeled safety bicyle, not much was said about the interactions within relevant social groups that led to this stabilization. Nor did I say much about the generation of new artifacts or the emergence of new relevant social groups. In various successive periods, the artifacts were analyzed by identifying the relevant social groups and then describing the artifacts. The interpretative flexibility of an artifact could then be used to illuminate specific puzzling aspects of bicycle history. Indeed, the focus on problems and solutions did suggest an account of the process of change, while it actually only made plausible the order of "scenes." How these various "scenes" had developed from one into another was still an open question. My story of the bicycle seems more like a slide show than a movie.

To make the descriptive model more dynamic, and to focus on the process of technical change rather than merely on a description of successive artifacts, we need additional concepts. The descriptive model casts technical artifacts in meanings attributed by relevant social groups. I already mentioned that this perspective places my project on the crossroads of social-interactionist and semiotic traditions. For if we want to analyze the artifacts' processes of change, we need to model these various attributions of meanings. And so, because of how these meanings are empirically found, we then need a concept to make sense of the interactions within relevant social groups. For that purpose the concept of technological frame was introduced.

I proposed five requirements to be met by a theoretical framework for the explanation of the development of technical artifacts: it should (1) be able to account for change in technology, (2) be able to explain constancy and lack of change in history, (3) be symmetrical with respect to success and failure, (4) encompass actors' strategies as well as structural constraints, and (5) avoid the implicit a priori assumption of various distinctions made by the actors themselves.

Technological frames provide the goals, the ideas, and the tools needed for action. They guide thinking and interaction. A technological frame offers both the central problems and the related strategies for solving them. But at the same time the building up of a technological frame will constrain the freedom of members of the relevant social group. A structure is being created by actions and interactions, which in turn will constrain further actions and interactions. Within a technological frame not everything is possible anymore (the structure and tradition aspect), but the remaining possibilities are relatively clearly and readily available to all members of the relevant social group (the actor and innovation aspect).

The concept of "technological frame" thus has an enabling as well as a constraining function. In this respect it is analogous to Giddens's (1984) structuration concept, in which social structure is viewed as being produced by, as well as acting back on, knowledgeable agents—actors who are the subjects of that structure and in turn "instantiate" it. This latter conception, however, unduly stresses the focus on an irreducibly active human agency.[169] In Giddens's (1984) framework it is not plausible to make reference to forms of collective agency such as organizations.

After introducing "technological frame" and demonstrating its empirical operationalization, I now shall discuss the concept in more detail—some of its characteristics, its origins, and its ontological status. In doing this, I am extending the discussion of the concepts "relevant social group" and "interpretative flexibility."

The existence of a celluloid technological frame was used to explain the stability of the "celluloid world." As such, the concept of technological frame functions in a similar way as Kuhn's paradigm, when used to explain the stability of normal science. The proposed conceptual framework, however, also has built-in incentives for change, related to the different degrees of inclusion in a technological frame. Partly working within the celluloid frame, but also partly in the frame of electrochemistry, Baekeland could become an agent of change. The concept of "technological frame" thus allows us to analyze both change and constancy in technological development—the first and second requirements for a theoretical framework.

Often it is ambiguous whether an artifact works or not. Whether an artifact works is determined in social interaction, and mirrored explicitly in the technological frame. Thus our conceptual framework allows for a symmetrical analysis of successful and failed technologies—the third requirement for a theory.

For discussing the fourth and fifth requirements, I need a longer excursion into the dynamics of technological frames. A technological frame does not reside in individuals—it is largely external to any individual, yet located at the level of a relevant social group. Thus a technological frame needs to be sustained continuously by actions and interactions. They are not fixed entities, but are built up as part of the stabilization process of an artifact. The building up of a technological frame mirrors the social construction of an exemplary artifact, just as much as it reflects the forming of a relevant social group. The social construction of an artifact (e.g., Celluloid), the forming of a relevant social group (e.g., celluloid engineers), and the emergence of a technological frame (e.g., the Celluloid frame) are linked processes.

In economic studies of technological change, various processes have been identified as contributing to this triad of processes. Most "heterodox" economists studying technological change recognize that, contrary to the assumptions of neoclassical economics, there is no freely accessible pool of technical knowledge from which innovations can be drawn. Thus, for example, learning processes play a crucial role in technological innovation. Economists have identified various forms of learning processes: "learning by doing," "learning by using," and "learning by interacting" (Dosi et al., 1988). Although these are primarily meant to describe innovators' and producers' activities, it is obvious that especially the latter two are directly applicable to the broader concept of technological frame.

In talking about these linked processes, I have until now implicitly assumed the existence of one-to-one relationships between a relevant social group, its technological frame, and the associated exemplary artifact (see figure 3.24). Though this served its purpose as an introduction, it is not an adequate representation of the conceptual apparatus. As depicted here, it would pose two problems. First, a group of actors (be they engineers, consumers, production people, or whoever) is of course working with a variety of artifacts—and not just the one at the starting point of the analysis that produced the relevant social groups (the high-wheeled bicycle in figure 2.13, for example). Strictly following the way in which the concept "relevant social group" was introduced in the second

Figure 3.24
Implicitly used relationship between relevant social group, its technological frame, and the artifact being analyzed.

chapter, each of these various artifacts (for example, a bicycle repair kit, a lighting generator, cycling trousers) would analytically entail an equally large set of relevant social groups. In practice these relevant social groups would of course largely overlap. In other words, these relevant social groups would be analytically different but empirically overlapping. Until now this problem was not acute because I was primarily concerned with the development of one artifact such as the safety bicycle or Bakelite; the other artifacts used by the relevant social groups did not play a prominent role. It is now time, however, to cast our nets wider, and to try to extend the analysis to the relationship between technology and society more generally. Hence this problem needs to be addressed.

The second problem of the conceptual scheme as depicted in figure 3.24 emerges when one artifact is used by different relevant social groups. In the analysis of chapter 2, this led to the demonstration of interpretative flexibility and the conclusion that there are as many artifacts as there are relevant social groups. That "sociological deconstruction" proved a fruitful starting point for investigating the subsequent social construction of the artifact; but now that we are interested in more than the construction of that one artifact and have widened our scope to include the continuing interplay between society and science, it will turn out that after the closure process different groups continue to use the increasingly stabilized artifact. Those groups would, following the analytical scheme of the previous chapter, be analytically identical because they attribute the same meaning to the artifact. In other words, the relevant social groups are in this case empirically different but analytically identical.

To address these two problems, it is necessary to recognize that the approach presented in this book has symbolic interactionist as well as semiotic traits. First I focused on relevant social groups and how they constitute, through social interaction, artifacts. Where the relevant social groups differed in their attribution of meaning, interpretative flexibility was observed—different artifacts were constituted. Such an interpretative flexibility could then be curtailed by arriving at consensus among the relevant social groups, that is, the process of closure. This is, I argued, a fairly straightforward social interactionist account.[170] But then there was the process of stabilization—an artifact acquired an increasing degree of stabilization, which can generally be traced by analyzing the language actors use to describe the artifact. Here, it is not the conflicts between relevant social groups that dominate—the main vehicle for change is a development of the artifact's semantic constitution. This process, I said,

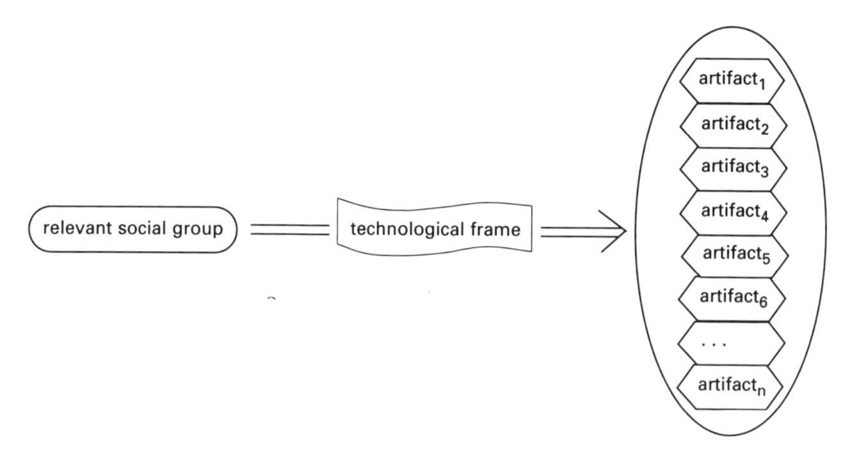

Figure 3.25
Social interactionist view of the technological frame.

should be interpreted from a semiotics perspective.[171] The concept of technological frame links the two types of analysis.[172]

Viewed from the social interactionist perspective, a technological frame provides the vocabulary for social interaction, the forming of social groups, and the constitution of a world. Artifacts are, as parts of such worlds, constituted by that interaction. Thus viewed, there is a one-to-one relationship between a relevant social group and its technological frame, resulting in the constitution of a world in which various artifacts may play a role (see figure 3.25). An example is the relevant social group of Celluloid engineers, the Celluloid frame, and the artifacts such as Celluloid, extrusion machines, presses, molds, etc. (Note that this variety of artifacts, as depicted in figure 3.25, is not the same as the one resulting from interpretative flexibility.) Viewed from the semiotic perspective, a technological frame mirrors technological development, provides the vocabulary for forming artifacts, and constitutes another type of world. Thus seen, there is a one-to-one relationship between artifact and technological frame, resulting in the forming of relevant social groups (see figure 3.26). The example here is Celluloid, a Celluloid technological frame, and then relevant social groups such as the Celluloid chemists, molders, pressing machine designers, etc. Combining these two partial perspectives, the special ontological status of technological frame and its crucial role as a hinge between the two world becomes clear (see figure 3.27). By rendering the two sides of the analysis—social groups and technical artifacts—into aspects of one world, "technological frame" will

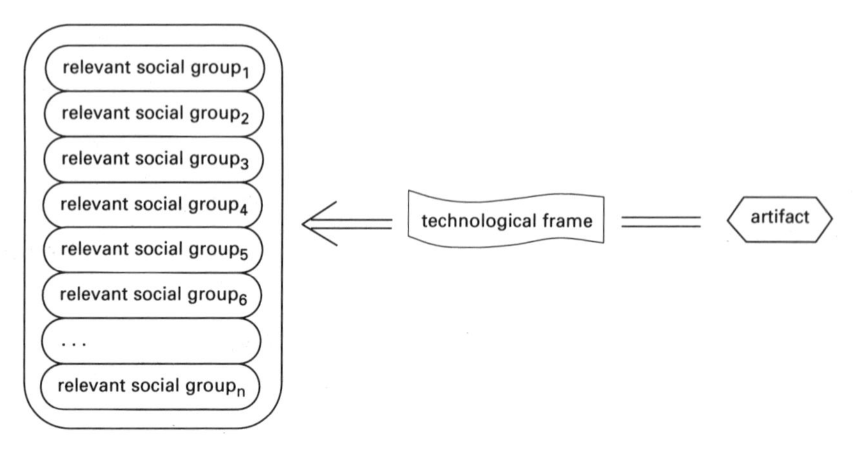

Figure 3.26
Semiotic view of the technological frame.

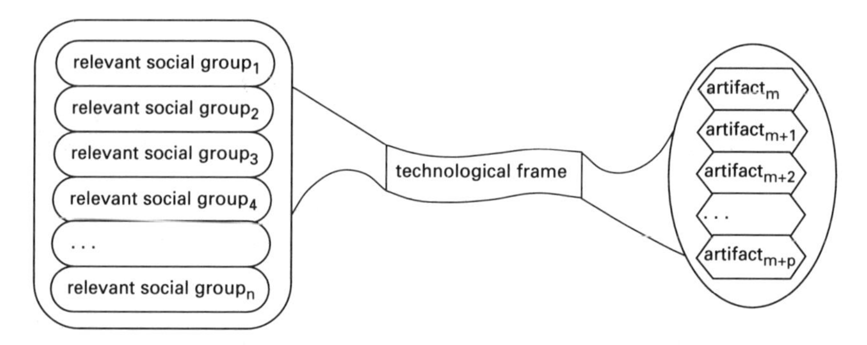

Figure 3.27
The concept of "technological frame" as a hinge between social-interactionist and semiotic views of technological development.

be helpful in transcending the distinction between hitherto irreconcilable opposites: the social shaping of technology and the technological impact on society, social determinism and technical determinism, society and technology.

3.11 Conclusion

As I have traced the social construction of Bakelite it has become clear, I hope, that my social-constructivist analysis does not diminish the greatness of Baekeland as an inventor. It does, however, provide a different slant on his individual abilities. To describe Baekeland, as I did implicitly, as a heterogeneous engineer (Law, 1987) or system builder (Hughes, 1983) does not reduce the development of Bakelite to the genius of this "grand duke, wizard, and bohemian." Rather, this characterization serves as the summary of all social processes in which Baekeland participated; his status as an inventing genius is the result of the social construction of Bakelite, not the cause.

Working with the descriptional model developed in chapter 2, I addressed the question of how Baekeland could succeed where other chemists had failed. To answer this question, I introduced the theoretical concepts of "technological frame" and "inclusion in a technological frame." With the previously developed notions of "relevant social group," "interpretative flexibility," "closure," and "stabilization" these were cast into one conceptual frame. In the last section I discussed how the origins of this conceptual frame can be traced to social-interactionism and semiotics.

We now have a conceptual framework that meets the criteria formulated in chapter 1. That is, it is able to account for change and constancy; it is symmetrical with respect to success and failure; it encompasses actors' strategies as well as structural constraints; and it avoids the implicit a priori assumption of distinctions. In the next chapter I will demonstrate how this framework can be put to work to analyze not only the development of artifacts, but also the development of society and social power.

4

The Majesty of Daylight: The Social Construction of Fluorescent Lighting

4.1 Introduction

In the case of the bicycle, the central role of such relevant social groups as women, "young men of means and nerve," and the anticyclists was fairly obvious. Technical artifacts like the Ordinary bicycle or the air tire fit nicely into the framework of the model of social construction. The bicycle's development was in fact quite a "social" affair. But it was reasonable at that point to ask what the SCOT model could contribute in the case of an invention that, in the accepted view, was credited to a single person. This was the crux of chapter 3, where I showed that, without diminishing Baekeland's crucial role as a heterogeneous engineer, the development of Bakelite could fruitfully be analyzed as, again, a process of social construction.

The bicycle and Bakelite are, in several obvious respects,[1] very different artifacts. First, they are from different fields of technology—mechanical and chemical engineering. Second, whereas new bicycle prototype might be constructed in a local blacksmith's workshop with a small amount of capital, the development of Bakelite required large capital investments. Third, the bicycle was a typical example of engineering without much "scientific" input, whereas Bakelite is generally considered an example of "scientific engineering." Fourth, the bicycle is typically a consumer product, while Bakelite was developed and produced as a molding material for industrial users. Fifth, the bicycle is (in the vocabulary of patent attorneys) a product invention, and Bakelite a process invention. Finally, they are from different periods: the bicycle underwent its complete development in the nineteenth century, while the construction of Bakelite occurred in the early twentieth century.

Nevertheless, in some sense the bicycle and Bakelite case studies are similar, for they both involve the construction of technical artifacts in their earliest stage of development—at the workbench. One might therefore conclude that the social constructivist form of analysis is limited to cases in which technology is actually under construction. This would seem to contradict my earlier claim that all relevant social groups—and not only the engineers and producers—are involved in social construction. To address this issue, I have chosen the fluorescent lamp as a third case for study.[2] I will show that the fluorescent lamp that most of us presently use was constructed during what the classical model of technology development would call its "diffusion stage." The engineers at their drawing boards had finished their work some months or even years before the "final" social construction of this lamp occurred in 1939—a social construction actually carried out by managers at a conference table.

The fluorescent lamp was developed in later than Bakelite; the war effort required from American industry during World War II will have important implications for this story. Though it is clearly a product and not a process invention, it is not as exclusively meant for the consumer market as the bicycle was. Economically speaking, the demand side of the market was not cohesive, and several segments can be delineated. The fluorescent lamp emerged, even more so than Bakelite, within a science-based industry. Also, it was developed in an industry that was, compared to the previous two cases, very capital-intensive. Indeed, the industrial setting of an oligopolistic market will allow us to examine the role of economic power, which was not nearly as relevant in the bicycle and Bakelite cases. Finally, having reviewed cases of mechanical and chemical engineering, we will now focus on an example of electrical engineering.

The central historical problem in this chapter will be to trace the social construction of the fluorescent lamp between 1938 and the early 1940s. The riddle of the case is to understand the discrepancy between the fluorescent tubes we now know (efficient daylight lamps for high-intensity illumination) and the lamps that were originally unveiled in 1938 by General Electric (low-intensity lamps for coloring purposes), although the two lamps are, in a naive sense, technically identical. A long prehistory of scientific research related to gas discharge lighting figures prominently in all traditional histories of fluorescent lighting, but this will be only briefly described here. The theoretical aim of the chapter is to address the economic and power dimensions of technological development. Reaping the fruits of this analysis, the chapter will provide means for understanding the different ways in which technology may acquire its characteristic hardness or obduracy.

Besides its interest for the overall argument of this book, the case is in itself fascinating for the view that it affords of the interplay among industries, government, and consumers and also among technology, science, and politics; in short, it encompasses all the key elements of complex modern societies and technological cultures.

4.2 Structure of the Electric Lamp Industry

If economic power relations ever played an important role in any technological development, then they certainly played one in the fluorescent lamp case.[3] The lamp was developed at a time when the economic power of General Electric in the electric lighting field was at its peak, though even then the company could not dictate the course of events because the market was oligopolistic rather than monopolistic. General Electric was active in many fields of electrical equipment, and its total annual sales were more than three times as large as the entire annual U.S. lamp production (Birr, 1957). The period of General Electric's establishment and consolidation coincided with the emergence of antitrust legislation and its enforcement.[4] Some of the most complex antitrust cases were fought between the United States and General Electric.[5] In addition, General Electric was a member of an important international cartel. The social group of Mazda companies, comprising General Electric and Westinghouse and named after their trademark "Mazda" for incandescent lamps,[6] thus is certainly relevant for analyzing any electric lighting development in the 1930s (see figure 4.1).

The electric lighting industry began with Thomas A. Edison, who built a technological system of electric power distribution in the early 1880s, with a new high-resistance lamp filament as its best-known component (Hughes, 1983). By the late 1880s three large companies had come to dominate the incandescent lamp industry: the Edison General Electric Company, Westinghouse, and the Thomson-Houston Electric Company.[7] Initially the Edison Company had a relatively strong position because of the patents it held, but this did not last long. Not only did the other companies acquire their own patents on slightly different versions of the important system components, but the patent situation generally grew less stable. There was a proliferation of patents for all sorts of technical artifacts, so that it became difficult to set up any electrical system without

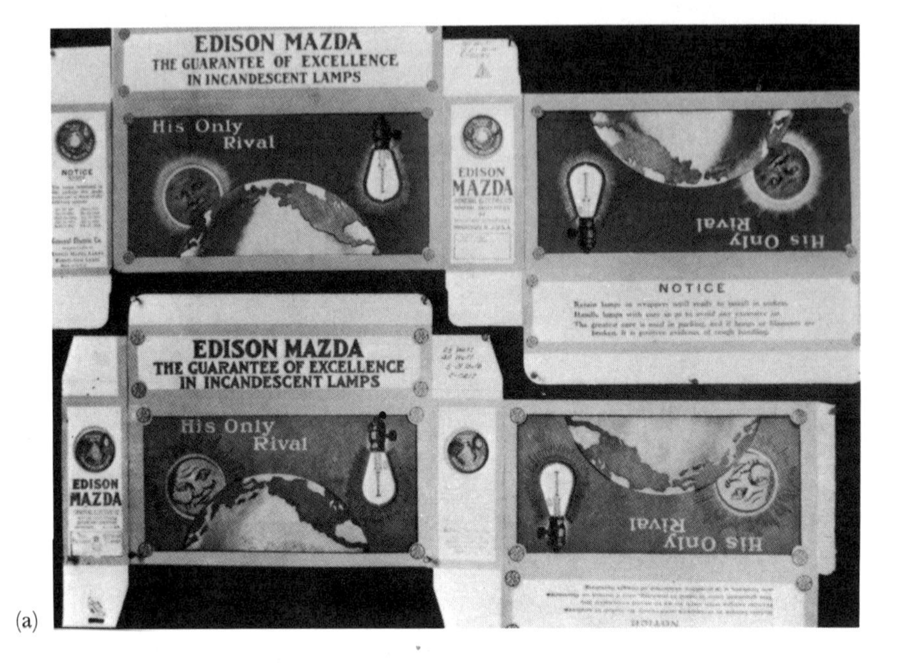

(a)

(b)

Figure 4.1
The Mazda trademark (a) and the General Electric symbol (b).

infringing upon one or more patents. Also, the value of many patents was quite unclear because of the large number of simultaneous, independent developments. Hence firms were reluctant to obtain patent licenses because they might not be of much value. Partly to surmount these problems, all three companies considered mergers with each of the other two. In 1892 the Edison and Thomson-Houston companies merged to form General Electric. Carlson (1991: 199–301) argues that this merger resulted from three other factors in addition to the patent problems: a desire to eliminate competition, the problem of raising sufficient capital, and the efforts of managers and investors to maintain their organizational capability. In 1896 General Electric and Westinghouse established a cross-licensing agreement allowing each access to the patents of the other. Thus two relevant social groups stand out clearly: the combination of General Electric and Westinghouse companies on the one hand and a group of small lamp producers on the other. General Electric had by that time 50 percent and Westinghouse 10 percent of the market, and the smaller firms shared the rest.

The small lamp manufacturers found it increasingly difficult to compete with General Electric. To solve this problem, several of them established the National Electric Lamp Company in 1901.[8] This was set up as a holding company, leaving the constituent firms relatively independent. General Electric provided much of the initial capital by purchasing 75 percent of the stock, but kept out of the day-to-day management. Publicly, the National Electric Lamp Company and General Electric presented themselves as competing, but they signed various patent licensing agreements, and most of the firms in the electric lamp business (including Westinghouse) participated in price and market-sharing agreements. All in all, General Electric thus indirectly controlled 97 percent of the electric lighting market in the United States. This situation led the federal Justice Department to bring equity proceedings under the Sherman Antitrust Act against General Electric and thirty-four other companies in 1911. As a result, General Electric was forced to take over the National Company.[9] By 1912 General Electric had a market share, under its own name, of 80 percent and licensed most of the remainder under its patents. The development of the ductile tungsten filament lamp, to which General Electric held all the crucial patents,[10] strengthened even further the company's hegemony. The tungsten filament lamp also dealt the final blow to gas and arc lighting. Between 1913 and 1945, lighting in the United States was incandescent lighting as manufactured by General Electric.

General Electric established a new two-class licensing system in 1916.[11] The class A license, granted only to Westinghouse, conferred the right to produce a fixed percentage (some 15–16 percent) of the lamp output of General Electric and permitted the use of the trademark "Mazda." Class B licensees were allowed to produce a smaller percentage (altogether less than 10 percent) of General Electric's own output and were not permitted usage of the "Mazda" trademark. Moreover, each class B licensee was allowed to make only one specific type of lamp, whereas Westinghouse could produce a broad range of lamps. Both classes of licensees agreed to extend royalty-free licenses to General Electric under all their patents relevant for the electric lamp business. The government brought an antitrust suit against General Electric and Westinghouse in 1924,[12] but its case was dismissed. The U.S. Supreme Court upheld this decision and thus sustained the legality of the licensing

Table 4.1
Approximate shares of U.S. market for large tungsten-filament lamps, 1937

Supplier	Lamps sold	License relationship
General Electric Co.	59.3	Licensor
Westinghouse Electric & Manufacturing Co.	19.0	Class-A
Hygrade Sylvania Corp.	4.4	Class-B
Consolidated Electric Lamp Co.	2.8	Class-B
Ken-Rad Tube and Lamp Corp.	1.1	Class-B
All other domestic firms (about 20)	8.8	Unlicensed
Imports	4.6	Unlicensed

Source: Bright and MacLaurin, 1943: 431.

system (Bowman, 1973). In 1927 the system was slightly adapted, allowing Westinghouse a larger share of production, but under tighter conditions.[13] Another revision of the agreement with Westinghouse, which would prove to be very important for the development of the fluorescent lamp, was the inclusion of vapor lamps. This was not done for the class B licensees. Many of the twenty-one class B licensees of 1917 went out of business before World War II (see table 4.1). Only the Hygrade Lamp Corporation survived as an important independent electric manufacturer, partly, as we will see, because of its particular strategy in the fluorescent lamp story.[14]

On the supply side, the market was thus relatively cohesive. General Electric and Westinghouse manufactured all types of lamps, while smaller companies had to limit themselves to a few types. On the demand side, to the contrary, the product market for electric lighting was quite segmented. For outdoor lighting, especially street lighting, incandescent lamps and neon-sodium discharge lamps were used. Colored outdoor advertising lamps formed another important market segment. Here high-voltage discharge lamps were employed. For the indoor market an important division existed between private households and large commercial buyers. Then there were the so-called miniature lamps (which were mostly automotive) and photographic lamps. Finally, a large variety of specialized lamps were employed for different purposes (Rogers, 1980). The geographic market for electric lighting relevant to this chapter is North America. After World War II the American market was quite isolated and not just a segment of the world market: the numbers of imported and exported lamps were very low (Rogers, 1980; Bright,

1949). They must have been even lower before 1945. This market was not internally segmented by geography: except for some very small companies, all manufacturers sold their lamps all over North America.

Though not of direct relevance for this case study, it is worthwhile to describe the international situation because it further shows the economic strength of General Electric (Stocking and Watkins, 1946). The international scene between 1918 and 1945 was dominated by a cartel of lamp producers, first only in Europe but then also in the Far East and South America. Following World War I the German Osram Company revitalized a cartel set up before the war.[15] It was registered in 1921 in Berlin as the Internationale Glühlampen-Preisvereinigung E.V., but it was commonly called the "Phoebus" cartel.[16] Though General Electric was never officially a member of this cartel, it played an important role in its formation by acting through its British subsidiary, General Electric Company Ltd. General Electric acted as if it were a member of the cartel and, through bilateral agreements between the U.S. company and the most important European lamp producers, Phoebus's and General Electric's spheres of influence were thus expanded to cover virtually the whole world. Moreover, General Electric owned stock interests in almost all European lamp-producing firms.

A final critical element for the introduction of the fluorescent lamp, at least in socioeconomic terms, was the relationships between the Mazda companies and relevant social groups other than the small lamp producers. Such groups included the sales agents, the central electric utilities, and the fixture manufacturers. There was, for example, a net of agents through which a large proportion of the Mazda lamps was sold. These dealers had to sell the lamps to the ultimate consumers at prices set by the Mazda companies. The marketing of the lamps was in this way much more strictly controlled than would have been the case under a normal jobbing procedure. Grocery, drug, and electrical stores throughout the country acted as retailers: more than 80,000 for General Electric and 30,000 for Westinghouse in 1939.[17] The two Mazda companies also had their own supply companies that provided lamps (and other electrical equipment) to large consumers. Second, and most important for events surrounding the social construction of the fluorescent lamp, there were the electricity-producing central stations—the relevant social group of utilities.

A crucial role in maintaining its almost absolute control of the lighting business by the relevant social group of Mazda companies was played by their intimate connections with the utilities. Many of the utility companies had originally been organized as licensees of the Edison, Thomson-Houston, or Westinghouse companies, and with the continued increase of the use of electrical appliances both relevant social groups stayed very much dependent on one another, even after direct financial ties had been broken.[18] The basis of this relationship was an understanding that each side would work toward the interests of the other. The utilities undertook to sell and promote Mazda lamps—and the appliances and other electrical apparatus of the Mazda manufacturers as well—and for their part, the Mazda manufacturers undertook to promote their products in such a way as to add to the amount of electricity consumed. The Mazda companies also supported and participated in campaigns and programs conducted by the utilities to increase the use of electricity. For example, in the 1930s a large number of utilities gave their customers free renewals of lamps of higher wattage to keep their sockets filled. The lamps used in these campaigns were Mazda. General Electric and Westinghouse supplied the lamps at reduced prices to the utilities giving free renewals. One of the things that most clearly emphasizes the intimacy of the relations between the Mazda companies and the utilities is that almost all of the utilities dealt only with Mazda lamps—and they not only sold Mazda lamps, but advertised them and promoted their use in every way.

Each utility was a private company, operating one or several central stations to generate and sell electricity. The utilities had a number of strong collective organizations and can be seen as acting, through these organizations, as one social group. The utilities, although ordinarily independent of each other, did act in concert in matters affecting their common interest. For instance, over 100 utilities belonged to the Edison Electric Institute (E.E.I.). Another large organization of utilities was the Association of Edison Illuminating Companies (A.E.I.C.). The operating area of each of these associations of utilities extended to every part of the country. The E.E.I. and A.E.I.C. were made up of committees and groups composed of representatives of the member utilities, who among other things handled policies for the industry. The policies were either determined at the meetings of the organizations as a whole, or they might be formulated by particular committees on the basis of their knowledge of the desires of the industry. Frequently this knowledge was derived from questionnaires sent out to all utilities (Anon., 1949).

Two important committees were the Lighting Sales Committee (E.E.I.) and the Lamp Committee (A.E.I.C.). These committees had over many years worked very closely with representatives of General Electric and Westinghouse in determining the policies to be pursued with respect to the manufacture, distribution, and use of (incandescent) lamps manufactured by the two Mazda companies, and to the promotion of such lamps by the utilities. Besides these two committees, the relevant social group of utilities also acted through the Electrical Testing Laboratories. This organization was owned by the utility companies and engaged in commercial testing work on electric lamps and other electrical equipment.[19]

A similarly close relation existed between the Mazda companies and the last relevant social group to be described: the fixture manufacturers. The Mazda companies produced mainly lamps. Sockets, reflectors, and other kinds of auxiliaries were produced by smaller companies. For incandescent lighting a system of specifications had been set up. This system was maintained by the RLM Standards Institute,[20] an association of the largest fixture manufacturers. The RLM Standards Institute established, in collaboration with the Mazda companies and the utilities, standards that the products of its members had to meet. Not surprisingly, it set criteria that favored Mazda lamps. The fixtures were tested by the Electrical Testing Laboratories, mentioned previously. As Bright (1949: 287) summarizes, "The testing and certifying arrangements among the 'Mazda' manufacturers, RLM and ETL enhanced the prestige of all three groups."

In summary, the new electrical industry could emerge only in the few nations with advanced capitalist societies: England, France, Germany, and the United States (Nye, 1985). In the case of the United States, the industry was closely connected with big private financiers that merged a unique combination of science and capital. As David Nye (1985: 13) notes, "there was nothing natural about such a finance system; it was a profoundly cultural creation, a system of superior abstraction that reduced a tremendous variety of goods and services to a single, flexible network of monetary values." General Electric formed a central node in this network, having become far more than a mere economic entity. Although for some purposes it can be treated as an individual actor, it wielded more power than some small cities. At the end of this chapter I shall return to this question of economic power.

More than socioeconomic terms are needed, however, to describe the circumstances under which fluorescent lighting was constructed; the broader background of the powerful mythic presence of modernity and progress in the 1930s must also be sketched.

4.3 Prehistory of the Fluorescent Lamp

I will now briefly outline the scientific and technological prehistory of the fluorescent (low-voltage) lamp. Besides sketching the technical background of the fluorescent lamp, this discussion will also serve the purpose of introducing some of the basic physics helpful to understanding the details of the case study.[21] Thus I will first discuss the phenomenon of electric discharge and the role of gas pressure, and then present the three architects of the fluorescent lamp—Moore, Hewitt, and Küch—with their respective lamps.

Early Electric Discharge Lamps

The first observations of light phenomena in "exhausted" glass vessels were already made with electrifying machines before 1700, a century before the voltaic cell (or battery) made it possible to do experiments with electric current. The first deliberate research was initiated by the German artist and glassblower Johann Geissler. He found in 1856 that when an alternating current was passed through a tube with air at a low pressure, light of low intensity could be seen. The phenomenon could only briefly be observed, because it was not possible to maintain the low pressure for more than a few moments. Further research indicated that all gases and vapors could carry current, and that some would emit light in doing so. Tubes filled with nitrogen and carbon dioxide even produced light with higher efficiency than incandescent lamps.[22] Different vapors produced light of various colors.

Between 1860 and 1890 there were several scattered efforts to put the light generated by a gas discharge to practical use for illumination. Some patents were granted for operating lighting buoys with Geissler tubes. These artifacts, however, did not stabilize. Around 1890 interest rose again. Edison experimented with the subject and took out a patent. The discovery by Roentgen of X-rays further stimulated interest in electric discharge phenomena.

A first physical explanation of electric discharge phenomena was established with the use of Joseph Thomson's electrons. When the voltage across two points (or the "tension between those points," as it was called in those days) is high enough, an electric current will pass between the points (called "electrodes")—even through air, which normally is an insulator. Electrons are pulled out of the negative electrode (called "cathode") and are accelerated by the electric tension between the cathode and the positive electrode (called "anode"). Collisions between these

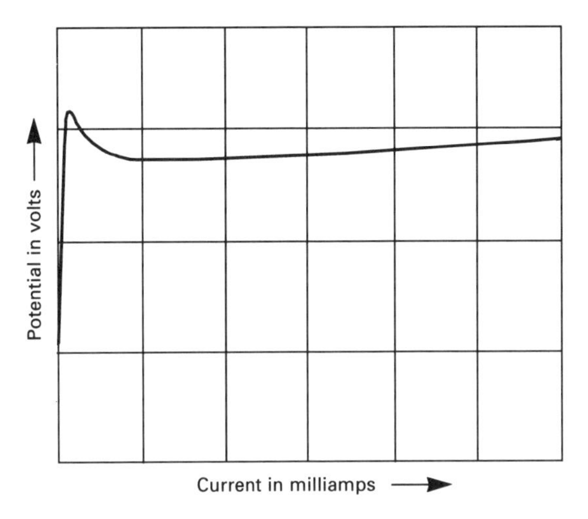

Figure 4.2
The voltage drops through a gas discharge tube.

speeding electrons and the gas atoms excite these atoms, which subsequently emit light. A second effect of the collisions is that atoms are ionized, so that new electrons are added to the avalanche and a stream of electrons starts to flow between the cathode and the anode, while a (much slower) stream of positive ions moves from anode to cathode: the electric discharge is established. To start this chain of events, a high voltage is necessary (see figure 4.2). Once the discharge has started, it can be maintained at a relatively low voltage. By heating the cathode, the voltage needed to emit electrons can be made very low, and the starting voltage may be almost as low as the operating voltage.[23]

It will be clear from the previous paragraph that gas pressure is an important variable for the operation of a discharge lamp. In a vacuum, ionization cannot take place at all and a discharge surge will not occur. At high pressures, the average unimpeded path of an electron is very short and electrons lose so much energy in elastic collisions that they will not acquire enough velocity to cause ionization; consequently, again, ignition of the discharge tube is very difficult. Hence it is not surprising that there is an optimum gas pressure for starting an electric discharge (see figure 4.3). But the gas pressure is not only an important variable for starting the discharge; it was, though much later, also understood to play a crucial role in the subsequent light emission of the gas.

The greatest problem in the 1890s was the short life of the discharge lamps, which made commercial use impossible. By now it was possible to

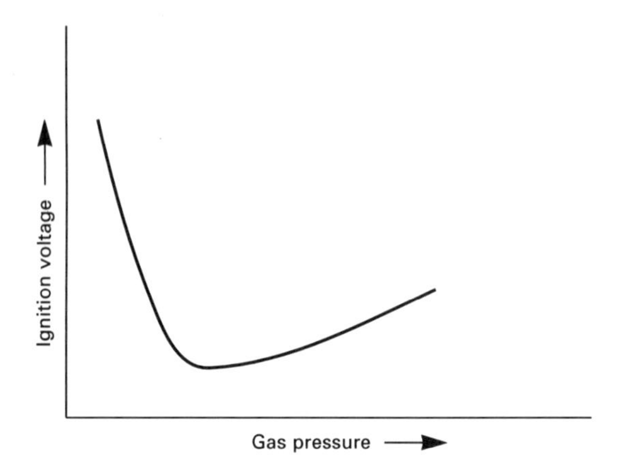

Figure 4.3
Ignition voltage as a function of the gas pressure (the so-called Paschen curve) (Elenbaas, 1959b).

seal glass tubes effectively and to maintain a low pressure for a long time. The problem now, however, was the reverse: because of chemical and physical interaction between the gas and the metal of the electrodes, the pressure slowly decreased until the electric resistance of the gas became too high. D. McFarlan Moore, a former employee of Edison, found a solution for this problem and thus made the first commercially useful gas discharge lamp. Moore's solution was a valve that opened automatically when the electric current changed (due to a change in electric resistance because of lower gas pressure) and then let new gas into the tube. Most of his tubes worked with normal air. Because the losses in the transformator (for stepping up the electric voltage to the required 16,000 volts) and the losses at the electrodes were fixed, one long, continuous tube was more efficient than a number of shorter ones. A typical Moore lamp had a diameter of 1.75 inches and a length of 180 feet. Despite the fact that the lamp and auxiliaries were expensive and difficult to install, its operating advantages were great enough to find some use for general lighting in offices and stores and also for advertising and decorative applications. The Moore lamp produced a nearly white light (Luckiesh et al., 1938). General Electric was worried enough to increase the efforts to find a better incandescent lamp (Hammond, 1941). With the stabilization of the tungsten filament lamp between 1905 and 1912, the advantage of the Moore lamp was annihilated and the lamp gradually disappeared. Moore returned to work for the Edison company. Some

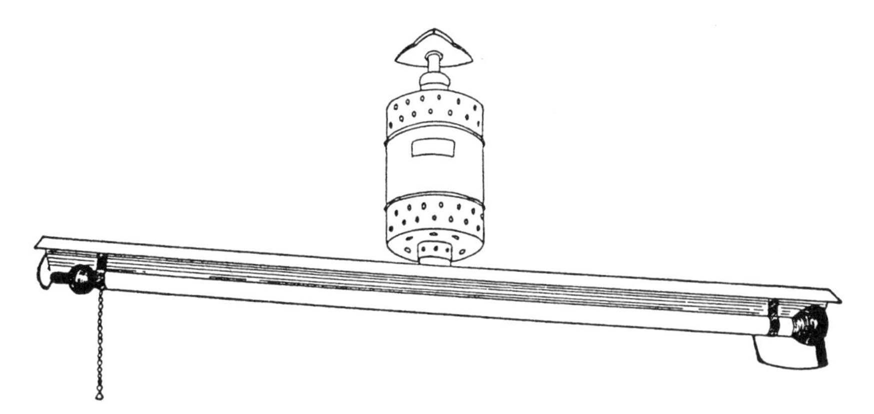

Figure 4.4
One form of the Hewitt lamp consisted of a tightly sealed tube, of which both ends were slightly enlarged so as to provide a container for the liquid mercury. In stable position, the lamp hung at an angle of 15 degrees with respect to the ceiling, so that all mercury was in the lower container of the tube. To switch the lamp on, one pulled the other side down, the mercury flowed to the other electrode and established a circuit between anode and cathode. The electric current then heated the mercury and vaporized it partially; when the lamp was let back into its original position, an arc was stroked and built up to extend the length of the tube. The lamp produced a green-blue light.

elements of his basic scheme would, however, be taken up by George Claude for his neon lamps.

At the same time that Moore was developing his lamp, another young independent American inventor worked on a gas discharge lamp. Peter Hewitt could build upon a long tradition of research on mercury lamps.[24] Many of these devices worked with liquid mercury in an inverted U-shaped tube: the passage of this stream vaporized some of the mercury, and the electric discharge through the mercury gas then produced a strong greenish light (Arons, 1892; Polak, 1907). These lamps, however, found only limited application in experimental physics. The commercially successful Hewitt lamps[25] employed different configurations. One rather spectacular form consisted of a glass tube—four feet long and one inch in diameter—that was switched on by tilting the lamp (see figure 4.4). This evidently posed practical problems and soon other means were developed to start the lamp, using automatic tilting mechanisms, induction coils, or extra cathodes (Hutter, 1988). To produce a less coldly colored light, fluorescent reflection screens were placed around the lamp and other gases were added to the mercury vapor, but these efforts failed. Another possibility was to combine the Hewitt lamp with a tungsten lamp; the combination of bluish and reddish light would then result in nearly white light (Grandy, 1933). These low-pressure mercury lamps did, however, occupy a small but stable part of the lamp market: in drawing rooms, factory halls, and for photographic purposes. The Cooper-Hewitt Electric Company was organized in 1902 with financial help from Westinghouse. In 1919 the company was bought by General Electric, its name was changed to the General Electric Vapor Lamp Company, and some twenty years later it merged with General Electric's lamp department.

The last lamp to be discussed here was developed by the German physicist R. Küch. He had developed a method of forming quartz glass tubes and reasoned that in an electric discharge lamp made of the stronger quartz, the gas could have a higher pressure and thus might yield a higher efficiency. He was right. The high-pressure mercury lamp that Küch developed, together with T. Retschinsky (Küch und Retschinsky, 1906, 1907), emitted much ultraviolet light because quartz is transparent for ultraviolet wavelengths, whereas normal glass absorbs them. The lamp could be used for special purposes such as sterilizing, but was—without extra shielding—harmful for human skin.[26]

Commercial Electric Discharge Lamps

Although the work by Moore, Hewitt, and Küch did not result in commercially viable lamps, it did produce a clear picture of the basic electricity physics involved. The next step was to tinker with the type of gas in the lamps and to integrate recent knowledge of atomic physics. This led to the first commercial gas discharge lamps. This success stimulated further research, and the first steps toward fluorescence were made.

The gas discharge lamps such as those developed by Moore used normal air, of which nitrogen and carbon dioxide (together comprising 99 percent of normal air) react chemically with the electrodes of the discharge device. Inert gases such as neon, argon, helium, krypton, and xenon (together only some 1 percent of normal air) would not engender that problem, but could not be extracted from air on a commercial basis in Moore's day. This achievement was, however, realized in 1907 when both George Claude and Carl von Linde, independently, improved techniques for the liquification of air. Once liquified, the components of air could be separated more easily, and a few years later the inert gases were available at moderate prices.[27]

In 1910 Claude made his first neon lamps.[28] In addition to an inert gas, the other important element in his lamp was a new electrode design. The deterioration of electrodes caused by the bombardment of gas ions (and called "sputtering" of the electrodes) had been a continuous problem for all engineers involved in electric discharge lighting. Claude patented an electrode design that specified that the electrodes required at least a surface of "1.5 square decimeters" per ampere (Claude, 1936). These cathodes were cold—the electrons were emitted by the voltage difference only, which thus needed to be relatively high. Another possibility would have been to use the hot cathodes I mentioned previously: the electrode is heated by a filament and the electrons acquire most of their energy to exit from the electrode in the form of heat (called "thermionic emission"), and the lamps could have been operated on lower voltages. Because this type of electrode required a more complex wiring and because the lamp engineers had less experience with them anyway, most "neon tubes" employed cold cathodes. These lamps were commonly called "neon tubes" as a generic term for all low-pressure cold-cathode electric discharge lamps, although neon was only used in lamps that were to give off red light. For other colors different gases had to be used: mercury vapor in a mixture of argon and neon for blue, and helium for cream. Other colors could be obtained by using tinted glass, of course with a corresponding lower efficiency, as when producing colored light with incandescent lamps. The Claude neon tubes were quite unambiguously a commercial success, especially in outdoor advertising.[29]

It was only during the first decade of the twentieth century that physicists started to acquire an understanding of light-emission phenomena. Their theory nicely explained the workings of Claude's neon tubes. When an atom is brought into a state of higher energy, for example by a collision with an electron, this state normally is not stable—the atom will fall back into its stable state of lower energy, and while doing so it will emit light. The energy difference between the two states determines the color of the emitted light. Each type of atom (or "chemical element") has a (large) number of possible states into which it may be excited. Accordingly, it will emit light of different colors upon falling back to its stable position. This set of colors (or "lines in the emission spectrum") is characteristic for the type of atom (see figure 4.5). Mercury, for example, has important spectrum lines in the invisible ultraviolet and in blue; sodium's most important line is yellow. The relative prominence of spectrum lines of a chemical element in a given electric discharge is dependent on the

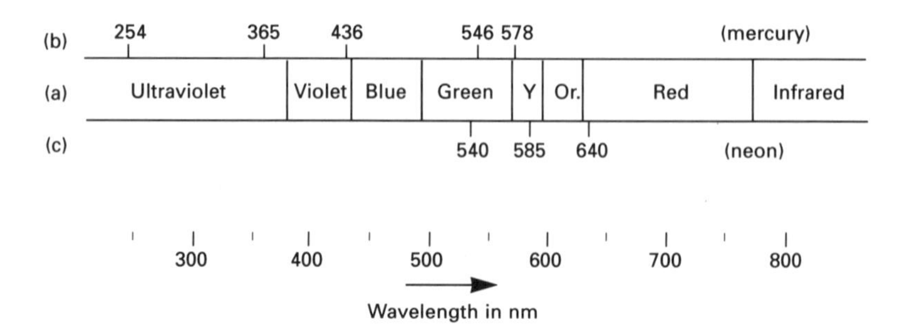

Figure 4.5
The light spectrum (a) and the emission spectra of mercury (b) and neon (c). Specific lines are designated by their wavelength (in nm (nanometer), 10-9 m).

energy with which the atoms are excited. Electric current density in the tube and vapor density are two important variables that determine that excitation energy.

On the basis of the Hewitt and Küch lamps, further research was devoted between 1910 and 1940 (especially by European firms) to the development of high-pressure mercury lamps for large-space lighting outdoor and in factories, and for film projection.[30] Although "high-pressure" at first only meant atmospheric pressure, by the end of the 1930s lamps were made (of 100 to 3000 Watt) with pressures of up to 100 atmospheres to increase the efficiency. An inner quartz tube and external water cooling tube were necessary for these high-intensity lamps (see figure 4.6). The high-pressure mercury lamps employed hot cathodes, which made it possible for many of them to be used on normal supply voltages. The color of the lamp was bluish white, but could be made better (that is, more yellow) by applying higher pressures, because then the emission lines with longer wavelengths would grow relatively more important.

The last commercially successful electric discharge lamp of the 1930s was the neon-sodium lamp. The luminous efficiency that such a lamp should be able to reach is obvious from a comparison of the emission spectra of sodium and incandescent lamps (figure 4.7). At normal temperatures sodium is a liquid and does not enable the establishment of an electric circuit by gas discharge. Therefore neon was added. The lamp is started by a discharge in the neon gas, helped by hot electrodes; after a few minutes the lamp is heated enough so as to evaporate the sodium, and the lamp then functions further as a sodium discharge device. These lamps were, and are, mostly used for street lighting, where high intensities

Figure 4.6
A water-cooled high-pressure mercury lamp, the Philips SP 500. Courtesy Philips Concern Archief.

are required and the monochromatic yellow light (and consequential lack of color discrimination) is no problem.

These successes in electric discharge lighting stimulated the search for a daylight discharge lamp. The color of electric lighting had been a continuous focus of interest, and now the gas discharge lamps suggested an escape from the "too small, too hot, and too red" characterization of the incandescent lamp.[31] Mercury lamps were combined with incandescent lamps, or other gases such as neon and cadmium were added to the mercury. An important line of research, first followed by Hewitt and later again by several European researchers, was the use of fluorescent materials.

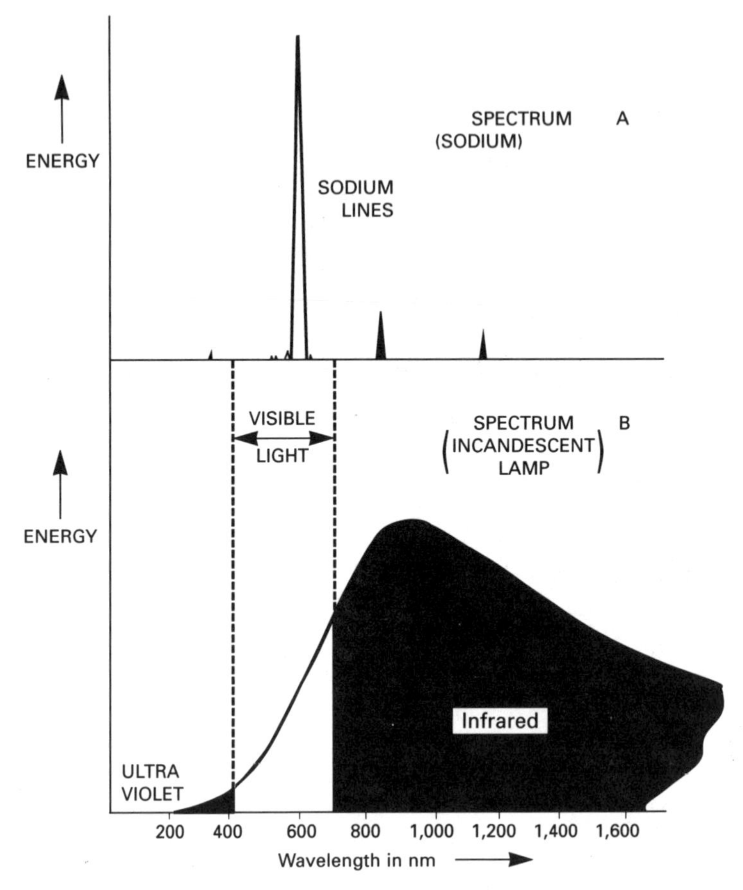

Figure 4.7
When sodium is heated (or an electric discharge occurs in sodium vapor), it emits all light as yellow (spectrum A). The emission spectrum of a standard incandescent lamp, however, extends for a large part into the invisible infrared: see spectrum B (figure is adapted from Wheatcroft, 1938: 257).

Fluorescence

The phenomenon of fluorescence had already been known for a long time, when the physicist George Stokes studied fluorite, a mineral that displays the effect strongly, and coined the term "fluorescence." The term is reserved for those illumination phenomena caused by some form of external radiation (for example, X-rays or ultraviolet light), as contrasted to illumination phenomena caused by heating (incandescence). Phosphorescence is fluorescence that continues for a while after the exciting external radiation has ceased. Confusingly, the word "phosphors" was generally used for all fluorescent materials. Stokes also formulated the law stating that the wavelength of the fluorescent light is never less than that of the light which caused it (Wood, 1911).

Becquerel (1857: 95) and Edison (in 1896) already had considered the possibility of employing fluorescent powders in electric discharge lamps (Claude, 1939), but this did not result in any stable artifact. Researchers in the United States and Europe tried various forms: reflectors behind the lamp,[32] internal or external coating of the lamp tube,[33] fillers baked into the glass of the lamp tube, or coatings on an external glass "balloon."[34] None of these resulted in a commercial lamp. In 1908 the assessment was that fluorescent lighting "est un mode d'éclairage ruineux."[35] European scientists, however, continued research and in the 1920s developed fluorescence lamps, which found some limited application in advertising.[36]

High-Voltage Fluorescent Lighting

The successful application of neon lamps in advertising led to the use of fluorescence to create a wider spectrum of more distinctive colors. The tubes were internally coated with the fluorescence powder. The ultraviolet light produced by the low-pressure mercury vapor could thus be transformed into almost any color desired. These fluorescence lamps used cold cathodes and high voltages, just like the other "neon tubes" described above.

The color of the light, the high voltage, and the expensive installation did not seem to make it useful for general indoor lighting. Nevertheless, Claude's company ("Société anonyme pour les applications de l'électricité et des gaz rares—Établissements Claude-Paz et Silva" in France) experimented with daylight lamps in the high-voltage range. In the 1930s André Claude (1936), George's nephew, and collaborators developed high-voltage neon tubes in combination with internally coated fluorescent tubes, and "from 1936 it has been possible to obtain 'white light' from a single tube. Presently the luminous efficiencies are much greater than those of the best incandescent lamps'' (Claude, 1939: 319). The efficiency, though, was not much greater than that of the incandescent lamps.

In Europe the high-voltage daylight fluorescence lamps were used for indoor lighting on a limited scale in situations where the expensive installation of high-voltage equipment would pay off, like in restaurants and warehouses. In the United States no such indoor usage developed. Claude's company had started conducting business in the United States and was quite successful there, but General Electric had managed to confine Claude's activities to the outdoor lighting field. General Electric and the Claude Company agreed not to enter each other's markets—until 1938 by an unwritten understanding, and since 30 December 1937 by a twenty-year licensing agreement.[37] General Electric would not enter outdoor fluorescent lighting, and Claude would not make indoor lighting installations. Claude continued to make (mostly colored) custom-designed electric discharge lamps for outdoor use. In Europe the Claude firm was actively developing daylight fluorescence lamps for indoor as well as outdoor lighting.

4.4 The Low-Voltage Fluorescent Lamp

Here the stream of events quickly becomes a cascade. General Electric started to work on fluorescent lighting, and soon a lamp was constructed. To understand the pressure on the Mazda companies to market this lamp very swiftly, we will make a short excursion into the technological culture of that time: the machine age in the United States. Demonstrating the lamp's interpretative flexibility, the controversy between the two relevant social groups of Mazda companies and utilities will be described. This controversy would erode the good relations existing between the Mazda companies and utilities, and end their almost total control of the electric lighting field in the United States. In telling this story, it will be necessary to follow the actors into many technical details. For these technicalities form the basic ingredients of the power plays among the various relevant social groups.

Fluorescent Lamp Research at General Electric

Although the General Electric Research Laboratory had looked into the possibility of fluorescent lamps before World War I, it was only in the 1930s that General Electric launched a serious effort in this direction, and this was not in the Research Laboratory but in the Lamp Development Laboratory at Nela Park (Birr, 1957). This laboratory operated almost as a separate company, although financially it was an integral part of General Electric (Bright and Maclaurin, 1943).

Before 1934 only scattered attention was given to the possibility of a fluorescent lamp. One problem was to find suitable phosphors. The earliest experiments were done, Inman (1943) recalls, with willemite ore, which was pulverized and then examined under ultraviolet light in a darkroom. The best bits were picked out with tweezers and used for the first lamp, which produced a weak green light. Many natural phosphors were tested in this way. In 1934 researchers at the General Electric Lamp Development Laboratory were inspired to step up this research by a letter from Arthur Compton, a Nobel Prize winner in physics who worked for General Electric as consultant. He reported about a relatively efficient fluorescent lamp in England.

Now progress was made swiftly. Within six weeks a fluorescent lamp was constructed that had significantly higher efficiency. In July 1935 this lamp was demonstrated at a closed meeting of General Electric officials and naval officers. Shortly afterward the lamp was tested aboard ship. Only three months later, a practical low-voltage fluorescent lamp was made public at the national convention of the Illuminating Engineering Society. It had hot cathodes, coated with a material that allowed better electron emission. For the glass tube production, the General Electric laboratory could draw on its expertise in producing tubular filament lamps (the Lumiline lamps).[38] A crucial result of the research into phosphors was a marked improvement of the efficiency of the fluorescent mechanism. A most remarkable aspect of fluorescence is that the spectral distribution of the emitted light is generally independent of the excitation mechanism. Willemite, for example, will always emit green light, whether it is excited by the mercury 254 nm ultraviolet, by the sodium 589 nm, or by X-rays. The efficiency with which the absorbed energy is converted into (visible) light is, however, not uniformly distributed. Phosphors are mainly excited by ultraviolet light,[39] and after General Electric researchers had compared the different emission spectra, their choice for a gas fell on mercury, which has an important emission line at 254 nm (see figure 4.8).[40] Further research led to the conclusion that this 254 nm line was most prominent when the discharge took place in low-pressure mercury vapor.[41] Concretely, this meant that in assembling the lamp a small drop of mercury was left in the vacuated tube. The thickness of the fluorescent layer influences the efficiency of the conversion of

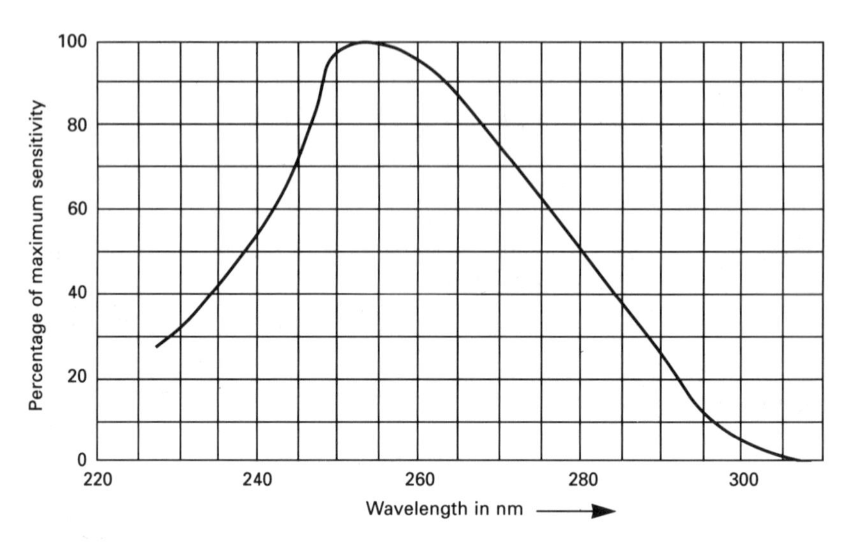

Figure 4.8
This curve shows the relative sensitivity of the zinc beryllium silicate phosphor to ultraviolet radiation of various wavelengths. There is a peak near the 254 nm of mercury's emission line. Many other phosphors have similar chracteristics (Inman and Thayer, 1938: 246).

ultraviolet radiation into visible light, and, as was reported in the first scientific article on the lamp, "the evolution of a process for obtaining smooth, uniform coatings ... has required much painstaking effort" (Inman and Thayer, 1938: 245). The fluorescent conversion was realized with an efficiency of 50 percent (Inman and Thayer, 1938).

Much research was devoted also to developing the necessary auxiliaries for starting the lamp. The earliest auxiliaries consisted of two parts in a single case: a choke to limit the current to the designed wattage of the lamp, and an automatic switch for starting the lamp (see figure 4.9). The choke presented no problems, but the switch did. Most automatic switches were of the thermal type: after a few seconds of electric current, a curved metal strip is heated so much that it stretches and breaks the contact.[42] Of course, the switch must be configured in such a way that it will not create a short circuit when cooling again (see figure 4.10). A practical application of the lamp was shown on 23 November 1936, at the 100th anniversary dinner for the U.S. Patent Office (Inman, 1954).

A pilot plant was installed and tentative production began. On 21 April 1938, General Electric announced the commercial availability of the low-voltage fluorescent lamp.

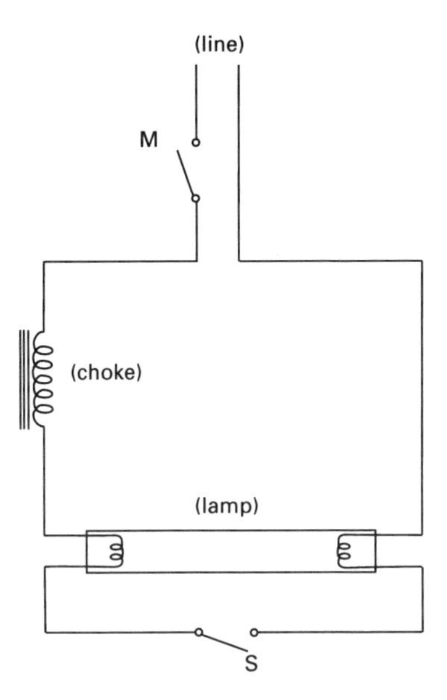

Figure 4.9
The elementary circuit of the first low-voltage fluorescent lamps. When the switch S is closed, the electrodes are heated. When, after two or three seconds the switch is opened (manually or automatically by a thermal switch), the induction coil (often called "choke" or "reactor," and after the power factor controversy—see below—called "ballast") will produce a voltage surge that starts the discharge (Inman and Thayer, 1938: 247). Switch M is the lamp's main switch.

Electrified America[13]

Before continuing the story of the lamps, a brief excursion into the technological culture of the United States between the two world wars is pertinent. In my theoretical framework, these wider cultural aspects come to bear on the social construction of technology via the technological frames of the relevant social groups. Because all groups in the lamp case share this culture, however, it is more straightforward to describe this background separately.

Despite the effects of the Great Depression of the 1930s, many Americans could see a unified period of science, technology, and industry. Technology in all its manifestations—as object, process, knowledge, and ultimately as symbol—became the fundamental fact of modernism. New materials such as Bakelite, new products like a range of electric household appliances, and new "arts" such as sound movies and the radio

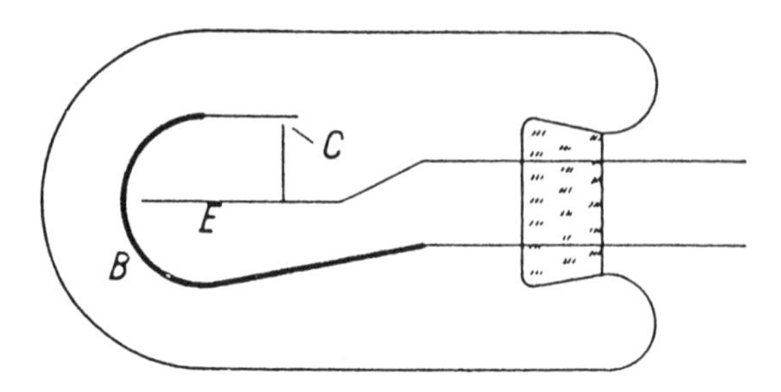

Figure 4.10
The glow switch: an automatic thermal switch (S in figure 4.9) to start the low-voltage fluorescent lamp. The switch is contained in a small gas discharge tube. The bimetal strip B forms one electrode; E is the other electrode. When the lamp is switched on (switch M), a "glow discharge" between B and E will occur, heating the bimetal strip. The strip will bend more, and make contact at C. Then a current will start through the lamp electrodes, which are heated to allow easy emission of electrons. After a few seconds, strip B cools again, and the contact C opens. Then the voltage surge should occur that starts the discharge in the lamp. If not, the whole cycle repeats. If the ignition voltage of the switch is higher than the normal working voltage of the lamp, the switch will stay dead when the lamp is operating. The ignition voltage of the switch should be lower than the mains voltage, so that it will discharge when the main switch M is closed (Kruithof, 1941: 67).

invaded society. The photographer Paul Strand (see figure 4.11) claimed that humans had "consummated a new creative act, a new Trinity: God the Machine, Materialistic Empiricism the Son, and Science the Holy Ghost."[44] Machines were everywhere, and their impact went beyond their physical existence to challenge perceptions of the self and the world, and thus to imply a whole new culture.

In the case of electricity, the same applies a fortiori. David Nye observes that there was never a time when Americans understood electricity in purely functional terms; they have always responded with a touch of wonder and ascribed symbolic dimensions to it. "As America electrified, in the imagination it became electrifying" (Nye, 1990: 382). Electrical terminology suffused popular language with metaphors of power, and in the Depression years, American leaders were expected to be "live wires" to recharge the American economy. These "human dynamos" could stage "electrifying performances" after having "powered up" in the morning with a strong cup of coffee, called "battery acid." If they were man and woman, their first meeting might "give off sparks" and one could "feel the electricity."

Figure 4.11
Paul Strand, "Lathe, Akeley Shop, 1923." Gelatin-silver print, 10″ × 8″. Courtesy the Paul Strand Foundation.

The New York World's Fair, "The World of Tomorrow," provides a typical image of the technological culture of the period.[45] To build the fair, city park commissioner Robert Moses turned the centrally located Flushing area—at that time some 1,200 acres of salt marsh land with waste and garbage dumps of monumental proportions—into the Flushing Meadows Park. Central in designing the fair were the same men who designed the machine age in the two previous decades: the industrial designers who had shaped the refrigerators, lamps, vacuum cleaners, and railway trains now incorporated the utopian ideas from technocracy, the architectural concepts of Lewis Mumford, and the esthetics of art deco in this model of the future world. Visitors were invited to live the world of tomorrow, staged in dramatized models and dioramas in a utopian landscape (see figure 4.12).

Figure 4.12
Constitution Mall, New York World's Fair, 1939. The Trylon and Perisphere, at the center of the Fair, were colored in dead white. From there the Constitution Mall, the main axis of the Fair, showed increasingly deeper reds, from rose to burgundy when moving from the center. The trees had soft individual lighting. Photograph courtesy Stichting Nederlands Foto- & Grafisch Centrum—Spaarnestad Fotoarchief.

General Motors' *Futurama* allowed visitors to see the world of 1960 as if they were hovering in a low-flying aircraft, sitting in comfortable chairs. The eighteen-story dome called Perisphere housed the miniature world *Democracity*—the "symbol of a perfectly integrated, futuristic metropolis pulsing with life and rhythm and music."[46] General Electric's *House of Magic* showed an electric motor running on solar light, suggesting the availability of limitless energy; a model train reacting to a human voice, intimating the easy control of technology by man; a metal bowl floating in the air on a magnetic field, suggesting the imminent conquest of gravity; and "visible sound" and "audible light," showing how technology extended human perception. Scientists and engineers had Promethean powers, the message clearly was, and were using these to create an ideal world with more pay and content for the workers, less hardship and more profit for the farmers, more comfort and less work for mother.[47]

But the New York World's Fair was not only important as an icon of technological culture and modernity; it also had a direct impact on the social construction of fluorescent lamps.

Commercial Release of the Low-Voltage Fluorescent Lamp by General Electric and Westinghouse

When the Mazda companies released the fluorescent lamp officially, it was explicitly aimed at "tint lighting."[48] The new lighting device could provide brighter and deeper colors of a wider variety than was previously possible with incandescent lamps. Because of their ability to produce "light in hitherto unobtainable pastel tints as well as pure colors,"[49] they were expected to find application in special lighting: "The new lamps are for use primarily in decorative lighting because they produce colored light efficiently in a variety of interesting shades" (Anon. 1938b).[50] Moreover, although their installation costs were higher, they were 30 to 200 times more efficient than incandescent lamps for color lighting: "Experimental 15-watt green fluorescent lamps have operated at 70 lumens per watt. As compared with about 0,3 lumen per watt for certain 60-watt green filament lamps, this represents an efficiency gain of about two hundredfold (Inman and Thayer, 1938: 247). Lighting applications mentioned ranged from theater interiors to ballrooms, from specialty shops to art galleries, from showcases to game machines, from railway cars to homes. Some of the latter applications show that the Mazda company executives were already thinking of general indoor lighting, but this is not yet very explicit.[51]

Also by the relevant social group of utilities, the lamp was considered for color lighting only. This is not surprising, as the utilities' knowledge of these lamps was rather limited and based almost exclusively on information provided by the Mazda companies. The new lighting device was introduced in a way that did not suggest any revolutionary change in lighting practice. Three utility men remembered the occasion: "Its presentation was as casual as developments in incandescent sources were wont to be. There was the usual amount of discussion, but the impression seemed to be that here was a light source rich in color and high in efficiency, but low in total light output, expensive, and generally suitable for only special applications."[52] Thus even when daylight lamps were discussed, this was done in the context of special purposes and tint lighting, as is clear from a memorandum of the Chairman of the A.E.I.C. Lamp Subcommittee: "The daylight tubes it is to be anticipated will have most utility. Because of the small wattages and small production of heat these lend themselves particularly well to showcase illumination. Because of the white light they should find large application for color matching purposes."[53] I will call this lamp the fluorescent *tint-lighting* lamp, thereby foreshadowing my demonstration of the fluorescent lamp's interpretative flexibility. The two other lamps to be revealed in this coming act of deconstruction are the high-*efficiency daylight* fluorescent lamp and the high-*intensity daylight* fluorescent lamp.

The origins of the fluorescent tint-lighting lamp can be traced to the 1939 New York World's Fair. Considering what we now know about the presently stabilized usage of fluorescent lamps (i.e., general indoor daylight lighting), it is intriguing why that first artifact was the fluorescent tint-lighting lamp and not immediately the other lamp that eventually stabilized: the high-intensity daylight fluorescent lamp. The fluorescent tint-lighting lamp seems to be a strange deviation from the (retrospectively apparent) linear path from the goal of general white indoor lighting to the modern fluorescent lamp for general lighting. As usually, the actors show how to understand this detour, in this case by guiding us to the New York World's Fair and its immediate predecessor, the Paris exposition "Arts et Techniques" in 1937.

Ward Harrison, Engineering Director of the Incandescent Lamp Department of General Electric and most prominent spokesperson for the Mazda companies in the early days of fluorescent lighting, admitted that "There were a couple of World's Fairs in the offing that were going to be lighted almost entirely with the high tension tube lighting if they were not supplied with some lamps of ours."[54] Other relevant social groups also saw the World's Fairs as the reason for dragging the fluorescent lamp "out of the research laboratories by a caesarian operation."[55] As the fixture manufacturers described this episode retrospectively, "the pressure of the demand for a new illuminant to be exploited at two World's Fairs was too much [for conservative judgment to prevail]. The 15- and 20-watt fluorescent lamps were produced for use at the Fairs—others wanted them—and a new illuminant, with a lot of unexplored implications, was launched."[56] This view is confirmed by the lighting engineers of the World's Fair themselves (Engelken, 1940). The Fair's context makes the emphasis on tint lighting understandable. Color schemes of architecture and artificial illumination played an important role in the planning of the New York Fair (see also figure 4.12):

> A zoning and color scheme adopted prior to the construction insured architectural unity, and harmony of plan, design, and treatment throughout the whole area.... The color scheme ... is coordinated with the physical layout. Starting with white at the Theme Center, color treatments of red, blue or gold radiate outward with progressively more saturated hues. Adjoining hues blend circumferentially along the avenues. The illumination was fitted to this scheme [so] as to maintain the basic pattern by night as well as by day, but with new and added interest and charm after sunset. (Engelken, 1940: 179)

Obviously, tint lighting was an important objective for the lighting engineers who were designing the first large-scale applications for these fluorescent lamps.

Load Controversy

But within half a year of the introduction of the fluorescent tint-lighting lamp, another artifact emerged: the high-efficiency daylight fluorescent lamp. A flood of advertising over the signatures of the major lamp companies streamed out, containing such statements as: "Three to 200 times as much light for the same wattage"; "Cold foot-candles"; "Amazing efficiency"; "Most economical"; and "Indoor daylight at last." Lamp engineers began to give more attention to developing daylight lamps. As Inman, a General Electric engineer, recalled: "Early white lamps at lower efficiencies were attractive for general lighting purposes, and they stole the show" (Inman, 1954: 37). Because different phosphors emitted various colors of light, it was possible to obtain white light by combining different fluorescent materials. The first white fluorescent lighting, installed in the General Electric Lamp Laboratory hallway, consisted of six different lamps—one green, two blue, and three pink (Inman, 1954). Soon single lamps were produced that were coated with a combination of different phosphors, such as cadmium borate (red), willemite (green), and magnesium wolframate (blue). In January 1939, the journal *General Electric Review*, aimed at engineers within and outside the company,[57] cheered: "The spectral quality of light from the fluorescent daylight lamp is the closest approach to natural daylight that it has ever been possible to produce by any artificial illuminant at an efficiency even approaching that of these lamps; and, similarly, the efficiencies with which these lamps produce colored light have never been approached" (Anon., 1939: 45).

The utilities started to fear that the high efficiency of the fluorescent lamp might reduce their electricity sales. As the utility executive Carl Bremicker of the Northern States Power Company said about his utility employees, "they had better get their white wing suits ready because very shortly General Electric and Westinghouse would have them out cleaning streets instead of selling lighting."[58] An internal Westinghouse memorandum lends support to the utilities' worries. It concluded that "the average utility lighting man sees in the rise of fluorescence a decrease in his relative importance."[59] The memorandum presented a comparison of the profits, based on a rate of 4¢ per kilowatt-hour and with equal costs to the user. It was presented as a conservative assessment: the design data were unfavorable to fluorescence and almost any other selection would have resulted in greater differences. The result of the comparison was that, to the utilities, fluorescence was only half as important as incandescence; to the lamp suppliers it was six times as important, to the equipment manufacturers three times as important, to the contractor 20 percent more important (see table 4.2).

Table 4.2

Comparison of profits to be gained by different relevant social groups in the cases of the fluorescent lamp and the incandescent lamp

For every dollar the user spends annually with	incandescence	fluorescence
The utility gets	80%	44%
The contractor gets	10%	12%
The equipment suppliers gets	6%	20%
The lamp suppliers gets	4%	24%

Source: R. G. Slauer (Westinghouse Lamp Division, Commercial Engineering Department to Westinghouse Lamp Division) to A. E. Snyder (Westinghouse Lamp Division, Executive Sales Manager), letter dated 12 July 1939 (Committee on Patents, 1942: 4818–4819; 4818).

Thus a controversy developed: the "load issue." It took the form of a battle over the two fluorescent lamp artifacts. The utilities, having been alerted by their discovery in January 1939 of the high-efficiency daylight fluorescent lamp, tried to keep the other artifact, the fluorescent tint-lighting lamp, presented in April 1938, in the forefront. The utilities argued that claims about high efficiency might be true, but only when fully qualified. And this, they claimed, was not done. For example, when the "three to 200 times as much light" statement was accompanied by the picture of an office, the customer might expect amazing efficiencies. But this, the utilities argued, would be true only if that customer were willing to have green or blue light in his office.[60] The utility lighting staff were irritated by this misleading publicity, and in trying to fill in the rest of the story found that they were immediately accused of excessive self-interest. They resented their position of apparently throwing cold water on fluorescent lighting, only because, as they would say themselves, they were trying to tell the complete story. Long and detailed arguments were given to point out that the high-efficiency daylight fluorescent lamp really did not exist, but that it was mistaken for the fluorescent tint-lighting lamp, which indeed was a valuable new lighting tool, but only for limited purposes.[61]

The principal spokespersons for the Mazda companies did not agree with the conclusion that the load on the electricity networks would fall, thus decreasing the profits for the utilities. And so they continued to push, albeit carefully, the high-efficiency lamp. Harrison, for example, was convinced that only in some instances would consumers experience reduced electricity costs, but that, on average, their electricity consumption would go up.[62] However, the Mazda companies had their own problems with respect to the high-efficiency lamp. Their main problem was that, at the moment of its commercial release, there was no known relation between life and efficiency in fluorescent lamps; in fact, the life of the lamp was not known. They knew that it was something more than 1,500 hours when the lamps were given their original rating, but it could work out to be 15,000 hours or much more. As Harrison said to an audience of utility executives, "Instead of having 93 per cent of our business in renewals in good times and bad, it may be that our first sale will be almost our last sale to a given customer."[63] Nevertheless, the Mazda companies were developing a more differentiated line of fluorescent lamps, because, as Harrison explained in 1940, "the effect of changes in the efficiency of fluorescent lamps, changes in their rated life and changes in price have radically affected their over-all operating costs, so that in twelve months . . . [these changes have] brought the lamp more seriously into the field of general lighting."[64] Obviously, the artifact he was describing was the high-efficiency daylight fluorescent lamp and not the fluorescent tint-lighting lamp.

Power Factor Controversy

The battle between the Mazda companies and the utilities would grow fiercer still. Besides the load controversy over the high-efficiency lamp and the tint-lighting lamp, a second casus belli soon emerged: the power factor controversy. The lines between the various parties were, however, less clearly drawn than in the load controversy.

The concept of "power factor" is intriguingly complex. Answering the chairman of the Senate Committee on Patents, who asked for an explanation, the antitrust investigator said that

> Power factor is a very, very difficult thing to describe. I know that the chief engineer of the lamp department of General Electric told me he had never heard two people describe it in the same way, but the effect of it is quite definite and can be stated—that is, that if electrical apparatus is operated at low power factor, it requires lines with greater capacity. . . . It is one of those things of which you may know what the effect is, but the thing itself is practically impossible to define, even though it is easy to understand.[65]

This phenomenon occurs in any electric device that is run on alternating current, and that is not a pure resistance—that is, any device that contains capacitors and/or induction coils. When the power factor of a device equals one (in the case of a pure resistance) the *useful power* dissipated in the device and the *total power* supplied by the mains are the same. When, however, the power factor is less than one, the useful power (in the form of, for example, light from a fluorescent lamp or work from an electric motor) is less than the total power.[66] The latter is provided by the electricity mains, thus causing an "extra" (with respect to what is usefully employed) useless load on the main system.

The fluorescent discharge tube itself has a high power factor of about 0.9, but the auxiliaries of choke and automatic switch result in much lower power factors of about 0.6.[67] The utilities were worried that the application of fluorescent lamps would take up large parts of the capacity of the distribution system, without the utilities being able to claim the selling of more (useful) electric power.

Although this may seem a rather straightforward electrotechnical concept, it is easy to show its interpretative flexibility; for different relevant social groups it constituted quite different facts. For the utilities it constituted a real problem, threatening to overload their distribution networks. Of the Mazda companies' reaction to this anxiety of the utilities, antitrust attorney Walker said:

> after fluorescent lighting had been on the market for over a year, the manager of the commercial engineering department of the lamp division of Westinghouse expressed the belief that, in view of the promotional policies of the Mazda manufacturers, there was "little justification for the anxiety expressed by the utility companies who, for years, have permitted and encouraged the sale of all sorts of appliances with even lower power factor than our first fluorescent equipment."[68]

The Mazda companies saw the low power factor as merely an additional argument of the utilities in their fight with the Mazda companies, rather than as a serious problem.

Several ways to solve the power factor controversy were tried. Already in the early stages of research at the General Electric Lamp Department, a combination of two fluorescent lamps was tried (Cleaver, 1940). One lamp was given an induction coil as ballast, while the other worked on a combination of coil and capacitor. Thus is was possible to achieve a 0.9 power factor. In 1939 such a twin-lamp auxiliary was made available under the name of "Tulamp." This was one example of solving the power factor problem "at the lamp." But that was not the only possibility.

Many utilities tried to regulate the type of electric appliances that might be connected to their distribution nets. Some used a rule prohibiting overall power factor from falling below a certain point. Others placed limitations on the power factor of equipment that might be connected to the lines.[69] Still another strategy was to let the customer connect whatever he wished, but for a fee:

> Studies are being made along another line which has very good public relations aspects. These are aimed at allowing the customer to have anything he wants as long as he is willing to pay for it. One method is based on inspection, and makes monthly power factor charges based on number and size of lamps. Another method entails finding an economical means of measuring power factor for the large mass of commercial and industrial customers so that suitable charges can be made. Some promise is offered so that the meter will measure the kw plus a given percentage of the reactive power.[70]

All these strategies were tried; the solution, however, would come in the course of 1940, from the enrollment of a new relevant social group, fixture manufacturers. Through them the production of certified auxiliaries that all had power factor correction incorporated could be secured. This will be described in the next section, after I have introduced some new relevant social groups.

Additional Relevant Social Groups

The load controversy, further heated by the power factor controversy, was fierce. Probably it was so acrimonious because the relevant social groups of Mazda companies and utilities both felt that their common control of the lighting market, as exerted in the incandescent era, was at risk. The utilities felt particularly threatened. They had come unscarred through the Great Depression because the service sector and the high-technology industries such as food processing, aircraft production, and chemical manufacturing had continued to expand. These growth areas all consumed large quantities of electricity. Also the domestic and rural sectors were particularly profitable because they paid the highest rates. During the first half of the Depression alone, 165 new central stations had to be built. Local communities, politicians, and consumers did not recognize the long-term benefit of this structural consolidation of the electricity distribution system, and accused the utilities of being rich because of exorbitant rates and monopoly control. As a result of President Roosevelt's New Deal, the government sometimes was virtually at war with the utilities (Nye, 1990). Under these circumstances, it is hardly surprising that the utilities were nervous about the prospect of losing support of the Mazda companies.

This threat of a crumbling control of the lighting market became especially acute when a third relevant social group entered the stage: the independents. Hygrade Sylvania Corporation announced its own line of low-voltage fluorescent lamps in 1938, Consolidated Electric Lamp Company followed in 1939, and still later a third firm, Duro Test Corporation, came on the market (Bright and Maclaurin, 1943). For the latter two companies, the fluorescent lamp was only an additional line to their incandescent lamps. For Hygrade it soon became more important.

Hygrade, restive under the quota restrictions of its B-class license, had long been seeking a way to grow in the lamp industry. It had bought as many other small lamp manufacturers as it could, but had not been able to expand its sales beyond 5 percent of the total market. Its engineers had been working on fluorescense, developing an effective method of coating glass tubes and making a first fluorescent lamp in 1934. This experimental lamp did not seem commercially attractive, and the project was abandoned until the Mazda companies introduced their lamp in

1938. Then Hygrade spurred on its research activities, gained control of some useful patents, and started its own low-voltage fluorescent lamp production. Indeed Hygrade felt so strong that, after considerable debate among company executives, it was decided not to accept the offer of a B-class license under General Electric's fluorescent patents (Bright and Maclaurin, 1943). (As I described earlier, the patent licensing system of 1927 did not cover discharge lamps.) This resulted in a patent infringement suit by General Electric against Hygrade, which was only decided after World War II.

Hygrade's fluorescent lamps signaled a potential break from the control of General Electric, and it prompted an aggressive promotional campaign. In late 1939 the Mazda people observed worriedly that Hygrade Sylvania's policy seemed to be successful: "There are figures which seem to indicate that the Hygrade Company is selling as many fluorescent lamps as General Electric and Westinghouse combined. Apparently, they are going out and 'beating the bushes,' so to speak, installing sockets in the smaller companies on main streets throughout the United States."[71] The aggressive sales policy employed by Hygrade Sylvania caused as much of a problem for the utilities as it did for the Mazda companies. The utilities sensed a realignment of forces taking place among lamp manufacturers. Hygrade claimed to have basic patents for the manufacture of fluorescent lamps and did not recognize the patents held by the Mazda companies. The utilities feared that this realignment of forces, together with the competitive situation that attended it, might lead to methods and activities that would disorganize the whole lighting market "to the detriment of the public and the utilities who were standing on the sidelines." Hygrade Sylvania was capturing a sizable portion of the market. This was claimed by them and admitted by the Mazda people.[72] Hygrade Sylvania clearly was advancing the high-efficiency daylight fluorescent lamp, although downplaying the economic risk for the utilities. For example, W. P. Lowell, a Hygrade executive, answering the question of why fluorescent lighting was demanded by the public, argued before an audience of utility and Mazda company executives:

> Why is it demanded? For many reasons: its daylight color, soft quality reduced shadows, novelty (it's new, modern, smart), real or imaginary economy. But don't worry too much about those who think they are saving money by using fluorescent lighting to save a few watts. If the overall value—combining the sheer dollars and cents with all other qualities—if the net value is not right, the product will fall of its own weight. You can't fool all the people all the time.[73]

Thus the activities of the relevant social group of independents, and Hygrade Sylvania especially, resulted in pouring oil on the controversy's fire.

Other relevant social groups came into play now as well. Until 1938 there was no public demand for anything like fluorescent lighting. The incandescent lighting experience seemed perfectly satisfactory to most people—a dollar spent on electric lighting was constantly buying more illumination. But after the lighting engineers of the World's Fair had used fluorescent tubing and the first lamps had been on the market for half a year, the general public was alerted. Bright and Maclaurin (1943: 439) adequately summarize the resulting situation by concluding that "fluorescent was bought, rather than sold." Now the general public had become a relevant social group, also in the eyes of the other relevant social groups. Hygrade, of course, enrolled this new relevant social group as an ally. In the words of one of its executives, "To be sure, we have advertised a little, but I think it is fair to say that without any real national advertising or great sale push, the public is buying fluorescent lighting faster than it can be supplied. Let's face facts—there is a genuine public demand for this better lighting tool and we must fill that demand if we are to serve the public properly."[74] Other relevant social groups were less positive about the role played by the public, as for example an executive of an auxiliaries manufacturer: "The public always clamoring for a novelty, brought increasing pressure upon all of us for the new equipment, often asking for ridiculous and impossible applications."[75] Within the relevant social group of the general public (which, for all actors, was synonymous with customers), the group of women was singled out (see also figure 4.13): "The widespread acceptance of fluorescent lighting in the home will depend directly upon the housewife, who is generally alert to new ideas that give comfort to her family and beautify her home, provided the cost does not exceed the family budget—and more important, provided she is made conscious of the advantages of the new equipment through national advertising and neighborly example."[76]

The last relevant social group that entered the fluorescent stage at this point was already mentioned in connection with the power factor controversy: fixture manufacturers. General Electric and Westinghouse produced only lamps and the necessary fixtures, and auxiliaries were manufactured by other companies. This was also what they hoped to do in the case of the fluorescent lamp, although the General Electric Lamp Department had already done a substantial part of the development of

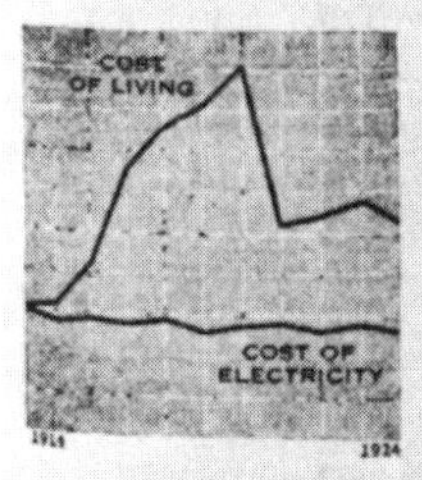

Figure 4.13

General Electric had a long tradition of specifically addressing women in their advertising campaigns. Here, in a 1925 advertisement, the implicit message is "The cost of electrical technology is so small that its price is irrelevant when compared with the value of children" (Nye, 1985: 130–131).

the key elements of the fluorescent lamp fixture (switch and ballast). This created, however, an important delay in the sales effort of the Mazda companies, as the fixture manufacturers had no experience with such a lamp. In contrast, Hygrade Sylvania produced its own fluorescent lamp fixtures and thus could market complete lamps. In the next section I shall describe how the Mazda companies and the utilities sought to end their controversy and stay in control of the lighting market by setting up specific relations with the relevant social group of fixture manufacturers.

At this moment of the story, the battle between the Mazda companies and the utilities is at its peak. My analysis of this controversy made use of the interpretative flexibility of the fluorescent lamp, yielding two different artifacts, the fluorescent tint-lighting lamp and the high-efficiency daylight fluorescent lamp. In the next section the fighting parties will arrive at a tentative ceasefire and then a more stable peace. As will become clear, a third fluorescent lamp was instrumental in, and resulted from, this closure in especially the load controversy. The power factor controversy was closed in the wake of the first agreement.

4.5 The Social Construction of Fluorescent Lighting

In the load controversy between the Mazda companies and the utilities closure was reached through the design of a third fluorescent lamp, the high-intensity daylight fluorescent lamp. The subsequent stabilization of this artifact was to a large measure further fostered by closing the power factor controversy, especially through an enrollment of the relevant social group of fixture manufacturers. To explain these processes, I will focus on the changes in technological frames of the Mazda companies and the utilities. Because the fluorescent technological frames of the Mazda companies and the utilities were quite similar (but for goals and the problem-solving strategies), their frames will be sketched by focusing on the dimensions of "goals" and "problem-solving strategies." In addition, the dimension of "current theories" will be discussed, because the closure that was reached resulted in important revisions of the frame.

Technological Frames

The utilities' main goal was to sell electricity, whereas the Mazda companies' goal, in the context of this study, was to sell lamps. Left at that, this would be a rather trivial observation. However, goals do not straightforwardly define the actions taken by the relevant social groups.

The respective technological frames influence, for example, the way these goals are translated into problem-solving strategies.

The theoretical base of the Mazda companies' fluorescent frame was formed by electricity and gas discharge physics, where the utilities obviously used, primarily, power electricity physics. Neither played an explicit role in the course of events I am describing. The utilities' frame was supplemented by what they called the "Science of Seeing," which focused on the quality of lighting, including such things as brightness, contrast, shadows, diffusion, and various kinds of glare.[77] This theoretical part of the utilities' frame did play a role: emphasis was placed on seeing and the prescription of lighting that would contribute maximum visibility to the task at hand. As the utility people said themselves, rather pretentiously, about the years of incandescent lighting: "a true Science of Seeing was born ... It was here that the Cooperative Better Light-Better Sight Movement was started, and lighting practice became firmly entrenched in the philosophy of 'results to customers.'"[78]

The last words in this quotation hint at an important element in the problem-solving strategy of the utilities: they pictured themselves as servants of the public, or even as teachers of that public.[79] Thus an important goal was to increase public confidence in lighting technology and to promote (the utilities' version of) knowledge about that technology. In this context, the utilities highly valued cooperation with the Mazda companies: "The lighting industry, based upon a sound Science of Seeing and united by the Better Light-Better Sight Movement, has presented a solid front to the public. This has captured the interest of strong professional groups, increased the customer confidence so important to future growth, and has proved successful commercially."[80] This is no devious strategy designed to mislead the public in order to boost electricity sales. As I have argued above, the promotion of electricity, of technology generally, and indeed of economic growth per se was deeply imbedded in American culture—it was modernity, it was "the American Way of Life." In this sense, the utilities' technological frame only mirrored the general American culture.[81] The utilities' frame was, however, specific in that it focused the strategies on promotional, advertising, and teaching activities.

The implication for the technological frame of the utilities is that, when confronted by a problem, their standard strategy was to reformulate that problem as an educational one—and hence to design better advertising strategies and sales methods. This is what happened in the case of the load problem. Talking about the public, which was thinking about lighting costs in terms of current costs instead of "true costs," they formulated as their task "to educate them properly to the true cost and value of adequate lighting [,which] is not an easy job."[82] It is important to see that this problem-solving strategy was not the only possible one. Another possible strategy would have been, for example, to define appropriate standards and to impose these upon other relevant social groups. The utilities were indeed going to follow this strategy, but this worked only as a "second choice" at a relatively late stage, when the Mazda companies had already proposed the certification scheme for fluorescent lamp fixtures.

After this brief characterization of the two technological frames, we will resume the story where we left it at the end of the previous section: in early 1939, when the load controversy took the form of a conflict between two competing artifacts, the fluorescent tint-lighting lamp and the high-efficiency daylight fluorescent lamp. During the first year after the commercial release of the fluorescent lamp, the tension between Mazda companies and utilities increased. A dissociation of the cooperation, established in the incandescent lighting era, seemed not unlikely. The utilities were accusing the Mazda companies of undermining this cooperation. Mueller, Sharp, and Skinner remembered: "The question was quickly asked . . . : could it be that the sound principles of the Science of Seeing so assiduously promoted were built upon sand, to be cast aside at the first gust of commercial expediency?"[83]

The Nela Park Conference: Rhetorical Closure

To settle this conflict, a conference between the representatives of the utilities and the Mazda companies was held on 24 and 25 April 1939, at the headquarters of the General Electric Lamp Department at Nela Park, Cleveland. The utility representatives referred to this meeting as "the fluorescent council of war."[84]

At this conference the idea emerged that fluorescent lighting might be reserved for high-level lighting only. At that meeting a third fluorescent lamp was designed—not on the drawing board or the laboratory bench, but at the conference table. This lamp was supposed to give light of daylight color, and with hitherto uncommonly high intensity: the high-intensity daylight fluorescent lamp. It came slowly into being during this meeting, as is apparent from the minutes:

there was considerable discussion on the outstanding features of fluorescent light with particular reference to daylight quality. Some thought that low foot-candles of daylight fluorescent lighting made a person appear sallow—on the other hand, 100 or more foot-candles in the Institute Round Table Room (previously inspected) seemed satisfactory to everyone. From the discussion, it was generally agreed that 50 to 100 foot-candles of fluorescent lighting could readily be installed without creating any impression of high level lighting. At least in some instances it was believed that 50 foot-candles of fluorescent lighting would appear like no more illumination than 25 foot-candles of filament lighting.[85]

What could be expected to happen to this idea? Considering the utilities' technological frame, it is understandable that the situation was perceived in terms of advertising. It was decided that the use of fluorescent lamps for general lighting would not be emphasized "until commendable equipment is available giving 50 to 100 foot-candles levels." This decision demonstrates clearly the effect of the specific problem-solving strategy in this technological frame. Instead of treating the problem primarily as one to be solved by advertising and educating, it would have been conceivable to treat it as, for example, a mainly technical problem, requiring the concentration of all effort on the development of lamps and fixtures that would provide high-intensity lighting. Indeed, quite the contrary happened.

In line with their technological frame, the utilities pressed the Mazda companies to adopt specific ways of advertising the fluorescent lamps; the utility representatives were quite satisfied with the outcome. After a difficult start of the meeting, the second day resulted in what was experienced by utility executives as "a most complete capitulation by the Mazda Companies."[86] Mueller, Manager of Commercial Sales, West Penn Power Company (and Chairman, E.E.I. Lighting Sales Committee), thought that he understood how this could have happened:

I think it was probably due to the fact that they realized they were definitely on unsound ground the way they had been operating, and they also knew . . . that the utilities realized it and were going to do something about it, and they knew that they really couldn't put across any lighting promotion without the help of the utilities. They were anxious to settle these matters with our group, because they thought that we were in the best position to get something in return for their capitulation.[87]

The large lamp companies issued statements of policy concerning the promotion of fluorescent lamps and tried to implement this new policy in all parts of their organization. For example, in the "statement of policy" by General Electric, issued officially on 1 May 1939, the company conceded that because the efficiency of fluorescent lamps is high, it might be assumed that the cost of lighting with them is less than with filament lamps; as often as not this conclusion is erroneous. The cost of lighting is made up of several items—cost of electricity consumed, cost of lamp renewals, and interest and depreciation on the investment in fixtures and their installation. All of these factors must be properly weighted to find the true cost of lighting in any given case. The fluorescent Mazda lamp should not be presented as a light source which will reduce lighting costs.[88]

Similarly, the Westinghouse statement read in part: "We will oppose the use of fluorescent lamps to reduce wattages."[89]

Mueller believed that one of the most important results of the conference was that the lamp companies now seemed inclined to take the utilities into their confidence, as part of the lighting industry, in the development of promotional plans, instead of "shooting the works" first and then letting them know about it.[90] The Mazda companies now clearly had the same ideas as the utilities about the need to reach an agreement. According to J. E. Kewley, manager of the Lamp Department of General Electric, "The ... statement of policy [was] issued particularly to allay the fears of the utility companies." And E. H. Robinson, a nother General Electric official, viewed the policy statement as a declaration by the lamp department signaling "Here's how we stand, boys, we'll play good ball with you central stations but we'll expect the same brand of ball from you too" (Committee on Patents, 1942: 4772). Thus, the agreement on the new high-intensity daylight fluorescent lamp not only solved the load controversy, but also bolstered cooperation between the two important relevant social groups of Mazda companies and utilities.

Stabilization of the High-Intensity Daylight Fluorescent Lamp

One would expect that this new lamp was quite successful, to have had such an impact on the two most powerful social groups in the U.S. electric lighting business. However, this was not so, at least not in any straightforward way. The lamp did not even exist physically; although closure had been reached, stabilization would take another year. According to Walker, the Antitrust Division attorney, there even was no immediate prospect of fluorescent lighting that would provide anything like 50 foot-candles levels for normal indoor use. Probably the average with incandescent lamps in 1939 was about 15 foot-candles, and there were no installations that emitted anything like 50 foot-candles of light. Nevertheless, the impact of this artifact—whose social construction started at the Nela Park Conference—was not small. Ironically, part of its impact at that conference may have been caused by its not yet being available, as Walker argued:

> The reason why the utilities did not want the fluorescent lamps promoted until they ... would give 50 to 100 foot-candles levels of lighting was that the utilities felt that if they could ever get fluorescent lamps of intensities that strong, fluorescent lamps would then use so much electricity that the utilities would not suffer as a result of the fluorescent lamps replacing the incandescent lamps (Committee on Patents, 1942: 4771).

The new statements of policy by General Electric and Westinghouse were not given out generally to the public. It is not difficult to guess why the public was not informed about the cancellation of the "high efficiency daylight fluorescent lamp" and the effort to sell the "high intensity daylight fluorescent lamp" instead.

Thus the utilities' technological frame (partly) shaped the fluorescent lamp. During the subsequent stabilization of the artifact, however, the technological frame is reshaped as well. In the case of the high-intensity daylight fluorescent lamp, this can be most clearly seen in an adaptation and further elaboration of the "theory" dimension of the technological frame of the utilities and in an adaptation of the fixture manufacturers' technological frame.

The Utilities' Technological Frame Adapted: A Science of "Daylight" Seeing

After reaching agreement at the Nela Park Conference, the utilities immediately started to elaborate on the idea of high-intensity lighting. Two days after conference, a note was written by the Electrical Testing Laboratories for the A.E.I.C. Lamp Subcommittee, where the point of daylight lighting was further argued by providing a theoretical evolutionary-biology justification:

> it will be noted that our eyes have evolved under the brilliant intensity of natural light in the daytime and under the dull flow of firelight in the evening. There is some reason to think that with light of daylight quality people will not be satisfied with the low intensity of illumination which is more or less acceptable in the case of light of warmer tone as that of tungsten filament lamps. Where the daylight lamps are to be used, the logical procedure is to work toward the equivalent of daylight illumination, which at once moves practice into higher ranges of illumination intensity.[91]

In addition to this kind of reasoning, research was boosted along the disciplinary boundaries of physics and biology.

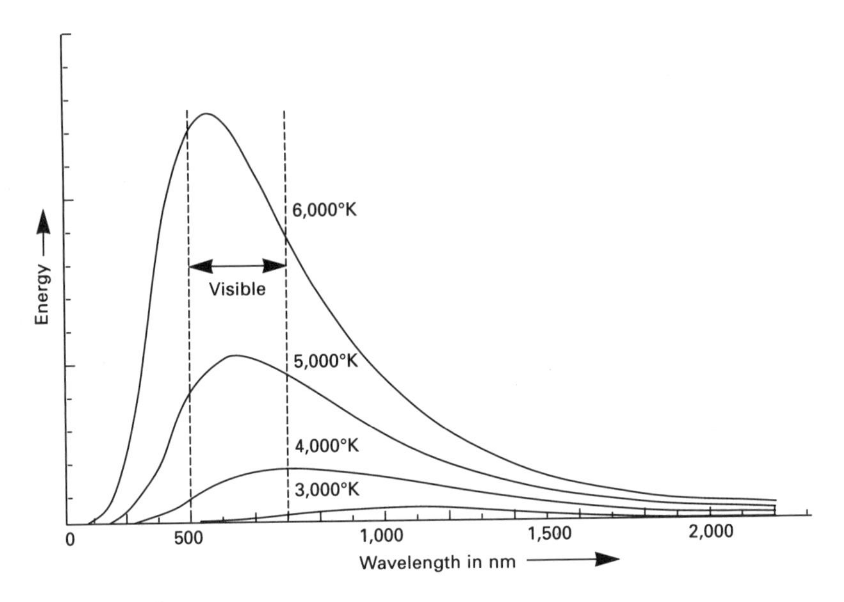

Figure 4.14
The spectral variation of light emitted by a black-body. Curves have been drawn for black-body temperatures of 3000°K, 4000°K, 5000°K, and 6000°K. The higher the temperature, the shorter the wavelength at which the black-body radiation has its maximum intensity. In other words: a relatively cold body emits much infrared, while a relatively hot body emits more visible and even ultraviolet light.

An important concept in these experiments and theoretical discussions is the "color temperature" of the light. The concept is founded in thermodynamics. One way to describe the color of light is to specify the light's wavelength, as I explained in connection with the early discharge lamps. Another way to characterize light is to compare it to the light that is emitted by a "black body"[92] of a given temperature. When the temperature is 0°K, there is no radiation emitted; when the temperature is, for example, 5000°K, the emitted light closely resembles sunlight. Black-body radiation has a spectrum of different wavelengths (see figure 4.14). The vocabulary of color temperatures offers a more effective way of comparing different "whites" than the wavelength vocabulary, as white light is, by definition, composed of a wide spectrum of wavelengths (see table 4.3).

Using the color temperatures framework, research was carried out to study the relation between the intensity and the color temperature of

Table 4.3
Color temperatures of natural white lights

Form of natural white light	Color temperature
Sunlight	5000°K
Light of overcast sky	6000°K
Light of blue sky	8000°K

Source: Kruithof, 1941.

comfortable lighting. Kruithof (1941), for example, described experiments to determine which combinations of lighting intensity and color are experienced as comfortable (see figure 4.15). The conclusion of such research was that with light of a daylight color-temperature, much higher intensities were needed than for light of lower (more red) color-temperatures.

In a 1939 report of the E.E.I., the biological argument was pushed even further. It was claimed that lighting research had indicated that the human eye functions more naturally above 100 foot-candles than under 15 to 50 foot-candles, which was considered the upper limit of most incandescent general lighting systems at the time. The ultimate advantage of fluorescent lighting to the consumer was therefore to be found in properly designed installations giving at least 100 foot-candles. In an E.E.I. memorandum, an elaborate argument was forwarded to explain why this leap to 100 foot-candles was not as big as it seemed—and indeed, was quite necessary for fluorescent lighting:

> lighting of substantially daylight quality, when appraised by the eye, appears to be much less than equivalent foot-candles of light from normal incandescent sources. The reasons for this are scientifically and psychologically obscure, but the fact remains that general satisfaction with lighting is based in large measure upon the user's appraisal of the amount available, and as such must be taken into account when applying light to large areas. Furthermore, the light from the "colder" tube appears blue and depressing at low intensities and produces an uncomplimentary effect upon goods or people in commercial or work areas. This effect disappears at levels of illumination above 100 foot-candles.[93]

Thus the utilities' technological frame was adapted to the new high-intensity daylight fluorescent lamp.

White light can be produced by combining different phosphors in the fluorescent coating of the lamp (see figure 4.16). Different combinations of the red, green, and blue will render different whites. The relation

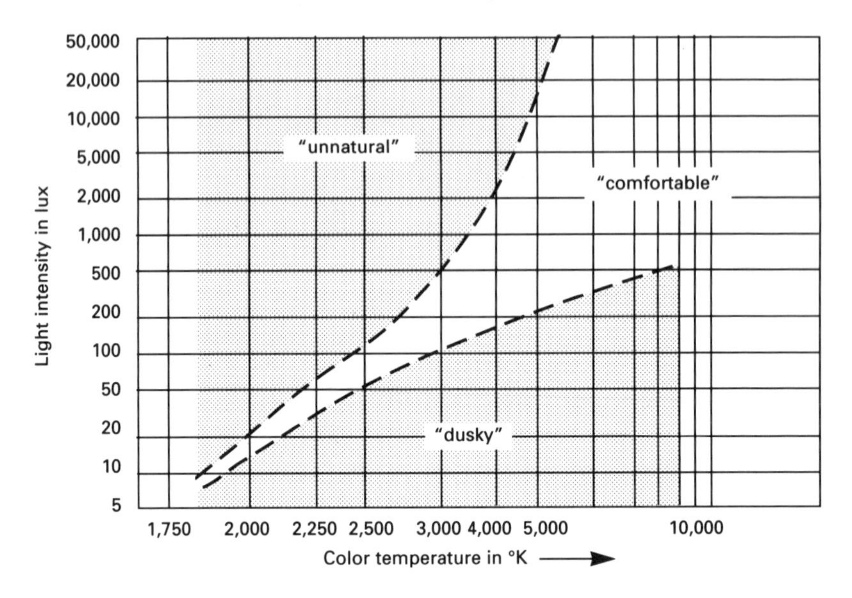

Figure 4.15
For each color temperature the intensity levels can be determined between which the lighting is considered comfortable. Above the maximum level the light was considered "unnatural," and below the minimum level it was "cold" or "dusky" (Kruithof, 1941: 69).

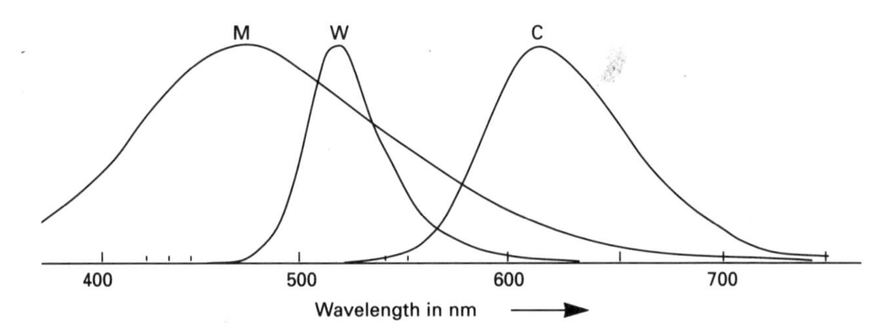

Figure 4.16
The emission spectra of the phosphors cadmium borate (C, red), willemite (W, green), and magnesium wolframate (M, blue).

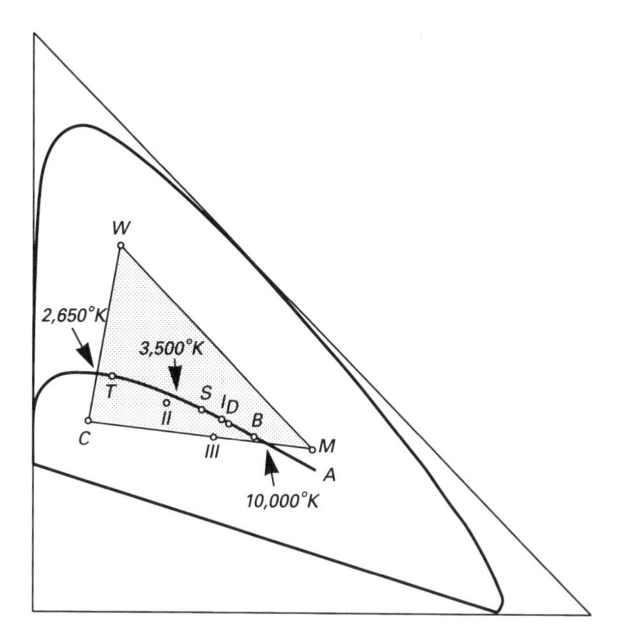

Figure 4.17
The chromaticity diagram. The colorpoints C, W, and M correspond to the light emitted by respectively the phosphors cadmium borate (red), willemite (green), and magnesium wolframate (blue). All colorpoints within the shaded area may be produced by combining these three phosphors in the appropriate proportions. The curve A shows the color points of black body radiation. Indicated are the color points of the incandescent lamp (T), sunlight (S), average daylight (D), blue sky light (B), and three daylight fluorescent lamps (I, II, III) (adapted from Kruithof, 1941: 69).

between the spectral and color-temperature characterizations of light can be depicted with the chromaticity diagram (see figure 4.17).[94] This diagram, in combination with data such as that presented in figure 4.16, formed an important element in the technological frame that was gradually built up around the fluorescent lamp: it provided a vocabulary with which to discuss various daylight colors and their possible renditions by mixing different phosphors (see figure 4.18). Thus, for example, lighting engineers in England [can] now identify

One of the most difficult problems which an artist has to overcome [which] is that his work may be done in the south-west corner of England, where 2000 hours of sunshine a year are recorded as compared with 900 in the industrial north; and then, if his work is accepted for one of the art galleries of the North,

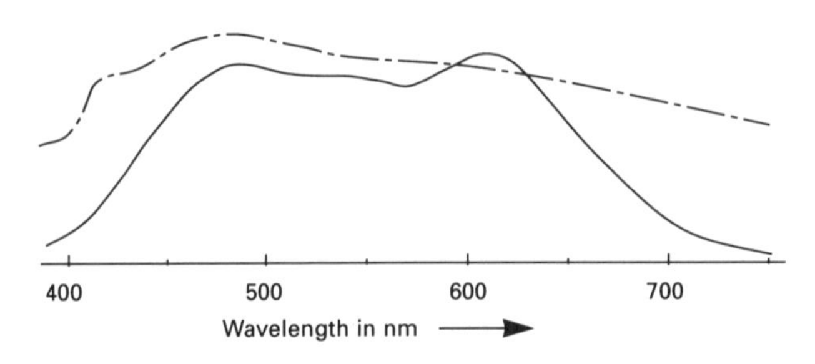

Figure 4.18
The emission spectrum of a fluorescent lamp with such a mix of phosphors in its coating that light point I in figure 4.17 is realized. The dotted curve represents the average daylight (adapted from Kruithof, 1941: 71).

where it is seen only by artificial light, the effect of the pigments which he has used is entirely lost. (Davies et al., 1942: 468)

And "Mrs. Housewife" was addressed again too. Because of presence of green and yellow "mercury lines in the spectrum of fluorescent lamps, overemphasis of these colors in the wall finishes, the draperies, upholstery, and the complexion often results in unattractive effects very disturbing to the lady of the house."[95] Even in the kitchen, where the lamp's shape and high efficiency were considered advantages, there were complaints about food's appearance under fluorescent lighting.

It is intriguing to see how the elaboration of illumination theory and practice, resulting in the ability to make varieties of white light, was used to develop the high-intensity fluorescent lamp. Instead, it would of course have been perfectly possible to develop a high-efficiency fluorescent lamp that produced light in colors similar to the old incandescent lamp—the first fluorescent lamp was, after all, a *tint*-lighting lamp.

Now closure had been reached; the interpretative flexibility of the fluorescent lamp had diminished greatly and the high-intensity daylight lamp had emerged as a new artifact with growing stabilization. It is to that increasing degree of stabilization—first within the relevant social groups of Mazda companies and utilities, but then also within other relevant social groups—that we will now turn our attention.

Certification Plan for Fixtures

An important role in the stabilization process of the new fluorescent lamp was played by the enrollment of a third relevant social group, the fixture manufacturers. I will now describe how the relevant social groups of Mazda companies and utilities succeeded in doing so and how this resulted in a decreasing number of modalities.

The group of fixture manufacturers already played an important role in the incandescent lamp era. The Mazda companies themselves did not produce much in the way of lamp fixtures. Instead, the Mazda companies and utilities had closely worked together with the RLM Standards Institute—an association of the most important fixture manufacturers—which maintained the standards for sockets, reflectors, switches, and other incandescent lamp auxiliaries. When the fluorescent lamp was developed, the Lamp Development Laboratory did research work on the switches and ballasts, but General Electric and Westinghouse never considered, in line with their practice in the incandescent era, the possibility of producing the fixtures themselves.

The introduction of the fluorescent lamp was so rushed that the Mazda companies had not been able to prepare the fixture manufacturers for their task in designing auxiliaries for the new fluorescent lamp. As a result, adequate fixtures were hardly available. Many individuals and small firms took advantage of the situation and started fluorescent lamp fixture businesses. These "tin-knockers" often produced quite objectionable installations, which were nevertheless so satisfactory to their customers that the demand for fluorescent lamps quickly increased further. On the other hand, this also boosted the power factor problem.

So it was not surprising that soon the Mazda companies, utilities, and fixture manufacturers came together. A number of meetings between representatives of these three relevant social groups were held, in Chicago, Cleveland, and New York.[96] For the Mazda companies and the utilities the enrollment of the fixture manufacturers seemed crucial—for the Mazda companies because adequate fixtures were a conditio sine qua non for the success of any of their lamps; and for the utilities because it was through the fixtures that the problem of low power factor had to be solved. And also the established fixture manufacturers could expect

to benefit from close collaboration: From a merchandising standpoint, we who make the fixture are in a rather precarious position. Although we fabricate only the secondary part of the lighting unit—the fixture or housing for the lamp—we are held responsible by the ultimate consumer for the performance of the complete lighting unit. Any dissatisfaction with it is promptly laid at our doors. Therefore, in self-protection, and in consideration of the clientele we serve, we must maintain a critical view toward the light sources we employ.[97]

In the meetings between representatives of the three relevant social groups it thus was quickly concluded that a kind of certification scheme had to be developed. Such a certification scheme would specify the technical requirements that all sorts of auxiliaries for fluorescent lamps had to meet.

The discussion that followed indicated that everyone involved favored some sort of certification test program. The Fleur-O-Lier association was established, sponsored by the Mazda companies. The Fleur-O-Lier specifications were to be developed by the lamp manufacturers. Progress was slow, however, and the initial specifications were not acceptable to the utility lighting interests. Even in September 1940, Mazda executives still sighed that "A good certification program has many advantages for utilities. It is hoped that they will give this subject more study so that certification will develop to a point where it will receive greater utility backing."[98] Again, it was Hygrade Sylvania that prompted the Mazda companies and utilities into action. Hygrade had started in 1939 to produce fixtures itself, which it had not done for its incandescent lamps. These "Miralumes" were well designed and enabled Hygrade to offer "a complete unit of light" to the highest standards. Its line of fluorescent fixtures included a wide variety of styles.

The fluorescent lamp did not yet have a high degree of stabilization; it still had many modalities. As in the case of the early stage of safety bicycle stabilization—when the type of drive, the size of the wheels, and the form of the frame still needed to be specified—a large number of technical details had to be specified when discussing a fluorescent lamp:

> You are all familiar with the catalog price quotation "complete except for lamps." Prices are also often quoted without a starter, without compensator, without ballast, without power factor correction, and so on. It sometimes appears that the words "without quality" would also be quite descriptive in some cases. Lack of familiarity with the required equipment makes it practically impossible for the average purchaser to know what he is buying and—believe it or not—the salesman himself most often does NOT know what he is selling."[99]

Consequently, that stabilization of the fluorescent lamp was still a long way off. The "dropping of modalities" had to be negotiated painstakingly, and this process was never carried out in social and economic isolation. As Harrison pointed out, "no one or two manufacturers can control the auxiliary business, and therefore can only recognize the desirability of providing power factor correction equipment and promote it to the greatest extent. Many equipments will, however, be purchased by the customer on a cost basis which will in many cases result in the use of uncorrected power factor equipment because it is less expensive."[100] More than a year of difficult negotiations followed the establishment of the Fleur-O-Lier association.[101]

The Mazda companies prepared specifications for the fixtures, ballasts, and starting switches to be made by the companies associated in the Fleur-O-Lier and the R.L.M., which then were to be tested by the Electrical Testing Laboratories. Utilities were given an opportunity to criticize these specifications during lamp conferences, and after a large number of revisions, sufficient consensus seemed to emerge by the end of 1940. They pertained to: lighting effectiveness, safety of operation, mechanical adequacy, and electrical requirements such as power factor correction. The arrangement also provided for cooperative national advertising by both manufacturers of fixtures and the Mazda lamp manufacturers, whereby a pro-rated contribution fund was being expended to promote certified fluorescent equipment under the trade name of Fleur-O-Lier (see figure 4.19).[102] By mid-1940, there were some forty fixture manufacturers who had contracts with the Fleur-O-Lier program, and another fifteen were still negotiating.[103]

The negotiations about the certification scheme show unambiguously that it was the artifact high-intensity daylight fluorescent lamp that was stabilizing. Antitrust attorney Walker concluded that "One of the reasons that the General Electric and Westinghouse and the utilities wanted to control the fixture manufacturers was to be able to compel them to make fixtures that would be equipped with this glass shielding so the lamps would not give as much light."[104] Thus the shielding would contribute to limiting the efficiency of the fluorescent lamp installation: "the glass shielding, such as that [indicating to a lamp], cuts off a certain amount of light, so that you have to use more electricity to get the amount of light that you would have if the glass shielding were not there."[105] In the case of the fluorescent lamp, however, such glass shielding is less necessary to protect the eyes from the glare that is so common in incandescent lamps.

This possibility of checking the light output finally got the relevant social group of utilities to fully cooperate with the certification scheme. Not only could their power factor problem be solved through this scheme, but they also realized that it was well suited to enhance the stabilization if the high-intensity lamp, while destabilizing the high-efficiency lamp. The utilities therefore sought to secure a key role in defining the certification standards. And indeed, the specifications were

Famous Electrical Testing Laboratories put FLEUR-O-LIER fixtures through exhaustive tests, which include such vital points as FLICKER CORRECTION, DURABILITY AND SAFETY, EASE OF MAINTENANCE, DEPENDABLE BALLASTS AND STARTERS, EFFICIENT LIGHTING PERFORMANCE, HIGH POWER FACTOR (85% OR OVER).

Based on these tests, Electrical Testing Laboratories gives the FLEUR-O-LIER manufacturer the right to affix the label of certification to all fixtures identical to the sample submitted. This label is your assurance that the fixture wearing it meets the 50 rigid specifications set up by MAZDA lamp manufacturers for better light, better service. You can buy Certified FLEUR-O-LIERS with confidence!

Certified FLEUR-O-LIERS are guaranteed by their manufacturers to be free from any defects in materials, workmanship or assembly for a period of 90 days. Note the complete guarantee with which every FLEUR-O-LIER Manufacturer complies.

IMPORTANT!
Before you buy fluorescent lighting, check with your local electric service company. They will be glad to give you expert advice on how to install fluorescent fitted to your specific needs.

WIDE VARIETY! *You can now get Certified FLEUR-O-LIERS in over 100 different designs and in a wide range of prices. Ask your electrical supplier.*

FLEUR-O-LIER
Manufacturers
Participation in the FLEUR-O-LIER MANUFACTURERS' program is open to any manufacturer who complies with FLEUR-O-LIER requirements

TEAR OUT AND MAIL
Fleur-O-Lier Manufacturers • 2126-8 Keith Bldg., Cleveland, Ohio
Please send me FREE new booklet "50 Standards for Satisfaction," together with list of Fleur-O-Lier manufacturers.
Name
Address
City State

ELECTRICAL WORLD • August 23, 1941 (613) 95

Figure 4.19
Advertisement for Fleur-O-Lier.

submitted by the Mazda lamp manufacturers to the utilities for their approval and revision before they were officially furnished to the Electrical Testing Laboratories. In Walker's words: "The specifications are prepared in accordance with the interests of the utility companies, both in that they provide that fixtures and ballasts will be power-factor corrected and in that they are designed to insure that fixtures complying with them will not be used to reduce electricity consumption—that is, they have a great deal of glass shielding on them."[106]

Once the three relevant social groups really started to collaborate on the certification scheme, they were very successful: it was estimated that by the end of 1940, 85 percent of all fluorescent lighting fixtures were made by companies associated with Fleur-O-Lier or the R.L.M. The collaboration then took on the familiar form of mutual support wherever possible. For example, General Electric and Westinghouse required fixture manufacturers in the Fleur-O-Lier and R.L.M. associations to affix to their fixtures labels that were worded to induce the use of fluorescent lamps manufactured by General Electric and Westinghouse. The Mazda companies sought to compel and cause all fixture, ballast, and starting-switch manufacturers, except those whose competition they wished to eliminate, to participate and join the associations.[107]

To conclude, the Fleur-O-Lier certification scheme furthered the stabilization of the fluorescent lamp artifact as it had emerged from the closure of the controversy between the relevant social groups of utilities and Mazda companies over the load problem. In doing so, it also contributed to the building up of new technological frames related to this new artifact. As an A.E.I.C. report summarized:

Certification does:

(1) Provide a method of dealing with manufacturers as a group.

(2) Offer a means of improving the quality of fluorescent fixtures.

(3) Provide a means for constructive national advertising.

(4) Help the utility to distinguish between fixtures of reasonably good quality and the large mass of slightly identified and locally produced fixtures on the market.

(5) Provide a design service for the smaller manufacturers, and a medium for prompt dissemination of news of new developments.

(6) Provide another means of insuring good-power-factor equipment.[108]

It reshaped the relevant social group of fixture manufacturers; it provided the technical specifications defining what fixtures and lamps should be like; it offered technical problem-solving strategies, through design services; and it provided economic and marketing strategies.

By the end of 1940, the fluorescent lamp (and by then it needed no explication that this was the high-intensity daylight fluorescent lamp) had a high degree of stabilization, and new technological frames had been built up around this artifact. To understand this building up of technological frames in more detail, we will now turn to the interaction within and between these relevant social groups.

Intragroup Interactions

Until now, I treated the relevant social groups of utilities, Mazda companies, and fixture manufacturers as monolithic entities. As we saw in the Bakelite case, it often is helpful to take a more differentiated view of relevant social groups; this was the reason to introduce the concept of "inclusion." Indeed also in the case of the utilities and the Mazda companies, the pressures from outside, caused by the processes of closing the load and power factor controversies and the stabilization of the fluorescent lamp, created tensions within these organizations. To complete the picture I have sketched of the development of the fluorescent lamp, I will now turn to some of these internal interactions.[109]

For example, within General Electric there was opposition to the Nela Park agreement. The Lamp Department, which had participated in the Nela Park Conference, experienced resistance within the large General Electric organization. When the General Electric Supply Corporation published a catalogue that listed and pictured fixtures with shielding, Harrison (of the Lamp Department) objected because "the repercussions from central stations are likely to be formidable."[110] The catalogue showed fixtures both bare and equipped with shields. However, all the listed prices applied to the bare lamp fixtures only and the shield was shown as an extra, requiring separate and additional catalogue numbers when ordered. Then the statement appeared that use of shields would produce 30 percent less light. It is evident that this way of presenting the fluorescent lamps would stimulate customers to buy the unshielded lamps, thus getting more light out of the lamp for the same amount of electricity. Harrison threatened: "Of course, it is up to the General Electric Supply Corporation . . . to formulate their own policies, but I do not think that a penny of Lamp Department money should be spent to support a campaign of this kind.[111] In its answer to Harrison's letter, the General Electric Supply Corporation justified the form of its advertising on the grounds that it was necessitated by the competition of Hygrade Sylvania.[112] Harrison had to protest because at the Nela Park Conference he had promised that fluorescent lamps would not be installed without "proper shielding."

A tension like the one within the General Electric organization is likely to occur among actors having different degrees of inclusion in one technological frame. The General Electric Supply Corporation was bound to have a relatively low inclusion as compared to the Lamp Department, because the latter was more intimately involved in the establishment of the new fluorescent frame of the Mazda companies, in which the selling of only the high-intensity daylight fluorescent lamp was the goal and which was aimed at encouraging collaboration with the utilities. For the sales people of the Supply Corporation, the old incandescent technological frame of simply selling as many lamps as possible, and thereby competing with other lamp manufacturers, was more prominent.

Similarly, such tensions can be observed within the group of utilities. The utility officials present at the Nela Park Conference were members of the E.E.I. Lighting Sales Committee, but did not act officially on behalf of the E.E.I. And though they generally had a high status within the E.E.I., they had to make quite an effort to get the rest of the utilities behind them. This was

> recognized by Mueller when he wrote to Harrison that he sensed a little feeling of disappointment on the part of a few when it was made clear that the utility men, in agreeing to the plan, were agreeing only for themselves as individuals, and not for the Edison Electric Institute or its Lighting Sales Committee.
>
> I believe I speak for the utility men present when I say that we are just as anxious as the lamp companies to officially get the utility industry at large behind the development of fluorescent lighting along the sound methods of promotion agreed to at the conference.[113]

This was not always easy. For example, Bremicker (of the Northern States Power Co.) wrote to Mueller, after having received a report on the Nela Park Conference, that this was not enough and that he wanted a specific reaction from the Mazda companies, in which it would state "that fluorescent lighting is not known to be applicable for any lighting purposes except colored or atmospheric lighting and certain phases of localized lighting such as wall cases, showcases, display niches."[114] Bremicker concluded that he did not want the utility companies to be "hoodwinked" into a cooperative program of promoting fluorescent lighting. The position of Bremicker was similar to that of the General Electric Supply Company, in that he did not attend the Nela Park meeting and, hence, was only marginal in the newly established technological frame.

Alerted by this reaction, Sharp proposed to Mueller (both were participants in the Nela Park meeting) not to send out the entire minutes of that Nela Park meeting. Instead, a letter with only a brief outline should be sent out, which "would indicate that the committee is still on the job, [and which would] serve to keep the utility group united, and give our committee some additional backing from the field, thereby making it harder for anybody to divide our forces."[115] Sharp evidently realized the potential tension between the participants of the Nela Park conference and the other utilities executives with a much lower inclusion. Such tensions might threaten the coherence of the relevant social group of utilities: "committee members should check with one another to be sure that there is substantial agreement before we go on record to manufacturers on matters pertaining to their promotion and publicity. No one has found a loophole in our armor yet and I think we want to maintain that record."[116] (This quote furnishes another illustration that actors have their own set of relevant social groups and thus will help the researcher to identify them.)

The intra-group interactions that I discussed in these pages were primarily meant to make explicit the internal differentiation of each relevant social group. This is necessary for a better understanding of the interactions among relevant social groups.

Intergroup Interactions

To analyze the intergroup interactions, I will focus on three issues: the carrying out of the Nela Park agreement, the making of a cost comparison method, and the antitrust and patent litigations. Through the latter the wider historical context of World War II will also enter the story.

At the Nela Park Conference the Mazda companies dropped the high-efficiency daylight fluorescent lamp and agreed to restrict themselves to making the high-intensity lamp. However, the construction of the high-intensity lamp certainly was not a complete victory for the utilities. Mueller clearly viewed the Nela Park agreement as a compromise, when he argued the need for the E.E.I. Lighting Sales Committee to make some additional concessions to the Mazda companies:

> Unless our committee does something now to give them [i.e., the Mazda companies] some publicity on their change of pace, and to get the utility industry as a whole interested in the promotion of fluorescent lighting along sound lines, I think they will drop us and either try to get action through some other body, or else come out with another "To Hell With The Utilities" campaign, and go it

alone, knowing that they have quite a strong customer appeal in their efficiency and novelty story.[117]

And so the utilities started slowly to adapt their policy toward advertising the fluorescent lamp. This resulted in the utilities switching from informing to selling in their fluorescent lighting presentations. Thus the conflict was solved by a piecemeal adaptation by both parties to the new situation.

This piecemeal adaptation did not come about smoothly. Neither party to the Nela Park agreement adhered to it without occasional lapses. Especially the utilities felt that the Mazda companies were regularly violating the agreement in their advertising. On 24 May 1939, Sharp wrote to Harrison that utility men had complained to him concerning a display in the General Electric Building at the New York World's Fair. This display purportedly consisted of a 20-watt fluorescent lamp and a 20-watt incandescent lamp with a foot-candle meter that showed how much more light was given by the fluorescent lamp. Objecting to General Electric having this display on exhibit in their building at the World's Fair, Sharp stated:

If this demonstration is as explained to me I think it does violate the spirit of the understanding that our group had in Cleveland. As a matter of fact, I would think it violated the fundamental concept of the lamp department that advances in the lighting art should not be at the expense of wattage but should give the customer more for the same money. I hope you can find a way to change this exhibit, so that it does not give misleading impressions to the crowd who will see it.[118]

Harrison replied to Sharp that the exhibit had not been intended to demonstrate the amount of electricity that could be saved by the use of fluorescent lamps, but that the exhibit was being withdrawn in any case.[119]

There were numerous incidents like these. In another letter Sharp commented on a draft article by a General Electric staff writer entitled "Choosing the Right Lamp for the Job"; almost all Sharp's comments boiled down to the suggestion of higher wattages of fluorescent lighting or the substitution of incandescent lighting for fluorescent lighting.[120] Several other utility managers complained about advertisements and other publications that, after the Nela Park Conference, still included the phrase "3 to 200 times as much light."[121] This kind of critical correspondence was directed by utility managers to whomever they happened to know inside the Mazda companies. It is fruitful, however, to focus for a while more closely on actors in specific positions within the two technological frames.

Let me investigate the position of an official like E. F. Strong, local sales executive of General Electric. Such a position is interesting because Strong was involved in two different technological frames: the Mazda frame and the utilities frame. In both technological frames, Strong had a relatively low inclusion. Within the Mazda companies, he had a marginal position when we take the Cleveland Laboratories as the center of fluorescent lamp construction and the building up of that technological frame. Because of his long collaboration with the local utility officials, he was, however, working partly in that technological frame too; but as he was no utility employee himself, there his inclusion was low too. What form do intergroup interactions take when they occur through an actor who has a low inclusion in both frames?

From the one side, a utility official exerted pressure on Strong by criticizing the newly issued policy statement by General Electric.[122] From the other side, the General Electric organization expected him to follow company policy. The tensions of Strong's boundary position between the two relevant social groups can be sensed from a letter to Harrison, in which he wrote: "I feel a very positive need of going along with the utilities in our territory in such manner as to impress upon them the fact, which is true, that we are attempting to use the fluorescent product to secure lighting effects through installations which really mean additions to the use of electricity in lighting."[123] Strong explicitly suggested that only the use of fluorescent lamps for "particularly efficient production of colors" would be advertised. He was prepared to avoid the use of fluorescent lamps for general illumination where they would replace tungsten filament lamps in present installations or where large areas are contemplated in new buildings or in the revamping of old buildings. As far as Strong was concerned, the fluorescent lamp was to be considered for lighting, or not at all. Strong concluded his letter to Harrison by asking for a revision of the General Electric policy statement.

The utility executives who participated in the Nela Park Conference also occupied a boundary position between the two relevant social groups. There are important differences, however. Executives such as Mueller and Sharp evidently had a high inclusion in the utility technological frame. They were in a position to negotiate and to make some concessions. I already described the tensions this generated inside the relevant social group of utilities. It also brought these utility officials closer to the technological frame of the Mazda companies. Indeed, Harrison and Cleaver, chief engineers of General Electric and Westinghouse respectively, proposed that "the more or less unofficial group which had gotten together in this conference in Cleveland be continued so as to act as a liaison body between the lamp companies and the utility companies."[124] It is exactly such a liaison group of actors, participating in the two technological frames, that Hygrade was missing.

Hygrade tried to make contact with the utilities, but was not very successful. Bremicker (of the Northern States Power Company in Minneapolis), for example, refused to exhibit Hygrade lamps. In a long letter in which he criticized Hygrade's recent promotional activities and spelled out some of the principles of advertising in the context of the "Better Light-Better Sight" movement, Bremicker concluded that "you will understand why we are not interested in spending any money for Hygrade demonstration equipment until we are satisfied that your company intends to cooperate with us in selling fluorescent lighting properly designed and properly applied."[125] Sharp, however, was slightly more open to Hygrade's advances and believed, for example, that the utilities could well afford Hygrade the courtesy of a conference, similar to the one they had had with the Mazda manufacturers.[126] Such a conference was indeed held on 11 June 1940, on the invitation of Hygrade Sylvania.

One of the ways in which the relevant social groups of utilities and Mazda companies interacted was by arguing over the "true" costs of fluorescent lighting as compared to incandescent lighting. This question soon was rephrased as how to establish a standard method for comparing the costs of incandescent and fluorescent lighting. Studying the debate about this cost comparison illuminates the differences between the technological frames of utilities and Mazda companies.

It was not easy to reach agreement on a standard method of cost comparison. In part, the cause of the problem was that this generation of lighting people had little experience with competitive illuminants. The incandescent lamp had been well nigh universal, so that lighting design mainly involved technical considerations, with relatively simple arithmetical calculations about equipment cost. Now that there was a light source as radically different as the fluorescent lamp, lighting design involved a more complicated cost comparison before it became clear which source would best meet specific requirements.[127]

However, an even more serious barrier to an agreement on standard cost comparisons was formed by differences in interests as the two parties perceived them. First, there was a difference between focusing on the costs of electricity versus focusing on the costs of the apparatus. For the Mazda companies, it was attractive to emphasize the low cost of electricity and to disregard the high price of the apparatus itself; for the utilities, it was the opposite. Second, the utilities' primary aim in developing a standard method for comparing lighting costs was to pursue the fight against the high-efficiency daylight fluorescent lamp. The Mazda companies, despite their "capitulation" at the Nela Park Conference, were of course not anxious to support the utilities in that fight.

Late in 1939, the E.E.I. Lighting Sales Committee proposed a standard method that it claimed to be universal in application and to ensure an evaluation of all factors. Utility lighting people seem to have been almost unanimous in their approval of this method, while manufacturers gave only lukewarm assent. Utility executives commented on this lack of enthusiasm of the Mazda companies:

> Their reluctance is founded on the fact that true cost calculations bring out the items of high fixed charges and expensive fluorescent lamp renewals. These are customarily slighted by manufacturers' representatives and jobbers in their eagerness to bring out unquestioned reductions in energy cost, foot-candle for foot-candle. Wide experience with the use of this method in investigating fields of fluorescent application have shown that no blanket statement as to cost can safely be made. As often as not, when a true cost comparison is made on a five- or six-year depreciation basis, the fluorescent installation is more expensive for the customer than filament incandescent lighting. This clearly points out that it is fallacious to sell fluorescent lighting on the basis that it is the most economical form of lighting.[128]

The utilities scorned the lamp manufacturers, because for years they had stressed the fact that electricity cost was a negligible factor in good lighting, so that "it seems a little ridiculous now to advertise appreciable reduction in this negligible cost, particularly if accompanying it there is retrogression with respect to already attained levels of illumination or absence of the raising of these levels."[129]

And indeed, apparently, there was not much choice open to the Mazda companies: some months later Mueller could come to the conclusion that "this method possibly cannot be dignified by being called an "industry standard," [but] it comes pretty close to that. It has also been endorsed by the lamp companies and is used by them."[130] Thus the development of this cost comparison method as a new element in the utilities' technological frame strengthened their struggle against the high-efficiency daylight fluorescent lamp and contributed to the stabilization of the high-intensity daylight fluorescent lamp.

But the utilities did not emerge unscathed from the battle over a standard cost comparison. Their image of "public servants" got stained. One example may illustrate this:

Two women operating a dress shop on the edge of a main shopping district called the local utility and asked if a man could come out to check their lighting as it didn't seem to be just right. Upon arrival he found an interior 10′ × 50′ lighted with two fixtures, each containing four 4-foot fluorescent lamps. Foot-candles ranged from sixteen to three. The windows each had two rows of bare lamps. The total job had cost them $300 cash; about six times the cost of an equivalent incandescent installation.... When asked why they had not consulted the utility first, they replied that they felt the company wouldn't render any help because the fluorescent lamps didn't use much energy.[131]

Not only the general public but also salesmen, dealers, and distributors criticized the utilities for being guided by fear for loss of revenue rather than by the desire to help customers. Also politicians scorned the utilities for their perceived hesitation over stimulating savings in electricity consumption.[132]

Another set of important intergroup interactions occurred in the U.S. courtrooms between General Electric, Hygrade, the Antitrust Division of the Department of Justice, and the War Department. When, in spring 1940, Hygrade refused an offer by General Electric to become a class-B licensee for the fluorescent lamp, General Electric instituted a patent infringement suit against Hygrade under two of General Electric's fluorescent lamp patents in May 1940. Hygrade reacted by instituting a countersuit against General Electric under three of its patents.[133] At the same time the Justice Department was investigating the electric lamp industry and decided in December 1942 to file a complaint against General Electric and eight other defendants under the antitrust laws.[134] General Electric's strategy in the first trial was to make Hygrade completely stop its production of fluorescent lamps. The Antitrust Division's objective in the second case was to demolish General Electric's licensing system. General Electric's lawyer tied to use the United States' increasing involvement in World War II to argue that the trial should be stopped because it "would necessarily involve detraction from an all-out war effort—because of the importance of the services of these defendants to national defense."[135] At first he seemed unsuccessful. Although he even got the District Judge, John Knox, on his side, the Department of Justice did not give in:

The Department deems it essential that conspiracies of the character found in your case be made hazardous. If such offenders are permitted to escape without penalty, provided merely that they stop when the discovery of the conspiracy is made, there will be no effective enforcement of the antitrust laws. Furthermore, any other attitude would encourage sabotage of the war effort by private groups having great economic power founded upon combinations in restraint of trade.[136]

Then, however, the help of the War Department was secured. The Secretary of War, Henry Stimson, addressed a request to the Attorney General, Francis Biddle, and the latter consented and overruled the Antitrust Division.[137] Further prosecution of the case was suspended for the duration of the war.

The resulting situation was quite extraordinary if one did not recognize that the government constituted, in this case, at least two different relevant social groups. The government's efforts to utilize the courts were suppressed by government officials in a suit by the government itself against General Electric. At the same time the suit of General Electric against Hygrade was not upheld. Because the production of fluorescent lamps for defense plants was an important contribution to the war effort and this suit might have resulted in stopping Hygrade from producing such lamps, this was quite remarkable, and can be taken as an illustration of G.E.'s economic power and lobbying effectiveness.

4.6 Power and the Construction of Technology

One point will be evident by now: the fluorescent lamp was developed in the midst of power games. Various exertions of power figured prominently, though not always identified as such: patent licensing, cartel forming, price setting, political pressure. I want to use the lamp case to discuss some general aspects of the way in which power seems to play a role in the context of technological development, and conversely, how technical artifacts are important in constituting power. This discussion will be limited to those aspects that are of direct relevance to a better understanding of the development of technological artifacts; it is not my intention to present a general theory of power.

What do we mean when we intuitively agree that the fluorescent lamp was developed in arenas of power? It seems prudent to distinguish again between intra- and intergroup relations. Within one relevant social group we saw the problem of enforcing a decision made "at the top"—did Harrison have enough power to make the Sales Department accept the policy statement about how to market the fluorescent lamp? Between different relevant social groups the clash of different technological frames can be stated in terms of power relations as well—was General Electric powerful enough with respect to the utilities to make them accept the high-efficiency daylight fluorescent lamp, or were the utilities powerful enough to stop that artifact's stabilization?

When we try to find a foothold for analyzing the role of power in technical development, we uncover an intriguing gap in existing research traditions. The few economists who explicitly address the role of technical change hardly ever refer to "power," even though their studies of market competition, corporate strategies, entry barriers, and state intervention, for example, pertain directly to matters commonly associated with economic power. On the other hand, in the large sociological literature on power there is no detailed analysis of technical development. When "technology" does play a role in power analyses—as for example in industrial sociology or labor theory—it appears as a macro concept and is not very useful for the purposes of this book, because the shaping of technology is not an important concern for such authors.[138] This missing link in power research is even more discomforting when we appreciate not just the theoretical interest of the relationship between technical change and power but the political importance of the issue as well.[139]

The previous paragraphs display a bouquet of different plain usages in which the term "power" may be encountered. "Power" sometimes is a quality ("Harrison had power"), a relation ("was G.E. powerful enough with respect to the utilities"), a domain ("arenas of power"), an outcome ("artifacts are important in constituting power"), or an agent ("power seems to play a role"). This variety of meanings suggests that the term "power" is not, in common language, very precise. I would indeed rather argue for abstaining from its usage completely. At best the term "power" can be a practical shorthand for more detailed and rich descriptions of situations, outcomes, relations, etc. This is, then, the purpose of the final section of this chapter: to take the rather imprecise usage of "power" as a shorthand indicating some important questions that can be raised in relation to the social shaping of technology and the technical shaping of society. During my analysis of power in this context, the term will be slowly written off into the background, while the constructivist framework developed in the course of this book will be extended.

First I will give an outline of the power conception that I will adopt. Then I will illustrate how technology—or more precisely, artifacts—can play a role in such an analysis of power. This discussion of power draws an end to the case study of the fluorescent lamp. It also lays the groundwork for the concluding chapter, in which I will show how a constructivist analysis of technology enables us to address wider political issues, partly by making sense of the hardness and obduracy of technology.

A Conception of Power

Few words have such a long and varied history in social sciences as "power," from Machiavelli and Hobbes, via Marx, Weber, Parsons, and Lukes, through Foucault, Giddens, and Barnes, to Mumford, Hughes, and Latour. I will not review this rich variety of different traditions, nor seek to contribute to the ongoing theoretical discussion in any general sense.[140] One observation can be made, however: all of these analysts have been quite vague about the relationships between power and the shaping of technology. I will try to outline the possible contribution of a constructivist analysis of technology to this end.

As a useful starting point I will take Giddens's (1979) definition of power as *the transformative capacity to harness the agency of others to comply with one's ends.*[141] Power thus is a relational concept that "concerns the capability of actors to secure outcomes where the realization of these outcomes depends upon the agency of others" (Giddens, 1979: 93). The emphasis on the *transformative* capacity to harness the agency of others stresses the relational aspect and thus avoids a "stuff" conception of power. For my purposes it will, of course, be necessary to extend "the agency of others" to include the agency of machines as well as of human actors, because technologies can also be instrumental in realizing certain goals. So power is a relational concept—it is exercised rather than possessed, and it is specific to these instances of exercise. Power also is ubiquitous and present in all relations and interactions. Taking power as a *capacity* will make it easier moreover, to analyze interactions as governed by more than only conscious strategies (skills, for example, play a role as well).

It seems natural to link the constructivist analysis of technology developed in this book with such an interactionist conception of power.[142] Just as an artifact is constituted in interaction rather than having an intrinsic context-independent meaning, so will power be analyzed as an instance of interaction. Just as the constructivist analysis of technology is symmetrical with respect to the working or non-working of a machine, so will this conception of power to be morally neutral. Just as a constructivist image of technology can be contrasted with what Latour (1987) called a diffusion image of technology, so can we contrast the interactionist concept of power with the causal "push and shove" image.

In addition to this interactionist view of power as a transformative capacity, we have also to distinguish the systemic or institutional aspects of power. We need a conceptualization of power that will allow us to combine an action and a structure perspective—one of the requirements for a theoretical framework as specified in the introduction to this book—and so overcome the traditional division between voluntaristic and structural notions of power. Giddens uses "domination" and "transformative capacity" as respectively the structure and action sides of his power coin. I will employ the terms "semiotic power" and "micropolitics of power" to forge a more direct link with my conceptual framework.

For the semiotic power conception, I draw on Laclau and Mouffe (1985) and Clegg (1989). To the extent that meanings become fixed or reified in certain forms, which then articulate particular facts, artifacts, agents, practices, and relations, this fixity is power. Power thus is the apparent order of taken-for-granted categories of existence, as they are fixed and represented in technological frames. This semiotic power forms the structural side of my power coin. The micropolitics of power describes the other side—how a variety of practices transforms and structures the actions of actors, thereby constituting a particular form of power. In Foucault's (1975) study of the development of discipline, this micropolitics of power results in producing obedient human bodies; in my framework the focus will be on producing technological frames. It is important to stress that the disciplining power of these micropolitics typically does not have a single center, and that these micropolitical interactions are not necessarily conscious strategies. It will be clear that semiotic power and micropolitical power are inextricably linked: micropolitics result in a specific semiotic structure, while the semiotic power in turn influences the micropolitics structures.

The semiotic and micropolitical aspects of this power conception can be directly linked to the closure and stabilization processes I have identified. The reaching of closure, whereby the interpretative flexibility of an artifact is diminished and its meaning fixed, can now be interpreted as a first step in constituting semiotic power, resulting from a multitude of micropolitics to fix meanings. In the subsequent stabilization process further interactions result in fixing more elements into the semiotic structure—enlisting more people in the relevant social group, enrolling new relevant social groups, elaborating the meaning of the artifact. A technological frame then constrains actions of its members and thus exerts power through the fixity of meanings of, among other elements, artifacts; this is the semiotic aspect of the new power conception. A technological frame also enables its members by providing problem-solving strategies, theories, and testing practices, for example, which forms the micropolitical aspect of power.

Power and the Development of Technology

Closure and stabilization result in a fixity of meanings. This fixity of meaning represents power. The role of this type of power in social interaction has been well documented, for example in Foucault's studies of discipline and punishment (1975) and of sexuality (1976), in Thompson's (1967) study of the time-structuring effects of technology in the emergence of capitalism, and in McNeill's (1982) study of military power. The architecture of prisons, contraceptive techniques, clocks, and guns figure as clear examples of how artifacts play various roles in the exertion of power.[143] I will now turn to the other, hitherto neglected question: how power, in this Machiavellian conception, may play a role in the shaping of technology.

The fixity of meanings affects the shaping of technology through technological frames. Technological frames specify, I have said, the way in which members of a relevant social group interact, and the way in which they think and act. In terms of the power discourse, one might say that technological frames represent how the discretion is distributed of who may do what, when, where and how, to whatever objects or actors. A technological frame is at the same time constituted by interactions of members of the relevant social group, and result in "disciplining" the members of that relevant social group. This may take on several forms.

Technological frames constrain freedom of choice in designing new technologies. The electricity distributed by the utilities—some 110 volts of alternating current—did not allow the use of high-voltage discharge lamps without extra transformers. Nor, the utilities argued, did it allow the use of low power factor devices. This latter examples demonstrates the need for a constructivist perspective in these analyses of power strategies. I have shown that it was not enough for the utilities to claim the technical impossibility of using fluorescent lamps because of their low power factor. Artifacts do not have intrinsic properties, but need to be socially constructed; similarly, actors do not have stored power to employ straightforwardly. General Electric, for example, partially owned many of the utilities; but this did not allow the G.E. Lamp Department to determine the actions of the utilities straightforwardly. Whether such an ownership relation is indeed transformed into a power relation depends both on the micropolitics employed and on the other elements in the semiotic power structure. In this case, this seems not to have happened. We have also seen how the exertion of power by the utilities was instantiated through a variety of micropolitical strategies, finally resulting in fixing the meaning of the fluorescent lamp as a harmful low power factor device.

Technological frames, on the other hand, are also enabling in addition to constraining the design work or—more generally—the interactions of actors. Artifacts are elements of a technological frame: they thus form part of the vocabulary in which interactions develop; they provide some of the resources on which actors draw for these interactions and for the transformative action by which they seek to harness the agency of other actors[144]; and finally they constitute an important part of the capabilities of actors. Often these capabilities take the form of routines.[145] Test procedures, standard theories, design rules-of-thumb, process control variable settings—these are all examples of routines that may form part of a technological frame. But an artifact such as the fluorescent tint-lighting lamp can provide a new routine, a different vocabulary for interaction. This particular lamp could have resulted, in due time, in another technological frame, with accordingly different forms of disciplining.

Patents are a particular form in which routines may appear. The patent system has been created to give to a firm a temporarily exclusive ownership of some specific routines. It is meant to present an incentive to firms to perform research and development internally, rather than buying it from outside.[146] In my analysis patents may have at least two different functions. First, they represent the routines and capabilities themselves, being elements of semiotic power; second, they may function as micropolitical devices in a broad spectrum of interactions such as negotiations about joint ventures, informal market agreements, or scientific claims.[147] A patent licensing system is one specific form of employing patents as a power strategy. By giving a patent license to another firm, General Electric provided that firm with the routines described in the patent. This was accompanied by General Electric specifying a quite narrow band within which the licensee had discretion to use the routines. As I described, General Electric not only limited the sales volume of the licensees (i.e., restriction of the discretion to use General Electric's routines), but the company also was entitled to use whatever improvements were made by the licensee to the existing technology (i.e., co-ownership of all the new routines that might be developed). The latter form of restriction is called a "grant-back license."[148] To have a patent does, in itself, not make an actor intrinsically powerful. As always, the crucial question is how the micropolitics of power will result in that patent being instrumental in transforming the capacity and routines of others. Baekeland already warned that one should not bother to take out a patent when unable to defend it in court. And when talking about "being able," he evidently had in mind "having enough money." Money, and capital, does lay a role in the micropolitics of power, though without unambiguously determining the course of events, as the case of G.E. and the utilities showed.

One other way in which this power analysis may shed light on the shaping of technology is by recognizing some artifacts as "obligatory passage points."[149] An example of such a point of obligatory passage is the certification scheme for fluorescent lamps. Artifacts may play the role of obligatory points of passage in a very concrete and physical sense, such as the plugs used to connect an electrical device to the mains.[150] In our lamp case, the starter to enable the electric discharge and the power correction device are other examples of artifacts as obligatory passage points. These represent instances of power, because they discipline the interactions of actors.

Finally, artifacts may represent specific interests. Interests, in our constructivist perspective, are not fixed attributes that can be imputed to relevant social groups on the basis of some theory of society. Rather, they are temporarily stabilized outcomes of interactions. This stabilization partly occurs in the form of artifacts. Whether the auxiliary condensers and switches for starting the fluorescent lamps are produced with the lamp or integrated into the fixture has implications, for example, for the interests of fixture manufacturers.

In concluding, it will be clear that the power conception that has been suggested does not add a completely new theoretical and explanatory level to the analysis. Rather, a description in terms of power strategies functions as a neat summary of processes that were otherwise described in terms of interactions, closure, stabilization, technological frames, and inclusion. It does, however, allow for some extra focus and sensitivity in addressing certain issues.

4.7 Conclusion

This chapter described the history of the fluorescent lamp. I made use of the conceptual framework introduced in the previous chapters to present an analysis of the social construction of the lamp. Closure was reached in the controversy between the utilities and the Mazda companies, resulting in the construction of the specific artifact we still have today, the high-intensity daylight fluorescent lamp. Then I described how, during its increasing stabilization, the various technological frames shaped this artifact. In doing so, a concept of "power" was introduced to account for the obvious differences in economic power between some of the relevant social groups.

After the closure and stabilization processes, a redistribution of power had occurred. The new artifact fixed some of the power relations; new technological frames embodied new power; new relations within and between relevant social groups mirrored changes in the distribution of power. History was rewritten accordingly. General Electric's executive Cleaver presented the development of the fluorescent lamp as a straightforward goal-oriented path toward high-intensity general illumination:

> New in theory, design, and appearance, the fluorescent lamp was introduced some two years ago to the public and the illuminating-engineering profession at almost the same instant, giving the latter little time to study and prepare essential data before the former began to insist upon its application to every lighting field. The past two years, therefore, have been a period of catching up—of constant improvement in the lamp and its auxiliary equipment, and of rapid assembling of data on its performance and its limitations as a practical and economical light source. (Cleaver, 1940: 261)

The early emphasis on color lighting, the controversies over load and power factor, the World Fair as external cause—this all had disappeared from the story. The new distribution of power is fixed by the artifact: daylight color, high intensity, auxiliaries according to a certification scheme. These elements together define the power relation between the relevant social groups of Mazda companies, utilities, independents, consumers, fixture manufacturers, and the government.

5

Conclusion: The Politics of Sociotechnical Change

We have had stories and theories; now it is time for politics. I have presented this book as an effort to turn into a main route what had, for me, started out as a detour. In the last three chapters I have told three stories while building a conceptual framework that allowed us to generalize beyond the confines of those stories. At the outset I argued that this project could lead to a new way of thinking about political issues involving society, technology, and science. The detour would thus be turned into a main route.

I will start this chapter by summarizing the central features of that conceptual framework. This will result in a suggestion that, rather than formulating the central STS problematic in terms of relations among three distinct domains, we should direct our research and politics to a new unit of analysis: sociotechnical ensembles. I will then use the power conception introduced at the end of the last chapter to focus attention on the obduracy of such sociotechnical ensembles, and on what that obduracy might mean. Finally, I will argue that analyses such as these suggest strategies for creating a more democratic technological culture.

5.1 Symmetry and Sociotechnology

The Argument So Far

I have developed my theoretical argument in three steps. First, using the bicycle case, I posited a need to analyze technical change as a social process, introducing along the way the concepts of "relevant social groups" and "interpretative flexibility." Then, in reviewing the history of Bakelite, I tried to show the usefulness of introducing the idea of a "technological frame." Finally, in the fluorescent lamp case, I proposed a concept of "power" that would fit a constructivist analysis of technology and society.

The bicycle case showed that the development of technical designs cannot be explained solely by referring to the intrinsic properties of artifacts. For example, the high-wheeled Ordinary was at once a dangerous machine, prone to failure in the marketplace, *and* a well-working machine that allowed highly skilled physical exercise, resulting in a commercial success. I showed that this double character could be clarified by looking at the alternative bicycle designs through the meanings attributed to them by relevant social groups.

Exploring the roles of relevant social groups gave us a basis for understanding the concept of interpretative flexibility with respect to artifacts. As noted in chapter 2, this concept finds its philosophical and methodological underpinning in the principle of symmetry formulated by Bloor (1973, 1976) for the sociology of scientific knowledge. Bloor argued that a useful understanding of scientific knowledge systems can be gained only if investigators are impartial to the supposed truth or falsity of (scientific) beliefs. To analyze true and false claims symmetrically, they must apply the same conceptual apparatus to each. Thus, one would not say that a claim that is presently considered true is explained by its better correspondence with "Nature," nor would one argue that a claim that is presently considered false is explained by the "social circumstances" surrounding its conception. In each case, instead, one needs to approach "Nature" as what is to be explained, not as an element of the explanation. "Nature," in other words, is not the cause of scientific beliefs, but the result.[1]

Pinch and Bijker (1984) argued that a similar principle should be applied to the analysis of technology—that we can understand technologies only if we analyze successful and unsuccessful machines symmetrically. Constructivist studies of technology strive not to consider the fact that a machine "works" as an explanation, but to address it as a subject requiring explanation. In this approach, machines "work" because they have been accepted by relevant social groups.[2]

Having demonstrated the flexibility inherent in the interpretation of artifacts, I argued that our next step should be to map the process by which artifacts attain or fail to attain a stable interpretation. In this descriptive model, an artifact does not suddenly appear as the result of a singular act of heroic invention; instead, it is gradually constructed in the social interactions between and within relevant social groups. I introduced the term *closure* to describe the process by which interpretative flexibility decreases, leaving the meanings attributed to the artifact less and less ambiguous; this process can also be described in terms of the artifact reaching higher levels of *stabilization.*

The concept of closure is also borrowed from the sociology of scientific knowledge, where it is used to describe how a scientific controversy ends with the emergence of consensus in the scientific community. Studies have shown that, following closure, the history of a controversy will be immediately rewritten. The interpretative flexibility of all scientific claims ceases to exist, and "Nature" is always invoked as the cause of consensus. Similarly, in the analysis of technology, closure results in one artifact—that is, one meaning as attributed by one social group—becoming dominant across all relevant social groups. In the case of the high-wheeled bicycle, closure resulted in the Macho Bicycle becoming obsolete and the Unsafe Bicycle becoming dominant. I did not trace in detail the latter's subsequent stabilization, but it can be appreciated from the change in names: The high-wheeled bicycle was no longer the Ordinary, but was henceforth nicknamed the Penny-farthing. After becoming dominant, the Unsafe Bicycle was actively improved and finally superseded by safer bicycles.

The process of closure is generally, but not absolutely, irreversible. Nowadays, for example, we find it strange that people once considered the Penny-farthing a well-working, comfortable machine. It is, in other words, very difficult to envisage the world as it existed before closure. This seeming irreversibility is, however, not exclusively, or even primarily, psychological, like a Gestalt that cannot be switched back. A specific technological frame, such as that of the solid tire or the air tire, comprises not only social-psychological elements, but artifacts, organizational constraints, and values as well. It is because of the heterogeneity of the frame that irreversibility is not absolute, and to formalize this idea I suggested that we consider *degrees* of stabilization. Each of the three stories revealed growing and diminishing degrees of stabilization. This perspective allowed us to understand the invention of the Safety Bicycle, for example, as an eighteen-year process (1879–1897), rather than as an isolated event occurring in 1884. We traced this process by noting the dropping of modalities in contemporaneous writings about the artifact.

In this way, the social-constructivist model highlights the contingency of technical development (by demonstrating the interpretative flexibility of artifacts), while describing how freedom of choice is narrowed by contextual constraints and alliances. Processes of social construction thus have a dual character: They include (almost) irreversible processes of closure, reflecting the steplike aspect of technical change, but they are also continuous between the steps, as indicated by changes in the degree of stabilization.

In chapter 3, I used the SCOT model to describe the development of Celluloid and Bakelite. Here again it proved difficult to label the inventor's work unambiguously as purely scientific, technical, social, or economic. Did Baekeland's condensation reaction result from a scientific fact (as he claimed himself)? Or was it the result of successful technical tinkering (as we may now think, knowing that Baekeland's theoretical explanation was superseded by macromolecular theories)? Or was it neither a scientific nor a technical accomplishment, but first of all a social and economic one—turning competitors into partners during patent litigation, and building networks of manufacturing companies that would use the new material? The explanatory concept of a technological frame, comprising knowledge, goals, and values as well as artifacts, was shown to mirror this heterogeneity.

After demonstrating the interpretative flexibility of the phenol-formaldehyde condensation product, I described the closure process and the subsequent social construction of Bakelite, during which the building up of a new technological frame could be seen. This constituted a new structural environment for further technical development. I must emphasize again that this structural interpretation is not meant to belittle the individual ingenuity, passion, and commitment of an inventor such as Baekeland. It does, however, place an individual's characteristics in a broader context, thereby rendering it subject to analysis by sociological as well as psychological tools.

In chapter 4, I used the technological frames of the utilities and the Mazda companies to explain the specific form of the high-intensity fluorescent lamp. I also showed how this artifact reshaped the technological frame, creating a new social order. This case introduced the distribution of *power* as a factor in shaping technology and society. The particular conception of power that I presented has two aspects: a semiotic aspect that emphasizes the importance of the fixation of an artifact's meanings, and a micropolitical aspect that focuses on the continuous interactions of relevant social groups in a technological frame. Power in this sense is not given a priori, nor is it an intrinsic property of actors; rather, it is itself an important explanandum. To explain a specific set of power relations—a semiotic power structure—and to reveal the micropolitics of power can thus be one cornerstone of an explanation of the development of a new order constituted by a particular combination of technology and society.

Beyond Technical Artifacts: From the Seamless Web to Sociotechnical Ensembles

In its crudest interpretation, the "seamless web" motif serves merely as a reminder that nontechnical factors are important for understanding the development of technology. It pulls us toward contextual approaches to the study of technology, as opposed to internalist analyses. A deeper interpretation, however, is that it is never clear a priori and independent of context whether a problem should be treated as technical or as social, and whether solutions should be sought in science, economics, or some other domain. Such an interpretation stresses that the activities of engineers and inventors are best described as heterogeneous system- or network-building, rather than as straightforward technical invention.[3] In my analysis I have taken an additional step beyond this interpretation by suggesting that it is not only engineers (even in their upgraded guise as system-builders or network-weavers) but all relevant social groups who contribute to the social construction of technology. In the case of the high-intensity fluorescent lamp, for example, the actual designers were not engineers at their drawing boards but managers at a business meeting.

Let us now take one further step and suggest a third interpretation of the seamless web. The "stuff" of the fluorescent lamp's invention was economics and politics as much as electricity and fluorescence. Let us call this stuff "sociotechnology." The relations I have analyzed in this book have been simultaneously social and technical. Purely social relations are to be found only in the imaginations of sociologists or among baboons,[4] and purely technical relations are to be found only in the wilder reaches of science fiction.[5] The technical is socially constructed, and the social is technically constructed. All stable ensembles are bound together as much by the technical as by the social. Social classes, occupational groups, firms, professions, machines—all are held in place by intimate social and technical links.[6]

The move from the domain of technical artifacts plus social relations into the domain of sociotechnology can be linked to Michel Callon's (1986) principle of general symmetry, which expands Bloor's principle of symmetry. While Bloor's principle advocates that true and false beliefs (or, in the case of technology, working and failing machines) are to be explained in the same terms, Callon's principle of general symmetry states that the construction of science and technology and the construction of society should be explained in the same terms. This principle outlaws both technical reductionism (in which society is explained as an outgrowth of technical development) and social reductionism (in which the technical is explained as a by-product of the social).

Linked to this principle was a proposal that symmetrical roles be assigned to human and nonhuman actors. (See table 5.1 for a summary of the various concepts.)[7] This proposal has been the subject of a heated debate that potentially divides students of sociotechnology into two camps, one Anglo-Saxon and the other French.[8] For the former camp, it amounts to a heresy against the Winchian (1958) tradition in the social sciences to allow machines as actors into the story. For the latter camp, the analyses within Bloor's symmetry scheme are hardly more than internal accounts and do not provide insight into such crucial questions as the relation between micro events and macrosocietal developments.[9]

In my conception, the sociotechnical is not to be treated merely as a combination of social and technical factors. It is sui generis. Instead of technical artifacts, our unit of analysis is now the "sociotechnical ensemble."[10] Each time "machine" or "artifact" is written as shorthand for "sociotechnical ensemble," we should, in principle, be able to sketch the (socially) constructed character of that machine. Each time "social institution" is written as shorthand for "sociotechnical ensemble," we should be able to spell out the technical relations that go into stabilizing that institution. Society is not determined by technology, nor is technology determined by society. Both emerge as two sides of the sociotechnical coin during the construction processes of artifacts, facts, and relevant social groups.

How might this extension of the principle of symmetry affect our work? One possible reaction might be to avoid the complexity of this new sociotechnical world by refraining from theoretical explanations. Only narratives are to be made, only "how questions" to be answered; no models are to be conjectured, no "why questions" to be asked. Clearly, this is not the route I would advocate. Complexity, for me, simply implies a challenge to develop more adequate conceptual frameworks.[11] The theory of sociotechnical change that I am developing, for example, must mirror the heterogeneity of this sociotechnical "stuff" without resorting to just "adding up" the social and the technical.

One way to avoid the twin horns of social and technical reductionism is to introduce other differentiations between explanandum and explanans, between dependent and independent variables, between foreground and background—differentiations that are not based on the distinction between the social and the technical.[12] This is exactly what

Table 5.1
Summary of concepts employed in the principle of symmetry (Bloor, 1973, 1976; Pinch and Bijker, 1984) and the principle of general symmetry (Callon, 1986) (analogous concepts are grouped in horizontal rows)

Symmetry (Bloor on science)	Symmetry (Pinch and Bijker on technology)	General symmetry (Callon on sociotechnology)
Impartial to a statement being true or false.	Impartial to a machine being a success or failure.	Impartial to an actor being human or nonhuman.
Symmetrical with respect to explaining truth and falsity.	Symmetrical with respect to explaining success and failure.	Symmetrical with respect to explaining the social world and the technical world.
"Nature" is the result, not the cause, of a statement becoming a true fact.	"Working" is the result, not the cause, of a machine becoming a successful artifact.	The distinction between the technical and the social is the result, not the cause, of the stabilization of sociotechnical ensembles.

the concept of "technological frame" is meant to do. Its heterogeneity should allow us to distinguish foregrounds and backdrops other than the technical and the social (or vice versa).

As a first illustration, consider the configuration model below, which will allow us to model sociotechnical change by mapping it onto different configurations characterized in terms of technological frames and degrees of inclusion.

A Configuration Model

Using the concepts of "technological frame" and "inclusion," I will now iron some pleats into the seamless web of sociotechnology. In place of differences between the social and the technical, I will distinguish among alternative *configurations*. A second step will be to use this foregrounding and backgrounding to build an explanatory model, generalizing beyond individual case studies by identifying processes that occur in specific configurations, irrespective of the particular case.

As a first-order analysis, three different configurations can be distinguished. In the first, *no* clearly dominant technological frame guides the interactions; in the second, *one* technological frame is dominant; and in the third, *two or more* technological frames are important for understanding interactions involving the artifact under study. Each of these configurations is characterized by different processes of technical change.

The first configuration occurs when there is no single dominant group and there is, as a result, no effective set of vested interests. The early history of the bicycle provides an example. Although there were many social groups involved, it is hard to characterize any of them as dominating the field and structuring the identification of problems and the problem-solving strategies to be used. Under such circumstances, and if the necessary resources are available to a range of actors, there will be many different innovations.

In the second configuration, one dominant group is able to insist upon its definition of both problems and appropriate solutions. This is probably the most common configuration—"normal sociotechnology," to paraphrase Kuhn. The period between 1880 and 1920 in the development of (semi)synthetic plastics provides an example, with the Celluloid technological frame being dominant. Under such monopolistic circumstances, innovations tend to be conventional.

In the third configuration, when there are two or more entrenched groups with divergent technological frames, arguments that carry weight in one of the frames will carry little weight in the other. Under such circumstances, criteria external to the frames in question may become important as appeals are made to third parties.

Having characterized three different configurations, our next task is to specify which processes of sociotechnical change can be expected to occur in each configuration. Without being in any sense complete, I will discuss several possibilities, drawing on work by other scholars to demonstrate the generalizing and integrating power of the configuration model. What I will present is thus a demonstration of the feasibility of developing theoretical models along these lines, not a full-blooded and comprehensive model of sociotechnical change.

When there is no dominant technological frame, the range of variants that can be put forward to solve a problem is relatively unconstrained. The alternative sociotechnical ensembles that are generated will tend to be *radical*, in the sense that they differ substantially from the pattern laid down in any given frame. This would apply to all sociotechnical ensembles, or their elements. For technical innovations this situation has been well described by Hughes (1987), but similar observations can be made about social innovations.

In the case of the bicycle, for example, radically different technical variants were proposed around 1880 to solve the safety problem, but there were also efforts to make radical changes in other elements of the sociotechnical ensemble "Ordinary + macho aristocrat + excluded Victorian ladies." In the American "Star" (figure 2.19), the small steering wheel was positioned ahead of the high wheel; Lawson's Bicyclette (figure 2.20) had a chain drive on the smaller rear wheel. Women's clothing was designed differently to accommodate the technical constraints of the bicycle and also to make an emancipationist statement about women's societal position. Thus "radically different" means that all aspects of this sociotechnical ensemble were subject to variation. Hardly any detail of the bicycle was taken for granted—not even the number of wheels (tri- and quatro-cycles were constructed) or the method of foot propulsion (besides moving cranks in a circular motion, various lever devices were constructed, requiring a linear vertical motion of the feet). Hardly any detail of the nontechnical aspects of the ensemble was taken for granted either: sex or class of the cyclist, purpose of the cyclist, societal function, and status of bicycling.

One of the most important stabilization processes in a configuration without a clearly dominant social group and technological frame is *enrollment* (Callon and Law, 1982). This describes the process by which a social group propagates its variant of solution by drawing in other groups to support its sociotechnical ensemble. More than in the other configurations, the success of an innovation will here depend upon the formation of a new constituency—a set of relevant social groups that adopts the emerging technological frame.[13] One way to do this is by the mechanism of *problem redefinition* (Pinch and Bijker, 1987). If an artifact such as the air tire offers a solution to a problem that is not taken seriously by other important social groups, then the problem may be redefined in such a way that it does appeal to them. Thus the problem for which the air tire was originally considered to be a solution (vibration) was redefined into a speed problem. Because speed was important to the racing cyclists, they were now enrolled.

In the second configuration type, when one technological frame is dominant, it is fruitful to distinguish between actors with high and low inclusion. Engineers with a relatively high inclusion in the technological frame will be sensitive to *functional failure* (Constant, 1980) as an incentive to generate variants. A functional failure may occur when an artifact is used under new, more stringent conditions. Celluloid's flammability presented such a functional failure when its use was extended to other applications besides dentures, such as photographic film. Actors with a high inclusion in the technological frame are bound to generate relatively conventional inventions such as improvements, optimalizations, and adaptations (Hughes, 1987). Thus, a large part of the innovative effort of the celluloid producers was directed toward rendering celluloid less flammable by finding another solvent.

Actors with a relatively low inclusion in the technological frame interact to a smaller extent in terms of that frame. A consequence may be, as in the case of Baekeland, that such actors draw less on the standard problem-solving strategies of that frame. Another consequence could be that such actors identify other problems than would actors with high inclusion. For example, the identification of *presumptive anomalies* is typically done by engineers with a relatively low inclusion in a particular frame. A presumptive anomaly "occurs in technology, not when the conventional system fails in any absolute or objective sense, but when assumptions derived from science indicate either that under some future conditions the conventional system will fail (or function badly) or that a radically different system will do a much better job" (Constant, 1980: 15). Constant cites the example of aerodynamic theory in the 1920s, which suggested a future failure of the conventional piston engine/propeller system of aircraft propulsion. It suggested that proper streamlining would allow aircraft speeds to be increased at least twofold, but that the propeller would probably not function at the near-sonic speed that would result. The theory also suggested the feasibility of highly efficient gas turbines. My contention is that young, recently trained engineers are in an especially good position to recognize and react to presumptive anomalies: they are trained within the current technological frame but have low enough inclusion to question the basic assumptions of that frame.

In the third configuration, more than one technological frame is dominant. For example, around 1890 electricity distribution systems based on both direct and alternating current had been commercialized, sometimes in the same town (Hughes, 1983). The closure process in a configuration like this can be quite erratic, particularly in comparison with the first configuration. Arguments, criteria, and considerations that are valid in one technological frame will not carry much weight in the other frames. In such circumstances, it seems reasonable to expect that criteria external to all technological frames will play an important part in closure and stabilization. This often makes *rhetoric* a significant closure mechanism (Pinch and Bijker, 1987). (Of course, rhetoric may also be a factor in the second configuration.) Hughes (1983) described such a rhetorical move in the "battle of currents." In a public demonstration, a dog was exposed to direct current of various voltages with no ill effects and was then electrocuted by quick exposure to alternating current. The object of the demonstration was to persuade citizens that direct current was the safer alternative. As Hughes observes, often in such a "battle of the systems"—a competition between powerful, equally dominant social groups with respective technological frames—no one wins a total victory. *Amalgamation of vested interests* is the closure process that often occurs.[14] Innovations that allow amalgamation will be sought, and such innovations (the construction of the high-intensity fluorescent lamp is an example) are, so to speak, doubly conventional because they have to lodge within both technological frames.

The configuration model gives a crude explanation of some of the processes of sociotechnical change that can be observed, but it does not yet make the link to a politics of technology. That is our next goal.

5.2 *Toward a Politics of Sociotechnology*

The detour into academia that I described at the beginning of this book can turn into a main route in several different ways. First, travelers might forget that their initial interest in STS issues was politically motivated and simply continue on the highway of *institutionalized academic work*. It should be abundantly clear by this point that this is not the route I have had in mind.

A second route does turn back to political concerns, but for purposes that are allied to the interests of a specific social group. Stimulated by recent successes in academic science and technology studies, encouraged by the demand for instruments to solve policy problems, and facilitated by a no-nonsense culture that allows critical activists of the 1970s to become management consultants in the 1990s, this route leads toward science and technology *policy studies*. It is not unlikely that the insights gained in the detour so far will indeed allow the development of concrete policy instruments, and that these instruments will be useful within the technological frame of policymakers. The route toward technology policy is, however, also not the main route I want to pursue.

I would like to argue for a third route leading toward a *politics of technology*. This politics will deal with questions of value-ladenness, of emancipatory and oppressive potentials, of democratization, and of the embeddedness of technology in modern culture.[15] In other words, it will be concerned with questions about sociotechnical ensembles and the semiotics and micropolitics of power. Such a main route would take the much-quoted dictum of Langdon Winner (1980) that "artifacts have politics" and link it to the epistemological insights of recent studies of the sociology of scientific knowledge and to the theoretical and empirical work of constructivist studies of technology. The politics of technology in this sense will not yield the concrete policy instruments that the second route promises to produce. It will be emancipationist rather than instrumental, it will politicize technological choices rather than pacify them, and it will problematize rather than absolve.

My argument for a politics of technology involves three steps. First, I will argue that a constructivist analysis, in some form, is a conditio sine qua non for such a politics. Such an analysis stresses the malleability of technology, the possibility for choice, the basic insight *that things could have been otherwise*. But technology is not always malleable; it can also be obdurate, hard, and very fixed. The second step, then, is to analyze this obduracy of sociotechnical ensembles, to see what limits it sets to our politics. And finally, as a sort of postscript, I want to comment generally on the possible role of STS studies in the politics of technology.

The constructivist perspective provides a rationale for a politics of technology and it exemplifies the possibility of a social analysis of technology. Demonstrating interpretative flexibility makes it clear that the stabilization of artifacts is a social process and hence subject to choices, interests, and value judgments—in short, to politics. If one does not accept interpretative flexibility, one is almost certainly bound to fall prey to determinist thinking.[16] Few students of technology would take a purely determinist stand nowadays,[17] but the issue remains crucial for any discussion of the political relevance of technology studies.

Determinism inhibits the development of democratic controls on technology because it suggests that all interventions are futile. This is as true for science as it is for technology (Bijker, 1985; Collins and Pinch, 1993). If scientific facts are dictated by Nature, rather than constructed by humans, then any given scientific controversy (for example, about the risk of radiation) will necessarily be resolved only when it is shown that one party to the debate is "right" and the other "wrong." If we accept the idea that the debate is only about Nature, then it is quite reasonable for citizens to react by standing on the sidelines and saying "let the scientists sort that out among themselves." Likewise, if we do not foster constructivist views of sociotechnical development, stressing the possibilities and the constraints of change and choice in technology, a large part of the public is bound to turn their backs on the possibility of participatory decisionmaking, with the result that technology will really slip out of control.

Without an understanding of the interpretative flexibility of sociotechnical ensembles, the analysis of technology and society is bound to reproduce only the stabilized meanings of technical artifacts and will miss many opportunities for intervention. But interpretative flexibility needs to be demonstrated in a rigorous way that goes beyond the simple level of observation that "technology is human-made and therefore subject to many societal influences." The constructivist argument is that the core of technology, *that which constitutes its working*, is socially constructed. This is a way to take up the challenge of unpacking the politics of artifacts and thereby overcoming the standard view in which "blaming the hardware appears even more foolish than blaming the victims when it comes to judging conditions of public life" (Winner, 1986: 20).

Arguing for the malleability of technology does not imply that we can forget about the solidity and momentum of sociotechnical ensembles. Overly optimistic expectations based on a false sense of infinite malleability can easily cause disillusionment. A politics and a theory of sociotechnology have to meet similar requirements in this regard: a balance between malleability and obduracy in politics, and a balance between actor and structure perspectives in theory.

In my discussion of semiotics, I argued that the fixing of meanings that occurs during the formation of a technological frame is a form of power. I will now further specify this argument in terms of the obduracy of sociotechnical ensembles. Artifacts have been described in previous chapters, albeit implicitly, as bundles of meanings, as exemplars, and as boundary objects. The descriptive model started by emphasizing the meanings attributed by relevant social groups. These meanings were said to constitute the technical artifact. Does this imply that such "bundles of meanings" have unbounded flexibility? Can relevant social groups fantasize whatever they want, without constraints? Of course, they cannot. Attributions of meaning are social processes and, as such, are bound by constraints. Previous meaning attributions limit the flexibility of later ones, structures are built up, artifacts stabilize, and ensembles become more obdurate. The concept of "technological frame" made sense of this process. Ongoing interactions with an artifact, within and between relevant social groups, results in the creation of a technological frame that bounds the attributions of meanings by relevant social groups.

An important element of any technological frame is the *exemplary artifact*, and this is the second way in which artifacts have been described in this book. An artifact in the role of exemplar (that is, after closure, when it is part of a technological frame) has become obdurate. The relevant social groups have, in building up the technological frame, invested so much in the artifact that its meaning has become quite fixed—it cannot be changed easily, and it forms part of a hardened network of practices, theories, and social institutions. From this time on it may indeed happen that, naively spoken, an artifact "determines" social development. As a sociotechnical ensemble, it is at the same time the result of micropolitical interaction processes and one of the elements of a semiotic power structure.

To analyze the obduracy of sociotechnical ensembles, it is also helpful to look at artifacts and ensembles in their role as *boundary objects*.[18] Part of the closure and stabilization process is the creation of inside/outside boundaries.[19] The concept of "inclusion" is used to describe the boundary area between inside and outside: Actors with a high degree of inclusion are more to the inside than actors with a lower degree of inclusion.[20]

As an example, actors willing to participate in the fluorescent lighting market were required by General Electric and the utilities to accept the high-intensity lamp as the only daylight fluorescent lamp in the game. A certification scheme for lamp fixtures and other auxiliaries was designed to give the exemplary high-intensity lamp this meaning. Creating the certification scheme as an obligatory passage point helped to establish the specific form of semiotic power by which General Electric tried to hold the other lamp producers in bondage. The actual high-intensity lamp was an obdurate element in this semiotic structure, helping to create a boundary between the inside and the outside of the fluorescent lamp technological frame. For an actor outside the frame that was being built up around the high-intensity lamp, it became obligatory to produce (or sell, or buy) high-intensity lamps; actors who would not comply were not allowed inside.

Typically a closure process results in one relevant social group's meaning becoming dominant. A new technological frame is formed, shared by several social groups that until then could be represented by separate frames. The formation of the Bakelite frame through the enrollment of automotive engineers and radio amateurs is an example. Bakelite had an increasingly fixed meaning when it was used in the negotiations between employees of the General Bakelite Corporation and engineers of the automobile, radio, and chemical instrument manufacturing industries. The artifact Bakelite thus functioned at the same time as an element in Baekeland's micropolitics of power and as a boundary artifact in the emerging semiotic power structure of the new Bakelite technological frame. We might say that Bakelite was a form of currency, allowing the negotiation of new relationships between previously unrelated social groups. The boundary artifact Bakelite helped to link, in an almost physical sense, the different social groups into one new semiotic structure.

The obduracy of technology can take on at least two different forms, one associated with the artifact as exemplar, the other with the artifact as boundary object. Artifacts as "currency boundary objects" are, by definition, primarily used by actors with a relatively low inclusion, although highly included actors may also use them strategically. Artifacts as exemplars are typically used by actors with a high inclusion.

For engineers of the Bakelite Corporation, the molding material was an important element in the technological frame. Because it had a high inclusion in that frame, the molding plastic set their research and design agenda with high specificity. The meaning of the artifact, as it resulted from the stabilization process, was quite unambiguous for these actors. This is, however, not the same as saying that such a sociotechnical ensemble has one monolithic, undifferentiated meaning. For actors with high inclusion, the exemplary artifact is *unambiguous* and *constrains* action, but it is also highly *differentiated* and enables them to do many things.

Bakelite engineers knew how many variables had to be controlled to produce specific forms of the artifact, they knew how many different shades could be produced, and they knew how tricky the manufacturing process could be.[21] Thus this artifact was "hard" for these actors with high inclusion, but in the very specific sense of enabling and constraining interaction and thinking. It was not hard in the sense of having one fixed, monolithic meaning. For actors with low inclusion, the situation is quite different: The sociotechnical ensemble, as an element of a technological frame, does not guide their interactions and thinking very strictly, but it does have a relatively undifferentiated, monolithic meaning.

For actors with low inclusion, the artifact presents a "take it or leave it" decision. They cannot modify the artifact if they "take" it, but life can go on quite well if they "leave" it. For highly included actors, on the contrary, there is no life without the exemplary artifact, but there is a lot of life within it. The obduracy of artifacts as boundary objects for actors with low inclusion consists in this "take it or leave it" character. For such actors, there is no flexibility and there is no differentiated insight; there is only technology, determining life to some extent and allowing at best an "all or nothing" choice. This is the obduracy of technology that most people know best, and it is what gives rise to technological determinism.

There is an interesting similarity between this view and Donald MacKenzie's analysis of uncertainty. MacKenzie (1990) observes a high uncertainty about technology and science among two groups of actors—those with high inclusion (my term) and those with low inclusion. In between, he conjectures, a "certainty trough" may exist (figure 5.1). My analysis suggests that the "uncertainty" at the right-hand end of the

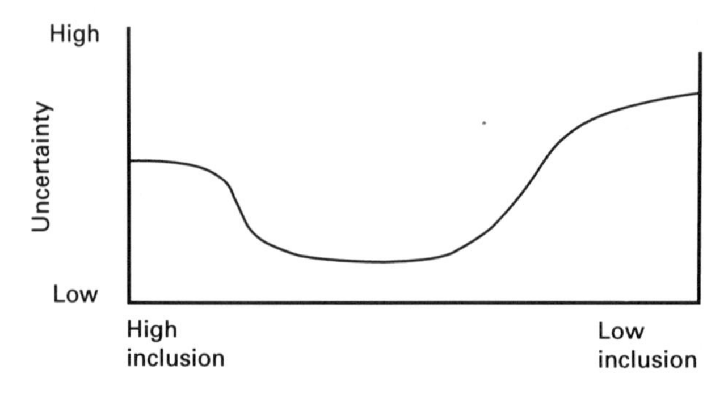

Figure 5.1
The "certainty trough" according to MacKenzie (1990: 372)

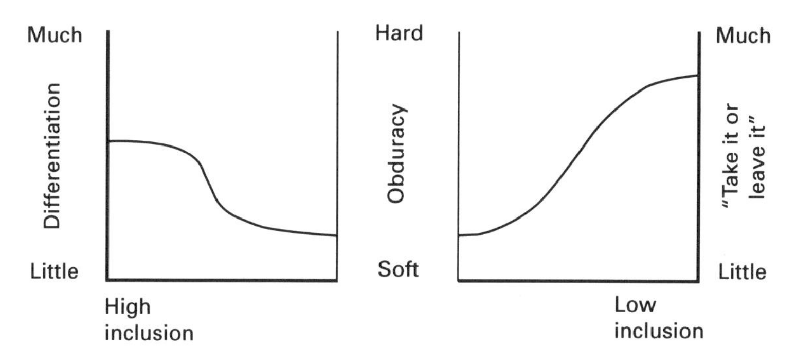

Figure 5.2
Two types of obduracy of technological ensembles, depending on the degree of inclusion

curve is due to the obduracy of the sociotechnical ensembles in the "take it or leave it" sense, while at the left-hand end it is due to their high differentiation for highly included actors (figure 5.2). Describing the differentiation of uncertainty as occurring along a single dimension erroneously suggests that the two types of uncertainty are the same. Consequently, I would argue that MacKenzie's trough is not a good representation of what actually occurs. Artifacts can have different shades of obduracy for actors with different degrees of inclusion; but when the boundary of a technological frame is passed, the character of this obduracy changes fundamentally.

Artifacts as boundary objects, in the form of obligatory passage points or currency, result in obduracy because they link different social groups to form a (new) semiotic power structure. Making a "take it" choice with respect to such an artifact leads to inclusion in the corresponding structure. One then becomes involved with power relations to which one would otherwise—in the case of a "leave it" choice—be immune. If you buy a car, for example, you become included in the semiotic structure of automobiling: cars-roads-rules-traffic-jams-gas prices-taxes. This will result in your exerting power, for example by choosing to use the car during rush hour and thereby contributing to a traffic jam, but it will also subject you to the exertion of power by others in that same jam. If you choose not to own a car, jams and oil prices simply do not matter.[22]

Artifacts as exemplars result in obduracy because they constitute to an important degree the world in which one lives. This also implies inclusion in a semiotic power structure, but with different possibilities and effects. There are now many alternative power interactions that involve the exemplary artifact. Leaving the car standing is a less likely option, but changing driving hours or routes (to beat the traffic jams), changing from gasoline to diesel or liquid gas (to beat the taxes), and changing to a smaller car (to beat the parking problem) are all real possibilities.

In the constructivist perspective, then, sociotechnical ensembles manifest at least two types of obduracy. More precisely, this obduracy, to an important degree, constitutes the semiotics of power, and it is within this semiotic structure that the micropolitics of power are staged.

The micropolitics of power can focus on the deconstruction of the powerful, on softening the obduracy of the sociotechnical ensembles that constitute the semiotic power structure. The bicycle case provides an example. The sociotechnical ensemble of Macho Bicycle, Victorian morals, and "young men of means and nerve" was part of a semiotic power structure that prevented women from using the bicycle. One important factor in unmaking this oppression was a softening of the obduracy of the Macho Bicycle. As I have described, this happened largely through the actions of racers on low-wheeled "pneumatic" bicycles—racers who were probably not more interested in furthering the cause of women's emancipation than were their brethren on the high-wheeled Ordinaries. But once the high-wheeled machine had lost its obduracy, the semiotic power that had excluded women from bicycling lost its coherence and force. It became feasible to make other elements that would lead to new ensembles: low-wheeled bicycles, women's frames, bloomers, different morals.

A similar, though much shorter, process was the softening of the high-efficiency daylight fluorescent lamp. This lamp had almost everything it needed to become the starting point of a new technological frame, but its deconstruction during the load controversy prevented the process of building that frame. Bakelite was also subjected to deconstruction processes, though without much effect. The most conscious deconstruction strategy was Lebach's, who tried to argue that Baekeland's patents were secondary to his own. Another example of the micropolitics of power, though less consciously strategic, was the negative image that plastics acquired during World War I. This could have resulted in a softening of Bakelite as an exemplary artifact, and it underlines the importance of having a conception of power that is not exclusively associated with the strategic actions of specific actors. Consumer behavior and product image may, under certain circumstances, turn out to be more important elements in the micropolitics of power than patent agreements, smoothly running production plants, or economic ownership relations.

Other micropolitics can result in a softening of the obduracy of boundary artifacts. In this case, it will be a matter of influencing the take-it-or-leave-it choice of actors with low inclusion. Ridiculing bicyclists and setting restrictive rules for bicycling in towns were micropolitical ways of discouraging people from cycling, but they left the bicycle as an exemplary artifact untouched. Nevertheless, if this strategy had been successful, the whole sociotechnical ensemble would have been deconstructed and the associated semiotic power would have crumbled. (This is exactly what had happened some decades earlier with the hobbyhorse craze.)

The refusal of some utilities to inform the public about the fluorescent lamp is another example of deconstructive micropolitics focused on an artifact as boundary object. Executives like Bremicker were not successful in tampering with the new lamp in its exemplary role, but their unwillingness to show it in their display rooms and to include it in their marketing campaigns posed a serious threat that might have softened the artifact as boundary object to such a degree that Bremicker's Northern States Power Co. could have "escaped" the semiotic power in which it was increasingly encapsulated by the Mazda companies.

Micropolitics can also lead to a bolstering of sociotechnical ensembles. Let me start with the contemporary example of routine ultrasound scanning.[23] Giving all pregnant women an ultrasound scan would effectively make the scan into a boundary device drawing women into the semiotic power structure of medical hospital technology. In the Netherlands, where a considerable number of women give birth at home, this process has not reached closure, and women can still choose *not* to have such a scan. "Not taking" this artifact would be a strategy for weakening medical power, while making women accept the artifact as "producing a nice picture," or as "a safe precaution," or as "a standard procedure" would result in the further extension of medical power, using the obduracy of the scan as boundary artifact.

Several of the processes that I identified in the discussion of the configuration model are also examples of micropolitics resulting in an increasing obduracy of artifacts. Enrolling the automobile and radio engineers into the Bakelite frame, negotiating a cost comparison method with the utilities, and selling a tricycle to the Queen all led to a bolstering of sociotechnical ensembles and their associated semiotic power structures.

Although some of the these examples do involve conscious strategies by particular actors, there is more to the politics of technology than explicitly planned, rationally decided, conscious action. This has been one of my reasons for introducing the structural semiotic aspect of power in the first place, and for emphasizing the need for balance between actor- and structure-oriented perspectives generally.[24] Technology is socially shaped and society is technically shaped, but there need not always be explicit "causal" links between specific artifacts and relevant social groups. The combinations of technological frames with actors and artifacts, and of the semiotics of power with the micropolitics of power, are meant to describe this process of developing sociotechnical ensembles.

Postscript: The Paradox of Sociotechnical Politics

This book started at the level of single artifacts and their social shaping and then moved to level of relevant social groups that are constrained and shaped by technology. Now, at the end, we have arrived at the highest level of aggregation—technological culture.[25] At this level the STS problematic with which I started can be given new formulations. Many of these have already been discussed: a constructivist view of artifacts and facts, the integration of the technical and the social into a new unit of analysis, a conception of power that combines structural and actor-oriented aspects, and finally an analysis of the hardness of technology. There is one final, general implication with which I would like to conclude, but I have chosen to do this by way of a postscript since it is more a suggestion for a research agenda than a well-polished conclusion.

One of the major implications of the analysis so far is that there are no actors or social groups that have special status. All relevant social groups contribute to the social construction of technology; all relevant artifacts contribute to the construction of social relations. This is mirrored in the power conception I have outlined. Both the semiotic and the micropolitical aspect of this conception stress that sociotechnical change cannot be understood as the product of one prominent actor, whether an inventor, a product champion, a firm, or a governmental body. The consequences of this are quite radical. Technology assessment methods, which are given a priority role by legislatures, and innovation management theories, which are given a similar role by firms, are equally inadequate. And even *my* attempts to describe micropolitical strategies to change technology and society will never yield a conceptual framework or suggest a set of policy instruments that are certain to guide sociotechnical change in a particular direction. This is a matter not of postmodern relativism but of recognizing that there will always be other actors who contribute to the construction of society and technology, actors that cannot be controlled.

One consequence of this observation is that we can no longer imagine that constructivist STS studies will principally or primarily benefit any specific social group, such as the less privileged or less powerful. One might attempt to argue that the sorts of STS studies I have discussed, by highlighting the constructed nature of facts, artifacts, social orders, and sociotechnical ensembles, will allow those who are kept hostage by the semiotic power structures involved—nonscientist citizens, consumers, patients, women, workers, neighbors to a chemical or nuclear plant, environmentalists—to sever these bonds and free themselves.[26] Although this may happen, there is no guarantee that it will always work out this way. First, science and technology may also be fruitfully employed by the less privileged. Environmentalists, for example, frequently use scientific data to argue their case, and the last thing they would want is to see those findings and arguments deconstructed.[27] And second, there is no reason why the powerful may not draw on the insights of the STS community or even hire constructivists to strengthen their micropolitical strategies. The relativizing force of constructivist STS studies thus prevents it from attaining a neat and simple political correctness.

STS work should not stop here. The observation that all citizens have access to insight might lead some to argue for a decentralization of control and responsibility—"local democracy," for example, in place of a national regulatory system. Such "free-market regulation" would leave larger semiotic power structures uncritically in place. The question of how a politics of technology can help establish institutional and structural ways of guaranteeing the democratic nature of technological culture thus reappears on the agenda. Its solution, however, necessarily takes the form of a paradox, recognizing that no principally prioritized group can exist in society—there is no Leviathan—while at the same time proposing some form of institutional regulatory system.

This open, and potentially paradoxical, agenda for future research mirrors the thrust of this book. Although I have wanted to argue for a politics of technology (rather than, for example, technology policy), and although I have tried to develop a constructivist theory of sociotechnical change (rather than, for example, an economic one), I have also wanted to tell my three stories with due respect to their own richness. Consequently, I have tried not to make this book into a closed world, which would leave no room for alternative interpretations of the stories. This, then, is my bottom line: a plea for combining empirical work with theoretical reflection to strengthen the links between academic STS studies and politically relevant action—a main route for STS studies out of what started as an academic detour.

"The time has come," the Walrus said,
"To talk of many things:
Of bikes—and bakelites— and bulbs—
Of theories—and kings . . ."[28]

Notes

Chapter 1

1. I will use the term "artifact" to encompass all products of technology: it denotes machines as well as technical processes, hardware as well as software. Thus artifacts include the bicycle, the chemical reaction patented by Baekeland, a complete lighting installation, the starter switch inside a fluorescent lamp auxiliary, and a method for cost comparison between incandescent and fluorescent lamps.

2. See Cutcliffe (1989) for a description of the emergence of American STS programs. Jasanoff et al. (1994) represent the state of the art in academic STS studies. The situation in the Netherlands is different from many other countries in that science and technology studies are the offspring of a marriage between two groups in particular: politically motivated STS academics and philosophers of science. Sociologists and historians did not play an important role in the early days of Dutch science studies. See Bijker (1988) for a review of the roots of Dutch science and technology studies.

3. See Bijker, Hughes, and Pinch (1987a) and Bijker and Law (1992a) for descriptions of this research program and for examples of empirical and theoretical work within it. See Bijker (1994) for a general review of sociohistorical technology studies.

4. Blume (1992) contributes to bridging the gap between economic studies of technical change and sociological work in the constructivist tradition. Winner (1991) argues for the need to complement recent empirical work in the constructivist tradition with philosophical and political analyses of technology.

5. Manuals describing resinous materials do mention Bakelite, but not with the amount of attention that, in retrospect, we would think justified. Professor Max Bottler (1924), for example, devoted only one page to Bakelite in his 228-page book on resins and the resin industry.

6. See Staudenmaier (1990) for a comprehensive review of the current state of the history of technology. Staudenmaier's monograph (1985) analyzes twenty

years of American history of technology as reflected in the journal *Technology and Culture.*

7. It would be silly to deny credibility as a scholar in technology studies, a priori, to anyone without "the right hardware credentials." On a personal level, though, I can only say that both my training as an engineer and my continued fascination with technology have been very important throughout this project. I have tried, however, to make the book readable for people without any technical background, and hope that it will convey some of this fascination.

8. Buchanan (1991) defends this position in a critical review of recent sociohistorical studies of technology. The two other participants in this illuminating debate are Law (1991b) and Scranton (1991).

9. Typical examples are Hughes's (1983) study of three electricity distribution systems and of Thomas A. Edison's role as a system builder, and Carlson's (1991) study of the creation of General Electric, focusing on the role of Elihu Thomson.

10. An example is Carlson and Gorman's work on Edison and Bell (Carlson and Gorman, 1990, 1992; Gorman and Carlson, 1990). See also Ferguson's (1992) and Hindle's (1981) studies of the role of visual thinking in creating new technologies.

11. Mumford (1964, 1967, 1970) did explicitly address the issue of power, but with a very different agenda. His concern was primarily with the role that technology played in the *corruption* of power and the emergence of an authoritarian society. Mumford's work is nevertheless quite relevant for issues pertaining to the role of the concept of power in the analysis of technology.

Early American sociologists of technology such as Ogburn (1933, 1945) and Gilfillan (1935a,b) focused on the impact of technology on society, but without addressing the power issue. More relevant is the work by German and English industrial sociologists who analyzed the role of technology in the capitalist production system. See Braverman's (1974) analysis of technology's "deskilling" effect on labor, whereby it enables management to hire less qualified and thus cheaper workers.

12. The project initially had three more case studies: aluminum, the transistor, and the Sulzer weaving machine. Research on these cases was completed and did contribute to the shaping of the theoretical framework presented here. Space limitations, however, prevented their inclusion.

13. This cross-fertilization between the sociologies of science and technology has been advocated by Pinch and Bijker (1984). Exemplary studies that demonstrate in various ways the contingent character of scientific development are: Collins (1985), Collins and Pinch (1982), Knorr-Cetina (1981), Knorr-Cetina and Mulkay (1983), Latour and Woolgar (1979), Pickering (1984), and Pinch (1986).

14. Earlier versions of this theoretical model (as in Pinch and Bijker, 1984) were more explicitly evolutionary. Evolutionary models, however, tend to reify that which is being modified through processes of variation and selection. To avoid

this connotation of a reified technology, which would be contradictory to my general constructivist approach, I have diminished this evolutionary element.

15. If this were the mechanism involved, it would be extremely difficult to keep a bicycle upright. For an observer who has such an interpretation implicitly in mind, the task seems quite impossible, and this is one explanation for the amazement of people in the 1860s when they saw bicyclists for the first time.

Chapter 2

1. The title of this chapter is taken from the bicycle lamps produced by Joseph Lucas around 1880 (Card, 1984).

2. The programmatic core of this chapter was published by Pinch and Bijker (1984, 1987).

3. Baudry de Saunier (1936), cited by Woodforde (1970: 7).

4. Some doubt whether the first machine, with the resting bar immediately linked to the front fork, was designed by Drais himself. They suggest that Drais only built the "running-machine" with the fixed resting bar and the separate steering handle (Plath, 1978).

5. The skills of riding a modern bicycle may be so tacit that many readers will not recognize readily that steering is the main balancing technique. Instead of giving a complex mechanical explanation, I suggest that those who doubt this carry out the experiment of trying to ride a bicycle whose steering wheel is firmly fixed (for example, by binding the mudguard with a rope to the frame).

6. Griffin (1877) describes the particulars of some fifty-one makes of high-wheeled bicycles.

7. For detailed figures, see Prest (1960), especially pp. 129–130.

8. Hounshell (1984: 188–215) has described a similar development in the United States, where Albert A. Pope contracted with the Weed Sewing Machine Company. The first high-wheeled Ordinary bicycles were produced by the Weed Company by using its sewing machine manufacturing equipment with only special fixtures and cutting tools added.

9. The following report is based on a contemporary account, found in the Starley family papers and quoted by Williamson (1966: 51–53).

10. For a list of early contests, see Rauck et al. (1979: 168–169).

11. For a detailed account, see Rauck et al. (1979: 169).

12. In his novel *Der Mann auf dem Hochrad*, the philosopher and novelist Uwe Timm gives a fascinating account of events in the small German town of Coburg when the local taxidermist, Schröder, started riding a high-wheeled bicycle. This

pioneering act upset the sleepy town and made social structures, values, and norms highly visible. The social democratic revolution, feminist actions, and technical progress all became closely linked.

13. "Bang, Bang, Bad, Always on his Head!" (Timm, 1984: 19).

14. If there were no children's bicycles these days, we might need riding schools again. I used to show visiting sociologists and historians of technology around the old town of Delft by taking them on a bicycle tour. Not all of them had had much cycling experience. I am happy to report that no accidents occurred, but in some cases a short refresher course in riding would have been useful.

15. C. Spencer, in *The Modern Bicycle,* quoted by Woodforde (1970: 112).

16. A. C. Pemberton, *The Complete Cyclist,* quoted by Woodforde (1970: 116).

17. That this was indeed an effect of the difference in wheel diameter can be appreciated without much physics, when the reader thinks of the difference in comfort between a baby carriage with large wheels and a stroller with smaller wheels.

18. This quotation is from an article in *The Spectator* of 1869, describing velocipede riding. It could just as well have been used to characterize cycling on Ordinaries. Quoted by Woodforde (1970: 24).

19. Schröder, in notes for a book he planned to write, quoted by Timm (1984: 149).

20. See Timm's (1984: 124–139) description of the antibicycle evening organized by the shoemakers of Coburg.

21. Also see Timm (1984: 83–84, 159, 187) and Wells (1896: 52).

22. Rev. L. Meadows White, *A Photographic Tour on Wheels,* quoted by Woodforde (1970: 49–50).

23. Ibid., p. 50.

24. Schumacher, *Das Recht des Radfahrers,* quoted by Rauck et al. (1979: 79).

25. Quoted from the December 1881 issue of *Wheel World* by Woodforde (1970: 52).

26. This article from the *Münchener Zeitung,* 1900, is quoted by Rauck et al. (1979: 76).

27. This prudence certainly was not restricted to the British isles; see the chapter on "Bicycle Etiquette" in the American book by Cooke (1896: 343–351).

28. Samuel Webb Thomas, British patent specification No. 361, 1870. Most likely, the velocipede was never built for actual use; the original Patent Office model is in the Science Museum, London. For a detailed description, see Caunter (1958: 6).

29. Reply to lady's letter in a magazine of 1885, quoted by Woodforde (1970: 122).

30. The Earl of Albemarle in G. Lacy Hillier, *Cycling*, quoted by Woodforde (1970: 45).

31. Quoted by Woodforde (1970: 47).

32. Here I draw on work by Collins (1981a). He used this sociological technique as an operational definition of the "core set"—the group of scientists centrally involved in a given controversy. Examples of controversy studies in the sociology of scientific knowledge are provided by Collins (1985), Collins and Pinch (1982), Pickering (1984), and Pinch (1986).

33. This limited set of actors was defined by Collins (1981a) as the "core set."

34. See Latour (1987) for more examples of "follow the actors" as an *adagium* for empirical research in science and technology studies.

35. Callon and Law (1989: 64) describe much the same process as "investments of form" in their network vocabulary: "An investment of form is the work undertaken by a translator to convert objects that are numerous, heterogeneous, and manipulable only with difficulty into a smaller number of more easily controlled and more homogeneous entities—entities which are nonetheless sufficiently representative of their heterogeneous and formless cousins that it becomes possible for the translator to manipulate the latter as well." They add that there is no reason that these investments of form would only be carried out to order the world of relevant social groups; the world of nonhuman actors is simplified in similar ways.

36. Also in ethnomethodology an important tactic is to disrupt the smooth flow of routine events. See Heritage (1984) on Garfinkel's (1967) "breaching" experiments.

37. The aim of the analysis is, of course, to show how differences between the various artifacts are constructed *by the actors*. Hence *the analyst* has to start his description by assuming no differences.

38. This is what Latour (1987) called the "diffusion model of technology": an unchanging artifact is "pushed" through time and space. Evolutionary models will often be sophisticated versions of the diffusion model: successive artifacts follow one another along "trajectories," as bullets through economic-technological space.

39. Many fantastic human-driven vehicles are described by Rauck et al. (1979): 8–11, 21.

40. For example, Rauck et al. (1979: 10–11) describe two such *Karossen* built by the Nürnberger fine metalworker Johann Hautsch for the Swedish crown prince (1649) and the Danish king (1663).

41. This is the story as told by Williamson (1966: 69–70). There is some discussion about whether Starley actually invented the device or adapted it from Aveling and Porter's road traction engine. Anyhow, he should be credited with recognizing its importance and applying it to the tricycle (Grew, 1921: 20).

42. James Starley is said to have written this letter in the house of the local agent, Mr. Roach, immediately after returning from his visit to the queen. It is cited by Williamson (1966: 78–79). It has been questioned whether the letter was actually written by Starley or only later by one of his relatives (Woodforde, 1970: 80). However, the core of the story is very probably true; the case of the watch carries an inscription, later added, that reads: "Watch presented to James Starley by Queen Victoria on the occasion of his supplying her with two tricycles—June 1881" (Williamson, 1966: 80).

43. Lord Albemarle in a letter to the Badminton Library's *Cycling,* 1886, quoted by Woodforde (1970: 80–81).

44. See, for example, Ritchie (1975: 112–114) and Woodforde (1970: 67).

45. In a report on the Coventry Machinist Company's Cycle Works, it is noted that "There are two machine shops, one exclusively for bicycles and the other chiefly for tricycles" (Engineer, 1886: 202).

46. This comprehensive catalogue was published in 1886 for the ninth year in succession and provides a rich source for detailed technical information on almost virtually all cycle designs of that period. The book was meant to be "a chronicle of the new inventions and improvements introduced each season, and a permanent record of the progress in the manufacture of bicycles and tricycles; designed to assist intending purchasers in the choice of a machine; and written from personal examination" (Griffin, 1886: ii). Besides sixty-one pages of bicycle descriptions and ninety pages of tricycles, it also contained an extra twenty pages with accessories, an index, and several advertisements. Its price was one shilling. See also Griffin (1889) and Roberts (1980).

47. Letter by "A Tricycler" in *Bicycling News,* 1878; quoted by Ritchie (1975: 112).

48. Lacy Hillier, a frequent contributor to *Cycling,* writing in 1886; quoted by Woodforde (1970: 125).

49. This accident which happened in 1904, was described by Richard Church (Church, 1955: 174–175).

50. This is a complicated optimization problem. With two equal-sized wheels, the most comfortable position is exactly between the wheels (think, for example, of a bus, where the most comfortable seat is halfway between front and rear). This would suggest that the saddle be shifted backward to a position between the wheels of the Ordinary. But the smaller rear wheel caused relatively more vibration. One could counter this by making the rear wheel bigger, but eventually this tactic would make it impossible to mount the bicycle. One could compensate by lowering the front wheel, but this lowered the bicycle's top speed as well.

51. These are mentioned by Palmer (1958: 75) and Rauck et al. (1979: 53). Griffin (1886) describes many other machines available on the market in 1886.

52. An advertisement in 1883, quoted by Ritchie (1975: 126).

53. From *The Cyclists' Touring Club Gazette,* November 1884, quoted by Ritchie (1975: 127).

54. Calling the Lawson and Likeman bicycle a "safety" may be an example of wishful retrospective distortion by Lawson himself, readily taken up by the bicycle historians. Ritchie reports a priority dispute in 1887 in the columns of *Bicycling News*, where F. Shearing rebuts the claim that Lawson had invented the "safety" in the 1870s. Shearing claims that he had invented, made, and ridden three different safeties "a half-score years ago" (Ritchie, 1975: 123).

55. Patent No. 285,821, issued on 2 October 1883 to William Klahr of Myerstown, Pennsylvania (Oliver and Berkebile, 1974: 53); P. A. Maigen, British Patent No. 4012 of 1880; W. S. Kelly, British Patent No. 8240 of 1885 (Caunter, 1958: 14).

56. See Palmer (1958: 77); also Oliver and Berkebile (1974: 54) and Woodforde (1970: 55).

57. The best description of this mechanism is given by Palmer (1958: 77–78).

58. Woodforde gives various examples of how, when skillfully used, the "Star" was an effective machine (1970: 55–58).

59. The "Star" was advertised in the Netherlands in 1884 (Hogenkamp, 1939: 58).

60. See Lawson (1879). One year earlier T. Shergold patented a bicycle with a chain drive on the rear wheel. It was of very crude construction and did not have any commercial significance. It can be viewed as a late survival of the tradition of homemade machines.

61. *The Cyclist,* 21 April 1880, quoted by Ritchie (1975: 124).

62. This race is described in great detail by Williamson (1966: 105–107).

63. British Patent 15342, issued on 21 November 1884.

64. Mudguards were introduced in this period (Caunter, 1955: 35).

65. In addition to its macho character, the Ordinary also allowed people to move around on the same level as horse riders—an aspect that surely appealed to the class-conscious British of the 1870s. I am grateful to Joep à Campo for this observation.

66. Nor should we, according to the strong program, use "reality," "rationality," or "truth."

67. I deliberately refrain from using typographical tricks to designate the two different types of artifact—the "real" artifacts and the sociologist's artifacts, or

the one artifact before demonstrating its interpretative flexibility and the several artifacts that result from this demonstration. I maintain that it is not possible to speak unambiguously about the first type of "real" artifacts; hence there is no ground for such distinction.

68. See Rauck et al. (1979: 105).

69. See Caunter (1958: 44). See also the report by Dunlop's daughter Jean, cited by Rauck et al. (1979: 106–107); it is not unlikely that this report suffers a little from retrospective and "heroic" distortion.

70. This account is by Grew (1921: 53), who was present at the event.

71. A tire according to this Michelin concept is still used on the lightest racing bicycles. Now it is called a "tubeless tire," an implicit reversion of inner and outer parts.

72. Advertisement by W. Edlin & Co., 33 Garfield Street, Belfast (Du Cros, 1938: 80).

73. Arthur Du Cros (1938: 51–53) relates how it took him and his father a number of months and quite a few races to be convinced that there was a "high-speed device" next to the "antivibration device." On June 1, 1889, riding on solid tires, he beat William Hume on pneumatic tires. Later in 1889 Du Cros tested the air tires extensively on both grass and hard track, and was then convinced of their efficacy in increasing the speed of bicycles.

74. See the personal memories by Du Cros (1938: 57–59).

75. Hence we have called this the "redefinition of problem" mechanism (Pinch and Bijker, 1984).

76. See Whitt and Wilson (1982) for more (ahistorical) technical analyses.

77. In the next chapter I will return to these two vastly different (and for some even irreconcilable) intellectual traditions, which are important for the conceptual framework developed in this book.

78. See Pinch and Bijker (1984) for a fuller discussion of the similarities between the studies of science and technology on this point. For an analysis of the closure of scientific controversies, see especially Collins (1985).

79. Pinch and Bijker (1984) have given an example of this closure mechanism at work in a scientific controversy. Although not labeled explicitly as "rhetorical closure," the paper by "Q" that effectively ended Weber's claims about the existence of high fluxes of gravitational waves provides another example (Collins, 1985: 92).

80. An advertisement of the "Facile Bicycle" in the *Illustrated London News,* 1880, cited by Woodforde (1970: 60).

81. Latour and Woolgar (1979: 176–177) employed this analysis of the decreasing number of modalities in their "splitting and inversion model" to describe the

phenomenon of an emerging reality that cannot be deconstructed once closure has been reached. At the beginning of the process,

> members of a laboratory are unable to determine whether statements are true or false (...); modalities are constantly added, dropped, inverted, or modified. Once the statement begins to stabilise, however, an important change takes place. *The statement becomes a split entity*. On the one hand, it is a set of words which represents a statement about an object. On the other hand it corresponds to an object in itself which takes on a life of its own. It is as if the original statement has projected a virtual image of itself which exists outside the stement (...). Before long, more and more reality is attributed to the object and less and less to the statement *about* the object. Consequently, an inversion takes place: the object becomes the reason why the statement was formulated in the first place.

82. To convince the reader of the gap between 1988 and 1888, I may mention the large number of bruises that I endured while learning to ride on a Penny-farthing in May 1988.

In the case of the air tire it is easier to see the interpretative flexibility, because the two devices actually coexist: pump your tire very hard and it is primarily a high-speed device, and with less pressure it is more effectively an antivibration device.

83. The quotation is part of the motivation for designing an ingenious device that would allow riders of low-wheeled bicycles to steer without having their hands on the handlebar (Engineer, 1889: 158). The device consisted of a pivoted saddle linked by a lever mechanism to the steering wheel. Moving one's body to one side would cause the saddle to tilt and accordingly move the steering wheel. It is difficult to imagine that engineers seriously attempted to design such intricate mechanisms to do something that most of us are now able to do simply by leaning our body a little to one side.

84. Discussions with Tom Misa helped to clarify this point.

85. The "Bantam" bicycle was produced by the Crypto Cycle Co. in about 1893. Its small front wheel was driven by a epicyclic gear train (Caunter, 1958: 18). The previously mentioned "Sun and Planet" employed a similar mechanism.

86. The roller chain was patented in 1864 by J. Slater (Caunter, 1958: 31). It consisted of extra rollers over the pins of the chain in order to reduce the friction and wear of the sprocket teeth.

87. See Miller's patent (1882), Caunter (1955: 37–38), and *The Engineer* (1897c: 569).

88. For example, W. and J. Lloyd and W. Priest in a 1897 patent (Caunter, 1958: 33), applied in the Quadrant bicycle of 1899 (Caunter, 1958: 2).

89. W. James, *The Book of Beauty* (circa 1895), quoted by Woodforde (1970: 92–93).

90. See also Mr. Hoopdriver's adventures with the "Lady in Grey" (Wells, 1896).

91. See Ritchie (1975: 162), who describes the discussions on the rational dress in detail.

92. Exact production figures are difficult to obtain, but Hounshell (1984: 192) estimates that the American bicycle industry produced, at its peak in 1896–1897, well over a million cycles per year.

93. About Holland it is said that "competition from this quarter must not be underrated, more especially when it is considered that these manufacturers are, in their own country, unfettered with patent monopolies, and have the advantage of cheap labour" (Engineer, 1890b: 138).

94. Williamson (1966: 116–119) describes the operations of the notorious Terah Hooley, who successively floated Raleigh, Singer, Humber, Swift, Trent, and Starley Brothers. The following may illustrate the exhilarated atmosphere in those days: Hooley obtained control over Dunlop's original patent (which the latter had sold in 1889 for £700) for £3 million and immediately resold it publicly for £5 million.

Chapter 3

1. Baekeland normally used his first two Christian names only, spelled as Leo Hendrik. For his complete name, see his birth certificate: Gent, 16 November 1863 (Gillis, 1965: fig. 3).

2. See, for example, Brandenburger (1938), Kaufman (1963), Ingenieursblad (1964), Gillis (1965), Jewkes et al. (1969), and Thinius (1976). These accounts all follow closely the pattern laid by Baekeland (1909a,b) himself in the two papers in which he reported the invention of his plastic.

3. A characterization reportedly given by one of Baekeland's friends; quoted by Kaufmann (1968: 4).

4. The dramatic events of the 1970s within the Baekeland family have only added to the heroic stature of self-made man Leo Baekeland. See Robins and Aronson (1985) for a semijournalistic account of the psychosocial events that supposedly formed the background for Baekeland's grandson killing his mother.

5. I found this quote in an anonymous biographical sketch in the Baekeland files at the National Museum of American History, Washington, D.C. This sketch is probably a chapter from a longer "company history" of the Union Carbide Corporation, into which Baekeland's "Bakelite Company" merged in 1958.

6. Baekeland in a letter to C. F. Chandler, 28 December 1915 (Baekeland Papers: CD II-16/box 8). See note 8.

7. In 1937 the Bakelite Corporation made the film "The Fourth Kingdom," in which the production and various applications of Bakelite were shown in much

detail. The film starts with a sonorous voice arguing along the following lines: "Mineral, Vegetable, Animal—the three Kingdoms of Nature. They served mankind for ages, but now our modern industrial society finds them insufficient to fill all needs. It has to turn elsewhere; it turns to the Fourth Kingdom—Plastics" (followed by a crescendo of symphonic music, of course). I am grateful to Robert Bud, Science Museum London, for showing me excerpts of this film.

8. Of course I did not do a chemical analysis of my pan's handles, so they may be produced from a different molding material. What is significant here is that the accompanying brochure *claims* that they are made of Bakelite.

9. The Baekeland Papers are kept at the Archives Center of the National Museum of American History, Smithsonian Institution, Washington, D.C. They are divided into two main parts: the "Collection Divisions" I through X (boxes 1–16) and the "Series" 1 through 8 (boxes 1–20). The Collection Divisions contain primarily personal and business correspondence and some printed sources, while the Series contain laboratory notebooks, diaries, and biographical and photographic materials. Carl B. Kaufmann's M.A. thesis (1968), also on file at the museum, offers a valuable guide to these papers; Kaufmann conducted interviews with relatives and former collaborators of Baekeland.

10. In saying this, I am implicitly following the common parlance of the early twentieth century in which "plastics" were divided into two categories: "natural plastics" such as shellac, horn, and rubber, and "synthetic plastics" such as Bakelite. Actors in the period I studied all used this classification. Systematic arguments (chemical or otherwise) do not unambiguously dictate whether materials such as shellac and rubber should be subsumed under the same class of materials as, for example, Bakelite. For a brief analysis of such arguments, see Friedel (1983: 24). The grouping of the two classes of materials into the category of plastics is probably a reflection of the goal for which most of the early synthetic materials were developed: to find a substitute for the expensive natural "plastics."

11. See Andes (1911) for a "modern" guide to the processing of natural plastics.

12. Shellac is unique among the natural plastics because of its animal origin. The cochineal-like insect *Tachardia lacca*, living in the lac tree *Butea frondosa*, excretes a resinous deposit that hardens into shellac. An important source was located some 500 kilometers south of Calcutta, India.

13. The vulcanization process was patented by Charles Goodyear (U.S.) and Thomas Hancock (U.K.) in the early 1840s.

14. William Hawes, Chairman of the Council of the Society of Arts, meeting on 20 December 1865, *Journal of the Society of Arts*, Vol. 14 (1865, No. 683): 85.

15. C. F. Schönbein in a letter to M. Faraday, 27 February 1846; cited by Kaufman (1963: 21).

16. Kaufman (1963) dedicated his history of the first century of plastics "to the memory of Alexander Parkes, 1813–1890" and gives many details about early

plastics history in Britain, especially as connected to the work of Parkes and Daniel Spill.

17. Parkes (1855): 1. Gutta-percha is similar to rubber, but has one crucial difference: It contains oxygen. This has at least two important consequences. First, it cannot be vulcanized and thus rendered more flexible and stronger. Second, it is resistant to corrosion and thus very suitable for coating electrical cables. Gutta-percha remained the most important insulation material for cables until the 1920s and 1930s.

18. These were mentioned in the explanatory leaflet alongside the Parkesine exhibits at the International Exhibition in 1862 (Class IV, Official Catalogue, No. 1112; reproduced in Kaufman, 1963: 22).

19. Alexander Parkes, testimony in *Spill v. Celluloid Manufacturing Company* (Circuit Court, Southern District of New York, 1880), cited by Friedel (1983: 29).

20. Quoted by Friedel (1983: 35).

21. Bowker (1992, 1994) analyzes patents on three levels—the courtroom battle, the company strategy, and the "official" technical history—to demonstrate that patents typically have two complementary roles: to describe a past reality (and impose that description) and to create a present one (and impose that creation). See also, for example, Hughes (1971, 1983), Hounshell and Smith (1988), and Carlson (1991). I will return to the role of patents in chapter 5.

22. *New York Times*, 19 September 1875, p. 7, mentioned by Friedel (1983: 97).

23. *Newark News*, 26 June 1949, mentioned by Friedel (1983: 96).

24. See also Vanderpoel (1914) for more anecdotal examples.

25. Baeyer (1872a,b,c). In subsequent years some of his collaborators further reported on similar reactions.

26. I have translated all German and French quotations.

27. The term "formaldehyde" is generally used in a broad sense. Chemically pure formaldehyde, CH_2O, is a gas at room temperature. The formaldehyde used by the chemists described in this study was a watery solution of formaldehyde that also contained variable proportions of methylalcohol. In this solution formaldehyde is present in various forms, for example, as methylenglycol and as methylal, and in various polymers of the same. The varied content of this formaldehyde solution is such that Baekeland considered it perfectly acceptable "that the work of Ad. Baeyer and his pupils on condensation products of phenols and formaldehyde was carried out by means of the latter's methylene representatives" (Baekeland, 1911b: 933).

28. The early history of formaldehyde and its applications is described by Bugge (1931, 1943). Some pro-German bias is not improbable; in his last paper the author sets out to give the industrial chemist Mercklin the recognition he deserves

because, according to Bugge, formaldehyde is so important for the plastics industry and other "*lebens- und kriegswichtigen Gebieten*" (life- and war-relevant domains).

29. This was reported to *Scientific American* (1892b) by Professor C. V. Boys, F.R.S., who continued his letter with an account of some additional experiments he had carried out. These showed that pieces of the celluloid buttons were ignited more easily than "a common wax match."

30. See also Claessen's (1905) patent, mentioning the difficulties of producing synthetic camphor and the high price of both natural and synthetic camphor.

31. Casein plastics were sold under names such as Galalith, Syrolith, Erinoid, Lactit, Alladinit, Zoolit, and Protoloit (Bonwitt, 1933). They were successful for a narrow range of applications. For example, by the early 1920s most buttons were manufactured from these casein plastics (Friedel, 1983).

32. Emile-Henri Fayolle added large amounts of glycerine to the reacting bodies and claimed that he had produce materials that could be substituted for rubber and gutta-percha (Fayolle, 1903, 1904). Fritz Henschke used alkalic condensation agents instead of acids as most of the others had done (Henschke, 1903).

33. Carlson and Gorman (1990) point, with Staudenmaier (1985), at the lacuna of the individual creative process of the inventor in historical studies of technology. Combining history of technology and cognitive psychology, Carlson and Gorman present a conceptual framework of mental models, heuristics, and mechanical representations to fill this lacuna. The difference between the proposed technological frame and their set of concepts is that the latter exclusively relates to the individual inventor, whereas the former is interactionist. For a review of cognitive psychology contributions to the art of invention, see Weber and Perkins (1989), who investigate simple inventions, such as the Stone Age knife, to extract principles of heuristics.

34. The analogy with Wittgenstein's analysis of following a rule is obvious. A technological frame constitutes a form-of-life.

35. Frames of meaning (Collins and Pinch, 1982) have cognitive and social but certainly no material elements. Jenkins's (1975) technological mind-set and the various concepts discussed in Laudan (1984a) all have a primarily cognitive character. The "mechanical representations" make the conceptual framework of Carlson and Gorman (1990) less purely cognitive, but this is too individualistic to be useful in connection with the concept of "relevant social group."

36. This aspect is an important difference between technological frame and the related concepts used by other students of technological development. Disciplinary matrix (Kuhn, 1970), frame of meaning (Collins and Pinch, 1982), technological paradigm (Dosi, 1982, 1984; Gutting, 1984; Van den Belt and Rip, 1987), focusing devices (Rosenberg, 1976), technological guide-posts (Sahal, 1981), technological style (Hughes, 1983, 1987), technological tradition (Constant, 1980, 1984; Laudan, 1984b), technological orientation complex (Weingart,

1984), technological mind-set (Jenkins, 1975), and technological regime (Nelson and Winter, 1977, 1982; Van den Belt and Rip, 1987) all apply exclusively to communities of scientists or engineers.

In addition, Tom Hughes's usage of the term "technological style" is primarily meant to account for national differences in technology, which places the concept on a much higher level of aggregation than the intended level for technological frame.

37. For most of this brief sketch, I have drawn on the extensive history of photography and the related (American) industry by Jenkins (1975).

38. For this episode, see Gillis (1965).

39. A copy of the legal document of the *Constitution de la société en commandite simple Dr. Baekelandt et Compagnie* is in the Baekeland Papers (CD VII, box 14).

40. See Kaufmann (1968).

41. Letter by Baekeland to Prof. A. Wagener, l'Administrateur-Inspecteur, 2 June 1889 (Gillis, 1965: 34).

42. Letter by Baekeland to l'Administrateur-Inspecteur, 23 or 24 June 1889 (Gillis, 1965: 36).

43. Letter from the Dean of the Faculty of Sciences to the Administrateur-Inspecteur, 22 July 1889 (Gillis, 1965: 38).

44. Letter by the Recteur to the Administrateur-Inspecteur, 24 July 1889 (Gillis, 1965: 39).

45. For example, the Administrateur-Inspecteur wrote on 6 August 1889 to the Minister that "the matter of Baekeland's wedding makes the date of his departure uncertain" (Gillis, 1965: 37).

46. See the wedding certificate reproduced by Gillis (1965: fig. 36).

47. Ministre de l'Intérieur et de l'Instruction Publique, 25 September 1889 (Gillis, 1965: 40).

48. For detailed information about the Anthony firm and its merging with the Scoville Company to form Ansco Company, see Jenkins (1975).

49. Letter by Baekeland to the Administrateur-Inspecteur, 5 November 1889 (Gillis, 1965: 40).

50. Letter from the Ministre de l'Intérieur et de l'Instruction Publique to the Administrateur-Inspecteur, 7 October 1889 (Gillis, 1965: 42); Arrest by the Ministre de l'Intérieur et de l'Instruction Publique, 30 November 1889 (Gillis, 1965: 43).

51. This research is reported in one of Baekeland's first notebooks (Kaufman, 1968: 31).

52. Jenkins (1975) refers to a testimony by F. A. Anthony in a patent suit, *Goodwin v. Eastman.*

53. For technical details of the various processes involved, and for the entrepreneurial developments, see Jenkins (1975).

54. Kaufmann (1968: 32), citing Baekeland family papers.

55. The relevant laboratory notebooks show that Baekeland prepared more than fifty different formulas. He evaluated almost all of them for smoothness, coatability, tonal qualities, sensitivity, shelf life, solubility, and permanence of image (Kaufmann, 1968).

56. Baekeland in *Photographic Journal,* November 1930, quoted by Kaufmann (1968: 43).

57. Quoted by Baekeland (1916: 185) without disclosing the author's name.

58. Another anonymous writer quoted by Baekeland (1916: 185).

59. See, for example, Baekeland (1897), in the Baekeland Papers, CD VI, box 13.

60. Letters by Baekeland to his Belgian friend Dr. Edward Remouchamps, 17 March 1899 and 30 May 1899 (Gillis, 1965: 63–64).

61. Letter by Baekeland to Remouchamps, 18 June 1899 (Gillis, 1965: 65).

62. See, for example, Kaufmann (1968).

63. Letter by Baekeland to Remouchamps, 18 June 1899 (Gillis, 1965: 65).

64. Letter by Baekeland to Remouchamps, 6 July 1899 (Gillis, 1965: 65).

65. Option Agreement between Eastman, Jacobi, Baekeland, and Hahn, 8 July 1899, quoted by Jenkins (1975: 201). Baekeland's allusions to Remouchamps about "several millions" can only be understood as referring to Belgian francs. This repeated mentioning of many millions, even by Baekeland's relatives in later years (Kaufmann, 1968: 48), probably induced the many popular histories of Baekeland's "one million sell to Eastman Kodak" (cf. the introduction to this chapter).

66. Baekeland in a letter to Remouchamps, 6 July 1899 (Gillis, 1965: 65).

67. Ibid.

68. The letters from Baekeland to Remouchamps show this shift in plans as the date of departure drew nearer. On 18 June 1899, the trip is still envisioned as "one or two years with my whole family"; on 6 July the length has become unspecified, but the plan still calls for an all-out holiday, except for a visit to the Paris Exhibition; on 18 July it is "only eight or ten months, provided that I do not want to return earlier"; on 29 July, Baekeland reported that "I have changed

my plans as to visit from December until May some German and Swiss laboratories, and only thereafter to make our small journey" (Gillis, 1965: 66–71). They probably stayed in Europe about half a year.

69. Letter by Baekeland to Remouchamps, 18 November 1901 (Gillis, 1965: 74)

70. See Haynes (1945b) for an overview of the early history of electrochemistry.

71. For more details on this case, see Hughes's biography of Sperry (1971: 89–98) and the rather straightforward history of Hooker's business by Thomas (1955). The more comprehensive company history in Haynes (1949: 210–215) devotes little attention to the early period.

72. Laboratory notes that document Baekeland's research in this period (between 26 April 1904 and 3 June 1904) are at the back of the Bakelite laboratory notebook "BKL III" (Baekeland Papers; S I/box 2).

73. See especially Kaufmann (1968) and Thomas (1955). In a festive speech on the occasion of Baekeland's receipt of the Perkin Medal, Hooker (1916) described in anecdotal form some of the problems encountered during the Brooklyn research.

74. A report of this presentation by Baekeland before the New York Section of the Society of Chemical Industry was published as "The New Electrolytic Alkali-Works at Niagara Falls" in *Electrochemical and Metallurgical Industry*, June 1907.

75. For a more detailed account of the research in this Brooklyn period, see the anonymous report "The New Electrolytic Alkali-Works at Niagara Falls" (1907).

76. This is an important difference between, for example, an actor with a low inclusion in a technological frame and the notorious "marginal scientist" as adequately criticized by Gieryn and Hirsch (1983). The marginality concepts they discuss, however, are one-dimensional. For example, in one study a scientist is considered marginal if he recently migrated from another field, whereas in another study "marginal" is operationalized as "being young." The different dimensions yield contradictory results: By tactically choosing the dimension to characterize a specific scientist, all ninety-eight scientists of Gieryn and Hirsch's sample turned out to be marginal in some sense.

77. See, for example, his (1907c) article in *Science* on "the danger of over-specialization," which was based on a talk he gave at the New York section of the American Chemical Society on 5 April 1907; see his diary #1, p. 4 and in later entries (Baekeland Papers; S8/box 1).

78. Baekeland (1907k) reported on this project in a series of articles in *The Horseless Age.* He provided detailed information about such different aspects as the choice of motorcar and accessories, the selection criteria for the handyman-driver who was to accompany the family, and the way to plan for fuel supplies at the various stopping places. The journey was described with a light touch and an eye for detail, and was illustrated with his own photographs.

79. Baekeland's diary #1, 17 April 1907, p. 10 (Baekeland Papers; S8/box 1).

80. This project was Baekeland's major interest in the spring of 1907, although still one of many. See his diary #1, 16 May 1907, pp. 23–25 and later entries (Baekeland Papers; SI/box 1)

81. See diary #1, 28 April and 18 May 1907, pp. 20 and 26 (Baekeland Papers; S8/box 1).

82. See Kaufmann (1968), referring to T. J. Fielding, *History of Bakelite Limited* (London, 1947).

83. See letter by Thurlow to Baekeland, 30 November 1904 (Baekeland Papers; CD II-1/box 7).

84. "Memorandum of agreement between L. H. Baekeland and Nathaniel Thurlow concerning process of manufacturing alcohol soluble gums," handwritten agreement, signed by Baekeland and Thurlow and countersigned by a notary public on 12 February 1907 (Baekeland Papers; CD VII-6/box 14).

85. For example, on 8 April 1907 a letter is received from Hooker, who objected to the smell of Novolak. Baekeland advised Thurlow to add specific oils to mask the smell. See his diary #1, p. 6 (Baekeland Papers; S8/box 1).

86. On 24 April 1907 Baekeland visited Berry Brothers in Detroit to see the upscaling to a commercial manufacturing process of Novolak. See his diary #1, pp. 15–16 (Baekeland Papers; S8/box 1).

87. In his diary, Baekeland wrote (18–21 June 1907: p. 47): "All this work has been carried out while Thurlow was in Detroit showing to Berry Brothers how to [?] Novolak. I am sure he will be surprised to hear about all what I have accomplished in so short a time" (Baekeland Papers; S8/box 1).

88. Previous research was carried out on, for example, 28 April. The research of 14–18 June is reported on the back pages of what is now labeled "Laboratory Notebook BKL II": pp. 5–7 (Baekeland Papers; SI/box2).

89. Baekeland's diary, 18–21 June 1907, pp. 45–46 (Baekeland Papers; S8/box 1). The name "Bakalite" was changed into "Bakelite" in 1908, following a suggestion by Baekeland's patent attorney Townsend. The patent he mentioned here was completed by early July.

90. Lab Notebook "BKL I," p. 10 (Baekeland Papers; SI/box 2).

91. Lab Notebook "BKL I," 19 June 1907, p. 10 (Baekeland Papers; SI/box 2).

92. Lab Notebook "BKL I," 19 June 1907, pp. 10–11 (Baekeland Papers; SI/box 2).

93. Lab Notebook "BKL I," 20 June 1907, p. 12 (Baekeland Papers; SI/box 2).

94. Lab Notebook "BKL I," 20 June 1907, p. 13 (Baekeland Papers; SI/box 2).

95. Baekeland's diary, 22 June 1907, p. 47 (Baekeland Papers; S8, box 1).

96. Baekeland's diary, 23 June 1907, p. 51 (Baekeland Papers; S8/box 1).

97. Baekeland's diary, 24 June 1907, p. 51 (Baekeland Papers; S8/box 1).

98. Lab Notebook "BKL I," 20 June 1907, p. 13 (Baekeland Papers; SI/box 2).

99. Lab Notebook "BKL I," entry dated 20 June 1907, p. 41 (Baekeland Papers; SI/box 2). This conclusion must have been written down on 21 June 1907.

100. Lab Notebook "BKL I," 20 June 1907, p. 43 (Baekeland Papers; SI/box 2). Again, this observation must have been made on 21 June 1907.

101. Lab Notebook "BKL I," 20 June 1907, p. 22 (Baekeland Papers; SI/box 2).

102. Lab Notebook "BKL I," 23 June 1907, p. 52 (Baekeland Papers; SI/box 2).

103. Lab Notebook "BKL I," 23 June 1907, pp. 54–55 (Baekeland Papers; SI/box 2).

104. This account is given in a seven-page untitled manuscript, dated January 1916, which is in the Baekeland Papers. It is a strange manuscript: the first two-and-a-half pages are written about Baekeland in the third person, but then the style switches to a personal account by "I," obviously Baekeland. The content of the manuscript, but for the switch in style, is almost literally identical to the section "Bakelite" in the Perkin Medal Presentation Address by Chandler (1916: 149–151). Baekeland thus had been prompting Chandler quite extensively, including the jubilant and heroic episodes of the story. And he was very successful: all stories about the "discovery of the first synthetic plastic" have followed this scheme.

105. Unpublished manuscript dated January 1916 (see note 77): p. 3.

106. Baekeland's diary, 7 April 1907, p. 5 (Baekeland Papers; S8/box 1).

107. Unpublished manuscript dated January 1916 (see note 77): p. 4.

108. Ibid., p. 3.

109. Lab Notebook "BKL I," 11 July 1907, p. 73 (Baekeland Papers; SI/box 2); Baekeland's diary, 11 July 1907, p. 95 (Baekeland Papers; S8/box 1).

110. Baekeland's diary, 11 July 1907, p. 95 (Baekeland Papers; S8/box 1).

111. Baekeland's diary, 11 July 1907, p. 96 (Baekeland Papers; S8/box 1).

112. Lab Notebook "BKL I," for example the entry dated 18 July 1907, p. 100 (Baekeland Papers; SI/box 2).

113. Lab Notebook "BKL I," 23 June 1907, p. 47 (Baekeland Papers; SI/box 2).

114. Lab Notebook "BKL I," 20 June 1907, pp. 30–32 (Baekeland Papers; SI/box 2). This must have been noted at the end of Friday 21 June.

115. Lab Notebook "BKL I," 18 July 1907, pp. 96–99, and 20 July, pp. 104–110 (Baekeland Papers; SI/box 2).

116. Baekeland's diary, 19 July 1907, p. 100 (Baekeland Papers; S8/box 1).

117. Lab Notebook "BKL I"; see for example the entry on 13 July 1907, pp. 75–82 (Baekeland Papers; SI/box 2).

118. Lab Notebook "BKL I," 14 July 1907, p. 83 (Baekeland Papers; SI/box 2).

119. Baekeland's diary, 25 July 1907, p. 105 (Baekeland Papers; S8/box 1); Lab Notebook "BKL I," 25 July 1907, p. 114 (Baekeland Papers; SI/box 2).

120. Lab Notebook "BKL II," 31 August 1907, p. 177 (Baekeland Papers; SI/box 2).

121. Baekeland's diary, 3 September 1907, p. 121 (Baekeland Papers; S8/box 1).

122. Lab Notebook "BKL II," 3 September 1907, p. 217 (Baekeland Papers; SI/box 2).

123. Baekeland's diary, 3 October 1907, p. 144 (Baekeland Papers; S8/box 1).

124. Baekeland's diary, 24 October 1907, pp. 153–154 (Baekeland Papers; S8/box 1).

125. See, for example, in Baekeland's diary, 17 October 1907, pp. 148–149; 19 October 1907, p. 151; 24 October 1907, p. 153 (Baekeland Papers; S8/box 1).

126. Baekeland's diary, 30 October–2 November 1907, pp. 160–167 (Baekeland Papers; S8/box 1).

127. Baekeland's diary, 2 December 1907, p. 179 (Baekeland Papers; S8/box 1).

128. See, for example, Baekeland's description of some test samples he prepared for the Norton Emeri Co.: "I send you a wooden slab on which I spread a mixture of 3 parts of fine abrasive and one part Bakalite. I took no pain whatever to make a smooth surface. I also send a small disc of Bakalite Emeri which was hardened on glass." Letter to Mr. Saunders of Norton Emeri Co., copied in handwriting in Baekeland's diary, 2 November 1907, pp. 163–167 (Baekeland Papers; S8/box 1): 164.

129. Baekeland's diary, 6 December 1907, p. 180 (Baekeland Papers; S8/box 1).

130. Ibid.

131. Ibid., pp. 180–181.

132. Baekeland's diary, 13 January 1908, p. 194 (Baekeland Papers; S8/box 1).

133. Baekeland's diary, 15 January 1908, p. 194 (Baekeland Papers; S8, box 1).

134. Account book (Baekeland Papers; SI, box 2).

135. Diary #2, 18 March 1908, pp. 35–36.

136. Ibid., pp. 37.

137. Baekeland's diary #3, 4 February 1909, p. 112 (Baekeland Papers; S8, box 1).

138. Baekeland's diary #3, 5 February 1909, p. 114 (Baekeland Papers; S8, box 1).

139. Baekeland based his account primarily on the patent literature, which explained his missing of Michael's purely academic work.

140. Work in the 1920s established the idea of macromolecular chemical substances, consisting of chains of thousands to millions of ordinary molecules, and identified the mechanisms of polymerization (Staudinger, 1926, 1961). In 1953 Hermann Staudinger would receive the Nobel Prize for this work.

141. Baekeland wrote in his diary: "More letters of requests about Bakelite came pouring in" and "Big mail again"; diary #3, 8–9 February 1909, pp. 124–125 (Baekeland Papars; S8, box 1).

142. The account book (Baekeland Papers; S1, box 2). See p. 23 for prices of the various products:

	More than 300 lbs.	More than 150 lbs.	Less than 50 lbs.
Solid A	31¢	40¢	50¢
Pasty A	31¢	40¢	50¢
Dissolved A	31¢	40¢	50¢
Liquid A	31¢	40¢	50¢
High-voltage insulating Bakelite	45¢	50¢	65¢

143. This fire broke out on the night of 1 to 2 March 1909. There was much material damage, but Baekeland was "glad my notes were in the house" (diary #3, 2 March 1909, p. 158).

144. This quotation is from an anonymous, undated piece, "Chapter II. The Bakelite Story and Dr. Baekeland," p. 21, which was found by Carl Kaufman in the files of the Union Carbide Corporation and transferred to the Baekeland Papers (CD VI-C-6, box 12).

145. Baekeland to Walter Damrosch, brother in-law-to Dr. Wiechmann, 25 November 1910 (Baekeland Papers, CD I, box 5, folder #I-109).

146. This shift was not made wholeheartedly, as he complained in public (Baekeland, 1916, 1932). His publications bear witness to the double character of his ambitions: In later papers he dealt with the theoretical issue of Bakelite's chemical identity (Baekeland, 1913; Baekeland and Bender, 1925), but all of his later patents merely described further applications of Bakelite.

147. Anonymous piece, "Project for the Formation of a Bakelite Company and Possible Sources of Income," undated but probably February 1910 (Baekeland Papers; CD VII-6, box 14).

148. Quotation from an anonymous, undated piece, "Chapter II. The Bakelite Story and Dr. Baekeland," p. 25; see note XX (Baekeland Papers; CD VI-C-6, box 12).

149. See, for example, Bayer (1907) and Helm (1907).

150. Baekeland (1909e: 2007); my translation.

151. The most detailed account of this patent struggle can be found in the anonymous "Chapter II. The Bakelite Story and Dr. Baekeland" (Baekeland Papers; CD VI-C-6, box 12).

152. The dry/wet variable was introduced by Redman et al. (1914) to distinguish their resin (Redmanol) from other synthetic resins.

153. The General Bakelite Company had filed a Bill of Complaint on 18 September 1917 against one of the users of Redmanol, the General Insulate Company. During the trial the Redmanol Company stood in fact behind the defendant. The trial started 31 March 1919 (in the U.S. District Court, Eastern District of New York) and lasted several weeks in open court; Judge Chatfield rendered his decision on 2 August 1921.

154. Editorial, *Journal for Chemical and Metallurgical Engineering*, Vol. 26, June 21, 1922: 1152.

155. Judge T. I. Chatfield, decision in the suit of the General Bakelite Company against General Insulate Company, U.S. District Court, Eastern District New York, 2 August 1922.

156. See the special issue of *Bakelite Review* (Vol. 7, No. 3, 1935), "Silver Anniversary Number, 1910–1935," pp. 19–22.

157. Lawrence Byck as quoted in "Chapter II. The Bakelite Story and Dr. Baekeland" (Baekeland Papers; CD VI-C-6, box 12): 26.

158. Lawrence Byck as quoted in "Chapter II. The Bakelite Story and Dr. Baekeland" (Baekeland Papers; CD VI-C-6, box 12): 27.

159. See tables 5 and 6 in Byck (1952: 21, 22).

160. See Meikle (1986). Nevertheless, also in America special campaigns were directed at improving the plastics image. See the next paragraph on industrial design.

161. See Haynes (1945b: 382); and Haynes (1945a), chapter 12.

162. This analysis can be only brief and will be restricted to a few remarks of immediate relevance to the development of Bakelite. For a more comprehensive

view of the fascinating subject of industrial design and the development of technology in this period, see Frankl (1930), Cheney and Candler Cheney (1936), Meikle (1979), and Wilson et al. (1986).

163. On the cover of *Plastics and Molded Products* 8 (Sept. 1932), cited by Meikle (1979: 80).

164. For a discussion of these styles, see Wilson et al. (1986).

165. The analysis in this paragraph is based primarily on the fascinating account by Meikle (1979).

166. Examples were the exhibition "The Machine Age in America, 1918–1941" at the Brooklyn Museum (Wilson et al., 1986), the Dutch exhibition "Industrie en Vormgeving" in the Amsterdam Stedelijk Museum (Stedelijk Museum Amsterdam, 1985), and the pioneering and comprehensive Bakelite exhibition "Bakeliet, techniek|vormgeving|gebruik" in the Rotterdam Boymans-Van Beuningen Museum (Kras et al., 1981).

167. The study was carried out in 1981 by the market survey bureau Intomart Qualitatief BV, Hilversum, the Netherlands, and is reported in Kras et al. (1981).

168. As already mentioned, on this point the situation was different in Germany as compared to the United States, where the word "plastic" was definitely better known than "Kunstharz" in Germany.

169. Clegg (1989) points at Giddens's implicit ontology, inherited from "social construction of reality" theorists such as Alfred Schutz and Berger and Luckmann (1966). Giddens's grand synthesis between structure and agency is not very successful—structuration theory is ultimately subjectivist, staying too close to the individualist and voluntarist side of the dualism of action and structure. See Hagendijk (1990) for a comparison between constructivism and structuration theory.

170. For an introduction to the social interactionist framework, see Blumer (1969) and Charon (1985).

171. For an introduction to semiotics, see Greimas (1987, 1990). Akrich (1992) and Latour (1992a) have applied this semiotic perspective to the analysis of technology.

172. Discussions with Jessica Mesman and Annemarie Mol have been helpful on this point.

Chapter 4

1. It is necessary to stress the "obvious," naive, and *a prima vista* character of these differences. After all, the argument in this book builds to exactly the opposite conclusion—that in an interesting way the cases are very similar, similar

enough to be understood with the same SCOT model. See Woolgar (1983, 1991) for a critical analysis of this kind of ironic argument.

2. The title for this chapter is derived from a conversation between Edison and D. M. Moore, a General Electric employee, as reported by Hammond (1941: 262). Moore researched gaseous discharge lamps because, he said, he hoped to produce an *imitation of daylight.* "What's the matter with my light?" Edison inquired. "Too small, too red, and too hot," replied Moore.

3. For a comprehensive historical overview of the electric lighting industry and technology, see Bright (1949). See also, especially for the European side of the story, Meinhardt (1932).

4. The Sherman Act was passed in 1890.

5. General Electric's president, Gerald Swope, launched an ambitious proposal for legitimizing a cartelized economy in 1931 (Mueller, 1983).

6. The trademark adopted in 1909 by General Electric was based on the Persian god of light, Ahura Mazda (Nye, 1985).

7. For detailed analyses of the early history of the American electric industry, and especially the origins of General Electric and Westinghouse, see Passer (1953), Carlson (1991), Chandler (1977), Hammond (1941), and Reich (1985).

8. This company often seems to be confused (Nye, 1985; Wise, 1985; Reich, 1985) with the National Electric Light Association (NELA), which was organized in 1885 to promote the interests of electric utilities (Bright, 1949: 146; Nye, 1990). NELA organized, often covertly, promotional campaigns and political lobbies (Nye, 1985: 135–147). Because revelations about some of these campaigns—in U.S. Senate hearings from 1928 to 1934—were detrimental to the utilities' public image, the organization was disbanded in 1933 and replaced by the Edison Electric Institute (Hirsch, 1989: 213; anon., 1949). The confusion may be caused by the fact that the two organizations had the same originator and manager, F. S. Terry. Moreover, the name NELA was also used for the General Electric research park where the research facilities of the National Electric Lamp Company had been located: Nela Park, Cleveland, Ohio.

9. The Justice Department contended that, because of the large amount of stock General Electric had in National, it "had combined and conspired to restrain commerce by concealed stock ownership of bogus independent companies" (quoted by Rogers, 1980: 96). The resulting decree was probably, quite contrary to the Justice Department's intention, anticompetitive in its effects.

10. The ductile tungsten filament was the outstanding achievement of the General Electric Research Laboratory. It solved the problem of the earlier (non-ductile) tungsten filament's extreme brittleness.

11. See Stocking and Watkins (1946), especially chapter 8, and Bright (1949).

12. This suit followed an investigation that the Justice Department had started at the request of General Electric, which wanted to vindicate itself publicly after

accusations were launched by several small companies that GE used unfair tactics, made exorbitant profits, and conducted legal harassment.

13. Stocking and Watkins (1946: 327–328) quote in full a revealing letter by the vice-president of General Electric to the president, in which plans are outlined for negotiating a new agreement with Westinghouse. See also Bright (1949: 258–259) for a summary of the principal provisions of the A and B licenses granted by General Electric after 1927.

14. Between 1917 and 1930 Hygrade acquired several other lamp firms, and after a merger in 1931, it changed its name to Hygrade Sylvania Corporation. In 1942 the name was changed to Sylvania Electric Products Inc.

15. The prewar cartel, registered in 1903 in Berlin as Verkaufsstelle Vereinigter Glühlampenfabriken G.m.b.H. (VVG), regulated the market shares for carbon filament lamps.

16. It soon became clear that during the war the situation had changed profoundly and that this cartel would not be able to control the market as effectively as its predecessor VVG had done. To regain control, the electric lamp industry established the General Patent and Development Agreement and the Phoebus S.A. Compagnie Industrielle pour le Développement de l'Éclairage. This new cartel also specified patent licensing and exchange of production experience and knowledge. See Meinhardt (1932) for more details.

17. The Mazda Companies kept strict control over their agents' network. They had, for example, a detailed price maintenance system with a minimum resale price for the lamps. See Bowman (1973) for a discussion of the history of the incandescent lamp price-fixing system and the 1926 lawsuit on this issue (*United States v. General Electric*). See Telser (1958) for a discussion of the rationale behind this policy.

18. For a brief account of the history of electric utilities in the United States before the Second World War, see Hirsch (1989), whose main concern, however, is the postwar era.

19. See "Stockholders of Electrical Testing Laboratories (November 15, 1940)" (Committee on Patents, 1942: 4942–4943).

20. RLM stands for Reflector and Lighting Equipment Manufacturers.

21. In doing this I will frequently put on my old hat of physics teacher and apply the vocabulary of "ready-made science" (Latour, 1987), because the demonstration of the socially constructed character of these scientific facts is not the aim of this book. For a comprehensive survey of the relevant physics, see Elenbaas (1959a).

22. This efficiency, the "overall efficiency" specifying the efficiency of a lamp in transforming electrical power input into light output, was measured in the units of lumens/watt or lightwatts/electric watt (Moon, 1936), but units of foot-

candles/watt were also used. The last was, strictly speaking, not correct, because footcandles are a measure of the illumination of surfaces, whereas lumens measure luminous flux from a lamp.

23. So the relevant social groups in this case used "low-voltage fluorescent lamp" as synonymous for "hot-cathode fluorescent lamp" and "high-voltage fluorescent lamp" as synonomous for "cold-cathode fluorescent lamp." The voltage vocabulary highlighted the possibility of installing the lamp on normal (low-voltage) mains, or the need for special installations. The cathode vocabulary focused on a crucial constructional detail of the lamp. The use of alternative names by specific actors in certain situations is of course not accidental and may be illuminating for the artifact they are working with. When not referring to specific historical situations, I will mostly use the the "high-voltage" appellation, because that will remind readers most directly of the application possibilities and problems of the lamps.

24. See Hutter (1988) for a history of mercury vapor lamps, especially at Philips, the Netherlands.

25. The lamp is often called the "Cooper-Hewitt lamp," using Hewitt's middle name.

26. Research along these lines continued well into the 1930s. See, for example, LeBel's patent (1929).

27. The German physicist Von Linde developed the countercurrent and cascade techniques to cool gases in 1896. Claude's work resulted in an industrial process for the liquification of air; see Claude (1909, 1913) for reviews. Kamerlingh Onnes achieved the liquefication of helium in 1908, for which he received the Nobel Prize for physics in 1913.

28. For a discussion of George Claude (1870–1960) as a colorful "scientist, industrialist, scientific popularizer and French politician," see Blondel (1985). Claude's (1957) scientific autobiography is also illuminating.

29. See, for example, such "handbooks for the lighting engineer" as Möbius (1932) and Sewig (1938).

30. See Hutter (1988) for a detailed account, especially for research by the Dutch Philips company.

31. General Electric had marketed in the early 1930s the "Mazda Daylight lamp," an incandescent lamp with a blue bulb that produced much whiter light than the ordinary tungsten filament lamp, but with the inevitable losses because of the filtering (Luckiesh et al., 1938).

32. See, for example, the patent issued to Ryde (1932), an employee of the British General Electric Company Ltd. For a brief and rather sketchy review of the gas discharge and fluorescent lamp research at the British GEC Research Laboratories, see Clayton and Algar (1989).

33. See, for example, the patents by Koch (1929, 1933), Fischer (1933), Jenkins (1935), Patent-Treuhand (1935), and Fritze and Rüttenauer (1935).

34. See, for example, another patent issued to Ryde (1934) and a patent by a German optical glass manufacturing firm (Jäckel, 1925).

35. A. Witz's conclusion after his efficiency measurements of fluorescent lighting, quoted by Berthier (1908: 186–187).

36. Much research was also devoted to the development of the fluorescent materials themselves; see, for example, the British General Electric patents by McKeag and Randall (1936, 1937) and Randall (1936), and the patents issued to Leverenz (1936), OSA (1936), and Cox (1937). Randall (1937) reviews this research. The French engineer Risler (1922a,b, 1923, 1925) patented some fluorescence discharge lamps, one of which had a hot cathode and was thus suitable for relatively low voltages. Some of the lamps had their fluorescence on the inside, others on the outside of the discharge tube. See Claude (1939) for a review of the French work.

37. See District Judge V. L. Leibell's decision not to grant the government's motion for leave to intervene in this combined patent suit, U.S. District Court, Southern District of New York, "General Electric Company, plaintiff, against Hygrade Sylvania Corporation, defendant, Raytheon Manufacturing Company, intervener-defendant, United States of America, applicant for intervention," 29 May 1942 (also published by Committee on Patents, 1942: 4929–4934); and Bright (1949).

38. During the very first months after its commercial release, the low-voltage fluorescent lamps were also called "new fluorescent Lumiline lamps" (anon., 1938b). Half a year later the new lamps were described as "tubular and somewhat resembling the familiar Lumiline lamps, with base contacts at both ends" (anon., 1939). But it soon stabilized enough to go without such reference to a familiar tubular lamp.

39. This explains why the fluorescent powder must be applied to the inner wall if normal glass is used, because ultraviolet radiation is absorbed by ordinary glass.

40. See Thayer and Barnes (1939) for a report on this research. Inert gases have their ultraviolet lines at much shorter wavelengths, which would result in the production of less visible light by the fluorescent powder (Elenbaas, 1959c). An additional advantage was, of course, that there was experience with mercury vapor lamps, especially at Westinghouse's laboratory, which also participated in the effort to develop the new lamp.

41. Retrospectively the engineers could now explain the low efficiency of Claude's fluorescent tubes: These tubes "used sulphide powders in conjunction with the mercury discharge, but since sulphide powders are excited mainly by long wave ultraviolet which is very weak in the low-pressure discharge, there was little or no gain in efficiency by their use" (Jenkins, 1942: 283).

42. The other type of automatic switch used a resonance circuit (Inman and Thayer, 1938; Cleaver, 1940; Kruithof, 1941; Elenbaas, 1959a).

43. For this sketch of American technological culture in the 1930s, I am mainly drawing on David Nye's (1990) *Electrifying America.* But see also, especially for the nonelectrical aspects, the beautiful exhibition catalogue by Wilson, Pilgrim, and Tashjian (1986).

44. Paul Strand, "Photography and the New God," *Broom* (November 3, 1922): 252; quoted by Wilson, Pilgrim, and Tashjian (1986: 23). Like many machine-age artists, Strand also was aware of the threats: "not only the new God but the whole Trinity must be humanized unless it in turn dehumanizes us" (ibid.).

45. See also Nye (1990) for a short description of the electricity-related elements of the fair; also Wurts et al. (1977) for a history of the fair's construction and a comprehensive series of beautiful photographs; Doctorow's (1985) novel for a vivid description of the fair's impact on a young spectator; and Engelken (1940) for a lighting engineer's detailed survey.

46. Quoted by Nye (1990: 371) from the *Official Guide Book* to the New York World's Fair.

47. See Schwartz Cowan's (1983) classic analysis of how the introduction of household technology has always included promises for more comfort, although often resulting in more work for mothers.

48. See the news item in *Electrical World* of April 23, 1938 (anon., 1938a).

49. Harrison and Hibben (1938: 1530); see also Inman and Thayer (1938).

50. See also the note in the spring issue of The Magazine of Light, Vol. 7 (1938): 44.

51. Inman and Thayer (1938) do mention that "fluorescence provides a means of producing for the first time a practical source of white light matching daylight in appearance," but there is no discussion of how to realize that. The photo at the end of the article shows a bank office hall with the caption: "One suggested use of the new tubular fluorescent lamps is for flooding ceilings with colored light" (Inman and Thayer, 1938: 248).

52. J. E. Mueller, H. M. Sharp, and M. E. Skinner, "Plain Talk About Fluorescent Lighting," confidential paper presented at the 55th Annual Meeting of the Association of Edison Illuminating Companies, held January 15–19, 1940 (Committee on Patents, 1942: 4802–4810; 4803). Mueller was Manager of Commercial Sales, West Penn Power Company and Chairman of the E.E.I. Lighting Sales Committee; Sharp was Manager of the Lighting Bureau of the Buffalo Niagara & Eastern Power Corporation; Skinner was Vice-President of the Buffalo Niagara & Eastern Power Corporation; Sharp and Skinner were members of the E.E.I. Lighting Sales Committee as well.

53. Millar, "Advance Memorandum for Meeting of Lamp Subcommittee, May 27, 1938" (Committee on Patents, 1942: 4820–4822; 4821). Millar was chairman of the A.E.I.C. Lamp Subcommittee.

54. W. Harrison, talk (confidential), probably to a E.E.I. Lamp Committee meeting in fall 1939 (Committee on Patents, 1942: 4943–46; 4945). He is referring to Claude's high-voltage lamps.

55. Wakefield, "The Objectives of the Fleur-O-Lier Association," paper presented at the Industrial Conference on Fluorescent Lighting, March 22, 1940, Chicago (Committee on Patents, 1942: 4900–4902; 4900). Wakefield was manager of the fixtures manufacturing company F. W. Wakefield Brass Co.

56. "The Object of the Fleur-O-Lier Association," synopsis of suggested presentation before E.E.I. Sales Meeting in Chicago, 5 March 1940 (Committee on Patents, 1942: 4961–4965; 4962).

57. See Nye (1985) for a fascinating discussion of how this journal and other communication means were used to construct a corporate image of General Electric among various relevant social groups.

58. Cited by D. W. Prideaux (Incandescent Lamp Department, General Electric Company) to A. B. Oday (Engineering Department, General Electric), letter dated February 1, 1940 (Committee on Patents, 1942: 4972–4973; 4973).

59. Slauer (Westinghouse Lamp Division, Commercial Engineering Department to Westinghouse Lamp Division) to A. E. Snyder (Westinghouse Lamp Division, Executive Sales Manager), letter dated July 12, 1939 (Committee on Patents, 1942: 4818–4819; 4818).

60. Mueller, Sharp, and Skinner, "Plain Talk About Fluorescent Lighting."

61. See, for example, J. L. McEachin (Commercial Sales Manager, Minnesota Power & Light Co.) to G. E. Nelson (Manager, Northern Division, Incandescent Lamp Department, General Electric), letter dated 15 December 1939 (Committee on Patents, 1942: 4920–4921); and H. Restofski (West Penn Power Company, Sales Promotion Manager, Pittsburgh) to J. Kernes, letter dated 7 May 1940 (Committee on Patents, 1942: 4953–4954).

62. W. Harrison, "The Need for More and Varied Types of Fluorescent Equipment," paper presented at the Industrial Conference on Fluorescent Lighting, March 22, 1940, Chicago (Committee on Patents, 1942: 4893–4996).

63. W. Harrison, talk, probably to a E.E.I. Lamp Committee meeting in fall 1939 (Committee on Patents, 1942: 4943–4946; 4945). See also Cleaver (1940), who reports about experiments designed to measure lamp life.

64. Harrison, "The Need for More and Varied Types of Fluorescent Equipment," p. 4894).

65. Testimony by J. W. Walker, attorney of the Antitrust Division, Department of Justice (Committee on Patents, 1942: 4753–4800; 4770).

66. In physics terms, the basis of the problem is the following. Purely resistive elements in an electric circuit only absorb electric energy. Capacitors store electric field energy while they are charged (during half of the alternating voltage cycle), but feed this energy back to the circuit in the other half of the cycle. Inductive elements (such as coils) store magnetic field energy when the current flows (during half of the cycle) but feed this energy back to the circuit when the current goes to zero. Thus under steady-state AC operation, capacitors and inductances alternatingly store and release energy, without dissipating energy. So, although such a circuit does not usefully dissipate electric energy, the main circuit is "feeling" the load of it.

In any real circuit, there will be resistive, capacitative, and inductive elements. The power dissipation of such a circuit is given by the expression

$$P = I_{eff} V_{eff} \cos \varphi,$$

where I_{eff} is the "effective current," V_{eff} is the "effective voltage," and $\cos \varphi$ is the power factor. The angle φ is the angle between the phases of current and voltage. An incandescent lamp has a power factor of 1; electric motors and high-voltage neon lamps may have power factors as low as 0.5.

67. Mueller, Sharp, and Skinner, "Plain Talk About Fluorescent Lighting."

68. Testimony by J. W. Walker, p. 4762.

69. Mueller, Sharp, and Skinner, "Plain Talk About Fluorescent Lighting."

70. J. E. Mueller, "Today's Fluorescent Lighting Situation," report prepared for the A.E.I.C. Sales Executives' Conference, Hot Springs, Virginia, 30 September to 3 October 1940 (Committee on Patents, 1942: 4810–4818; 4812).

71. O. P. Cleaver (Westinghouse Electric and Manufacturing Company, Commercial Engineering Department), internal memorandum dated 25 April 1940 (Committee on Patents, 1942: 4855–4957; 4955).

72. Mueller, "Today's Fluorescent Lighting Situation."

73. Lowell Jr. (Hygrade Sylvia Corp.), "Industrial Applications of Fluorescent Lighting," paper presented at the Industrial Conference on Fluorescent Lighting, March 22, 1940, Chicago (Committee on Patents, 1942: 4899–4900; 4899).

74. Ibid.

75. Wakefield, "The Objective of the Fleur-O-Lier Association," p. 4900.

76. O. P. Cleaver, Westinghouse Lamp Division, Westinghouse Elec. & Mfg. Co., "Fluorescent Lighting in the Home Field," paper presented at the Industrial Conference on Fluorescent Lighting, Chicago, 22 March 1940 (Patent Commission, 1942: 4903–4904; 4903).

77. See Luckiesh and Moss (1937) and Alger (1939).

78. Mueller, Sharp, and Skinner, "Plain Talk About Fluorescent Lighting," p. 4803.

79. See Murdock (1977) for an analysis of this "duty to serve" and its enforcement by administrative bodies that are generally known as "public utility commissions."

80. Mueller, Sharp, Skinner, "Plain Talk About Fluorescent Lighting," p. 4802.

81. Maher (1977) studied the values endorsed by the executives of two electric power utilities in the 1960s and 1970s. Hirsch (1989) shows that it was no longer just a matter of values: The utilities' business practice of "grow and build"—growing by erecting larger power plants for higher thermal efficiency and then building consumer demand—had resulted by the end of the 1930s in large amounts of capital sunk in huge plants. This compelled them to frantically build load.

82. Mueller, Sharp, and Skinner, "Plain Talk About Fluorescent Lighting," p. 4807.

83. Ibid., p. 4804.

84. H. M. Sharp to W. J. Amos (executive of the New Orleans Public Service Co.), letter dated 6 April 1939 (Committee on Patents, 1942: 4848).

85. Draft of Detail Minutes of the Nela Park Conference, April 24–25, 1939 (Committee on Patents, 1942: 4843–4847; 4846).

86. Jim Amoss, quoted by J. E. Mueller to H. M. Sharp, letter dated 29 May 1939 (Committee on Patents, 1942: 4857–4859; 4858).

87. J. E. Mueller to H. M. Sharp, letter dated 29 May 1939 (Committee on Patents, 1942: 4857–4859; 4858).

88. "Statement of Policy Pertaining to Fluorescent Mazda Lamps," published by the General Electric Company Incandescent Lamp Department in *Lamp Letter No. S-E-21A* (Superseding S-E-21), 1 May 1939 (Committee on Patents, 1942: 4849).

89. Westinghouse Lamp Division, Westinghouse Electric & Manufacturing Co., "Statement of Policy on the Use of Fluorescent Mazda Lamps. Our Policy" (Committee on Patents, 1942: 4848–4849).

90. J. E. Mueller to H. E. Dexter (Commercial Manager, Central Hudson Gas & Electric Corp.), letter dated 11 May 1939 (Committee on Patents, 1942: 4855).

91. "Notes from the Electrical Testing Laboratories," prepared for the A.E.I.C. Lamp Subcommittee Meeting of 18 May 1939, dated 28 April 1939 (Committee on Patents, 1942: 4991–4992; 4991).

92. In physics a "black body" is one that absorbs all the light that falls on it. "Black-body radiation" is the (partly invisible) light that is radiated by a black body because of its temperature.

93. Lighting Sales Committee E.E.I., "Recent Developments in Fluorescent Lighting and Recommendations for the Immediate Future. Supplemental to Report of April 1939" (Committee on Patents, 1942: 4868–4877; 4872).

94. For a more detailed understanding of the chromaticity diagram, which is not needed for the present argument, see Kruithof (1959) or physics textbooks.

95. Cleaver, "Fluorescent Lighting in the Home Field," p. 4903.

96. Mueller, Sharp, and Skinner, "Plain Talk About Fluorescent Lighting."

97. Blitzer (Vice-President, The Lightolier Company; Member Board of Governors, American Lighting Equipment Association), "Prospects for Residential Fluorescent Fixtures," paper presented at the Industrial Conference on Fluorescent Lighting, 22 March 1940, Chicago (Committee on Patents, 1942: 4804–4906; 4905).

98. Mueller, "Today's Fluorescent Lighting Situation," p. 4815.

99. W. J. Amos (New Orleans Public Service Inc.), "The Value of Certification," paper presented at the Industrial Conference on Fluorescent Lighting, 22 March 1940, Chicago (Committee on Patents, 1942: 4890–4893; 4892).

100. Draft of Detail Minutes of the Nela Park Conference, April 24–25, 1939 (Committee on Patents, 1942: 4843–4847; 4844).

101. See, for example, the series of letters exchanged by Sharp, Mueller, and Harrison, April–July 1939 (Committee on Patents, 1942: 4988–4990, 4993–4995); "Notes Upon Informal Conference Concerning Certification of Fleur-O-Liers, August 3, 1939" (Committee on Patents, 1942: 4986–4988).

102. Wakefield, "The Objective of the Fleur-O-Lier Association."

103. Lamp Committee, "Fluorescent Lamps and Lighting" (Committee on Patents, 1942: 4950–4953).

104. Testimony by J. W. Walker, p. 4791.

105. Ibid.

106. Ibid., p. 4792.

107. Ibid., p. 4793.

108. Mueller, "Today's Fluorescent Lighting Situation," p. 4815.

109. Nye (1985) describes how different groups within General Electric are approached by management, each in their own particular way. The explicit aim is to create an effective corporate image, but one that is different in the various relevant social groups within General Electric. Nye's analysis thus demonstrates the interpretative flexibility of General Electric as a corporation.

110. W. Harrison to N. H. Boynton and E. E. Potter, letter dated 20 May 1940 (Committee on Patents, 1942: 4917–4918; 4917).

111. Ibid., p. 4918.

112. W. Booth (Manager Lighting Sales, General Electric Supply Corporation) to W. Harrison, letter dated 28 May 1940 (Committee on Patents, 1942: 4818).

113. J. E. Mueller (Manager Commercial Sales, West Penn Power Company) to W. Harrison (General Electric Co.), letter dated 2 May 1939 (Committee on Patents, 1942: 4853–4854; 4854).

114. C. T. Bremicker (Manager Lighting Sales Department, Northern States Power Co.) to J. E. Mueller, letter dated 16 May 1939 (Committee on Patents, 1942: 4855–4856; 4856).

115. H. M. Sharp to J. E. Mueller, letter dated 22 May 1939 (Committee on Patents, 1942: 4860).

116. H. W. Sharp (member Lighting Sales Committee, E.E.I.) to J. E. Mueller (Chairman, Lighting Sales Committee, E.E.I.), letter dated 6 May 1940 (Committee on Patents, 1942: 4954–4955; 4954).

117. J. E. Mueller to H. M. Sharp, latter dated 29 May 1939 (Committee on Patents, 1942: 4857–4858; 4858).

118. Sharp to W. Harrison, letter dated 24 May 1939 (Committee on Patents, 1942: 4916–4917).

119. Harrison to H. Sharp, letter dated 1 June 1939 (Committee on Patents, 1942: 4917).

120. H. W. Sharp (Manager, Lighting Bureau) to D. R. Grandy (Lamp Department, General Electric Company), letter dated 9 April 1940.

121. See, for example, J. L. McEachin to G. E. Nelson (Manager, Northern Division, Incandescent Lamp Department, General Electric), letter dated 15 December 1939 (Committee on Patents, 1942: 4920–4921); J. E. Lynch (Vice-President, Minnesota Power & Light Company) to R. H. Fite, Jr. (Evasco Services Inc.), letter dated 18 December 1939 (Committee on Patents, 1942: 4921–4922); H. M. Sharp (Sales Manager, Buffalo, Niagara & Eastern Power Co.) to A. L. Powell (Supervising Engineer, Lamp Department, General Electric Company), letter dated 3 January 1940 (Committee on Patents, 1942: 4923); H. G. Isley (Sales Manager, Carolina Power & Light Company) to O. P. Cleaver (Commercial Engineering Department, Westinghouse Electric & Manufacturing Co.), letter dated 22 January 1940 (Committee on Patents, 1942: 4924–4925); L. B. Paist (Supervisor, Lighting Sales Department, Northern States Power Co.) to the editor of *Electrical World* (25 January 1941: 367), about an article by W. J. Leemhuis (General Electric Company) who had compared an incandescent and a fluorescent lighting installation under the heading "Eight Times Illumination with Equal Wattage."

122. Vice-President of Buffalo, Niagara & Eastern Power Corporation to E. F. Strong (General Electric), letter dated 7 March 1939 (Committee on Patents, 1942: 4825).

123. E. F. Strong to W. Harrison, letter dated 12 April 1939 (Committee on Patents, 1942: 4827).

124. J. E. Mueller (Manager, West Penn Power Company), "Some Items of the Fluorescent Conference Requiring More Emphasis," internal memorandum dated 5 May 1939 (Committee on Patents, 1942: 4853).

125. Bremicker (Manager, Lighting Sales Department, Northerm States Power Company) to R. W. Seamans (Hygrade Sylvania Corporation), letter dated 6 April 1940 (Committee on Patents, 1942: 4926–4927; 4927).

126. H. M. Sharp (Buffalo, Niagara & Eastern Power Corporation) to J. E. Mueller (West Penn Power Corporation), letter dated 6 May 1940 (Committee on Patents, 1942: 4954–4955; 4955).

127. Mueller, Sharp, and Skinner, "Plain Talk About Fluorescent Lighting."

128. Ibid., pp. 4806–4807.

129. G. E. Whitwell (Vice-President, Philadelphia Electric Co.), "Keynote Message by Chairman" to the Industrial Conference on Fluorescent Lighting, 22 March 1940, Chicago (Committee on Patents, 1942: 4881–4883; 4883).

130. Mueller, "The Economics of Fluorescent Lighting," paper presented at the Industrial Conference on Fluorescent Lighting, 22 March 1940, Chicago (Committee on Patents, 1942: 4883–4890; 4884).

131. Mueller, Sharp, and Skinner, "Plain Talk About Fluorescent Lighting," pp. 4807–4808.

132. See, for example, Senator H. T. Bone, as Chairman of the Hearings Committee interrogating J. W. Walker (Committee on Patents, 1942: 4753–4800; especially 4760–4761, 4764–4766, 4790.)

133. District Court, Southern District of New York, No. 9–35 (civil), "General Electric Company, plaintiff, against Hygrade Sylvania Corporation, defendant, Raytheon Manufacturing Company, intervener-defendant." Court hearings started in 1942 and took several months; On March 30, 1944, Judge V. L. Leibell expressed as his opinion that the General Electric patents were valid and infringed, and the Hygrade patents invalid—a sweeping victory for General Electric. The decision was handed down only in 1954, when the other suits in which General Electric was involved were sorted out (Bright, 1949; Rogers, 1980).

134. District Court of the United States for the Southern District of New York, No. 110–412 (criminal), December 9, 1942, *United States of America v. General Electric Company, International General Electric Company, Inc., et al.* Previous indictments in the same case were returned August 30, 1940, and October 21, 1941. Final judgment was filed on March 26, 1954, as Civil Action No. 2590. In 1940, however, this outcome could not be foreseen, and GE felt quite threatened.

135. Letter from W. G. Merritt, counsel for the defendants in the case *U.S. v. General Electric Co. et al.*, to Hon. T. Arnold, Assistant Attorney General, Department of Justice, December 22, 1941 (Committee on Patents, 1942: 5015).

136. Letter by T. Arnold, Assistant Attorney General, to W. G. Merritt, counsel to General Electric Co. et al., January 19, 1942.

137. Letter from H. L. Stimson, Secretary of War, to F. Biddle, Attorney General, April 20, 1942; response letter April 25, 1942 (Committee on Patents, 1942: 5030).

138. The *locus classicus* is Braverman (1974).

139. John Mathews (1989a,b) provides the exception to the rule: his work explicitly seeks to link issues of power, control, and democracy to questions of economic and technical change, drawing on recent studies in political science, the economics of technical change, and the sociology of technology. His main objective, however, is not academic but political. Two authors who do contribute to an analysis of technology and power are Winner (1977, 1986) and Noble (1984).

140. Classical texts are Machiavelli (1958), Hobbes (1962), Marx (1867), Weber (1947), Parsons (1967), Lukes (1974), Foucault (1975), Giddens (1979), Barnes (1988), Mumford (1967, 1970), Hughes (1983), and Latour (1986a). See Clegg (1989) for a comprehensive and illuminating review.

141. This is indeed only a starting point. I am not trying to contribute to a "Giddensian" power theory, nor indeed to any comprehensive power theory. See Giddens (1984) for an elaboration of his social theory.

142. Authors working in the recent constructivist tradition of science studies have followed a similar line. See, for example, the seminal work by Callon (1986) and Latour (1986b, 1987) and the volumes edited by John Law (1986, 1991a).

143. In a more traditional perspective, the fluorescent lamp can be said to have had power implications through its economic effects. See Bright (1944) for a discussion of these effects. He concludes that the lamp resulted in "a definite net increase in national income and a definite net increase in employment."

144. See Giddens (1979: 92) for his particular notion of "resources" as "the media whereby transformative capacity is employed as power in the routine course of social interaction [although] they are at the same time structural elements of social systems as systems, reconstituted through their utilisation in social interaction."

145. See Barnes (1988) for an analysis of the role of routines in power interactions.

146. For a comprehensive introduction, with bibliography, to the economics of the patent system, see Kaufer (1989). Economic studies of the patent system typically do not focus on the power issues I am discussing, but are motivated by such

long-term questions as how the allocation of scarce societal resources is optimally directed to the benefit of consumers. See also Bowman (1973) and Coombs et al. (1987).

147. Bowker (1992) describes how the oil prospecting company Schlumberger succeeded in its goal of becoming a crucial and inevitable actor in the oil fields by *losing* a patent battle in which they tried to defend two patents that did not really work. See also Bowker (1994) and Misa (1992).

148. Kaufer (1989) and especially Bowman (1973) discuss other forms of restriction employed in patent license systems.

149. The concept of an "obligatory passage point" has been proposed by Callon (1986). See also Law and Callon (1992).

150. Those who have traveled in countries with different standards for electric plugs have experienced the disciplining power of this technology. Like all strategies for exerting power, however, these may be circumvented—by using, for example, two single copper wires (not recommended), by constructing an adaptor cable, or by using a fancy "universal traveler's plug."

Chapter 5

1. For a discussion on the principle of symmetry, see Laudan (1981) and Bloor (1981). See also Collins (1985) and Latour (1987).

2. It may be necessary at this point to stress once more my stance with respect to the need for a balance between *narrative* and *theory* for fruitful STS studies. It is just for candor's and clarity's sake that I assume a strict tone in prescribing *explananda.*

3. See Hughes (1986) for a discussion of the "seamless web" approach to the study of technology and science.

4. Bruno Latour has made baboons into a classic illustration of this point (Strum and Latour, 1984).

5. For a description of an imaginary world that is constituted by purely technical and natural relations, see Stanislaw Lem's novel *The Invincible.*

6. MacKenzie (1990: 409–417) has argued along the same lines in his study of nuclear missile guidance.

7. See also Latour (1992a).

8. See, for example, Collins and Yearly (1992a,b) and Callon and Latour (1992).

9. I will not follow Callon, Latour, and Law as far as assuming the equivalence of human and nonhuman actants. Although this does raise interesting questions about philosophical and ontological issues related to (Kantian) modernity, I do

not think that it is helpful for studying sociotechnical change. Here it seems more fruitful to use the principle of general symmetry in a less ontological sense—to issue a methodological warning against producing a priori distinctions that are to be studied as constructed rather than as given. As numerous examples in this book have demonstrated, this is much more than a trivial plea "to take all relevant aspects into consideration."

10. I prefer the word "ensemble" to other candidates such as "network" or "system" because it neatly conveys the a priori unstructured character I am seeking to describe. The advantage, for some, of the term "system" is that it suggests a hierarchical structure, while others prefer "network" because it conveys an unboundedness.

11. Thus I disagree with Latour's (1984) plea for "irreductionism."

12. See Law and Bijker (1992) for a discussion of various ways to make such a distinction between foreground and backdrop.

13. See also Staudenmaier's (1985) use of different constituencies in his interpretative model for the history of technology.

14. I adapted this mechanism from Hughes's (1987) "amortization of vested interests" mechanism, but now prefer Misa's (1992) term "amalgamation" because it has a less economic connotation.

15. My usage of the term "politics of technology" does not include party and state politics in the narrow sense of the word. If relevant at all, it would subsume such politics under the rubric of technology policy. See Elzinga and Jamison (1994) for a comprehensive and enlightening review of policy and politics studies in science and technology.

16. See Bimber (1990) and Bijker (1994) for discussions of the debate on technological determinism. MacKenzie's (1990) historical sociology of nuclear missile guidance provides the most convincing demonstration so far of the fallacy of technological determinism.

17. See, however, Misa (1988) for an interesting discussion of how elements of technological determinism may still play a role in modern technology studies. He argues that studies on a micro level will generally deconstruct technological determinism, while analyses on a macro level tend to produce technologically deterministic accounts. Misa concludes with a plea for meso level studies of technology.

18. The term "boundary object" was introduced by Star (1988) as "objects which are both plastic enough to adapt to local needs and constraints of the several parties employing them, yet robust enough to maintain a common identity across sites." Star discusses various types of boundary objects: repositories (for example, a museum's collection), platonic objects (for example, a geographical map), standardized labels (for example, anamnestic protocols for the intake of epileptic patients).

19. See Law and Bijker (1992) for a discussion of how this view of creating inside/outside boundaries relates to various approaches in recent technology studies.

20. It is difficult to give a universal criterion for identifying the "final" boundary between low included actors on the inside and others on the outside. Here the actors' and analysts' concepts of "relevant social group" should be distinguished. For actors it will often be clear whether somebody is "in" or "out." Crozier and Friedberg (1977) have used the concept of "minimal membership" to describe actors' intuitions about what is minimally needed to be a member of an informal group. Analysts can try to follow the actors, but at some point they must make their own choices as researchers.

21. This is very similar to Fleck's (1935) analysis of the differences between esoteric and exoteric circles. See also Collins's (1987) observation that it is the scientists in the core set who are most aware of the uncertainty of scientific facts. Consequently, certainty about natural phenomena as well as about technical artifacts tends to vary inversely with the degree of inclusion.

22. I realize that this is a particularly Dutch example. In the Netherlands it still is feasible *not* to have a car, and to use a combination of bicycle and public transport instead.

23. See Blume (1992) and Pasveer (1992) for histories of ultrasound.

24. MacKenzie (1990) demonstrates the same by disaggregating "the state" and other important actors in the guided missile history, as well as disaggregating every important decision.

25. The classical methodological vocabulary of micro, meso, and macro levels of analysis is not adequate for the perspective that I have tried to develop. Sociotechnical ensembles transcend these levels almost by definition. For a more extended argument along the same lines, but focusing on networks rather than ensembles, see Callon and Latour (1981).

26. Discussions with Trevor Pinch and Brian Wynne have been very helpful on this point.

27. See, for example, Brian Wynne's (1982, 1991) work and recent work by Steven Yearley (1992, 1994).

28. Adapted from Carroll (1872: 75).

References

(anon.) 1907. "The New Electrolytic Alkali-Works at Niagara Falls." *Electrochemical and Metallurgical Industry* (June 1907).

(anon.) 1938a. "Fluorescent Produces Tint Tone Lighting with High Efficiency." *Electrical World* 110 (23 April 1938): 1394.

(anon.) 1938b. "Lampholder for Lumiline Lamps." *General Electric Review* 41: 509.

(anon.) 1939. "Illumination." *General Electric Review* 42: 44–49.

(anon.) 1949. "Edison Electric Institute." *Electrical World* 131: 565–566.

Akrich, M. 1992. "The De-Scription of Technical Objects." In Bijker and Law (1992a): 205–224.

Andes, L. E. 1911. *Verarbeitung des Hornes, Elfenbeins, Schildpatts, der Knochen und der Perlmutter (Abstammung und Eigenschaften dieser Rohstoffe, ihre Zubereitung, Färbung und Verwendung)*. Vienna: Hartleben's Verlag.

Appelbaum, S. 1977. See Wurts (1977).

Arons, L. 1892. "Ueber einen Quecksilberlichtbogen." *Annalen der Physik und Chemie* 47: 767–771.

Aylsworth, J. W. 1909. "Fusible Phenol Resin and Method of Forming Same." U.S. Patent No. 1,029,737, filed 14 May 1909.

Aylsworth, J. W. 1911a. "Improvements in Phenolic Condensation Products and Method of Preparing the Same." British Patent No. 3496, filed 11 February 1911.

Aylsworth, J. W. 1911b. "An Improved Plastic Composition Applicable for Use in Coating Surfaces." British Patent No. 3497, filed 11 February 1911.

Aylsworth, J. W. 1911c. "Phenolic Condensation Products and Process for Manufacturing the Same." British Patent No. 3498, filed 11 February 1911.

Aylsworth, J. W. 1911d. "Plastic Phenolic Condensation Product." British Patent No. 9559, filed 19 April 1911.

Aylsworth, J. W. 1911e. "Improvements in the Production of Printing and Embossing Surfaces." British Patent No. 12,659, filed 25 May 1911.

Aylsworth, J. W. 1911f. "Plastic Phenolic Condensation Product and Process of Forming Same." U.S. Patent No. 1,020,594, filed 2 June 1911.

Aylsworth, J. W. 1911g. "Improvements in Plastic Compositions and Method of Preparing the Same." British Patent No. 24,124, filed 31 October 1911.

Aylsworth, J. W. 1911h. "Improvements in and Relating to Phenolic Condensation Products." British Patent No. 26,029, filed 21 November 1911.

Aylsworth, J. W. 1912. "Phenolic Condensation Product and Method of Preparing Same." U.S. Patent No. 1,020,593, filed 11 February 1910.

Aylsworth, J. W. 1913, "Cement for Pipe Connections." U.S. Patent No. 1,065,495, filed 11 February 1910.

Aylsworth, J. W. 1915, "Varnish, Enamel, or Lacquer Composition." U.S. Patent No. 1,137,374, filed 11 February 1910.

Baekeland, L. H. 1887a. "Des plaques photographiques développables dans l'eau." Belgian Patent No. 78957, filed 22 July 1887.

Baekeland, L. H. 1887b. "Photographische Trockenplatten, welche durch Eintauchen in Wasser entwickelt werden." German Patent No. 43521, filed 18 October 1887.

Baekeland, L. H. 1897. "Some Pecularities of Velox Paper." *Wilson's Photographic Magazine* (February).

Baekeland, L. H. 1906a. "Apparatus for Regenerating Electrolytes." U.S. Patent No. 844,314, filed 16 February 1906.

Baekeland, L. H. 1906b. "Electrolytic Diaphragm and Method of Making Same." U.S. Patent No. 855,221, filed 10 March 1906.

Baekeland, L. H. 1907a. "Method of Making Insoluble Products of Phenol and Formaldehyde." U.S. Patent No. 942,699, filed 13 July 1907.

Baekeland, L. H. 1907b. "Method of Indurating Fibrous and Cellular Material." U.S. Patent No. 949,671, filed 18 February 1907.

Baekeland, L. H. 1907c. "Indurated Product and Method of Preparing Same." U.S. Patent No. 942,852, filed 13 July 1907.

Baekeland, L. H. 1907d. "Varnish." U.S. Patent No. 954,666, filed 15 October 1907.

Baekeland, L. H. 1907e. "Condensation Product and Method of Making the Same." U.S. Patent No. 942,809, filed 15 October 1907.

Baekeland, L. H. 1907f. "Abrasive Composition and Method of Making Same." U.S. Patent No. 942,808, filed 26 October 1907.

Baekeland, L. H. 1907i. "Condensation Product of Phenol and Formaldehyde and Method of Making the Same." U.S. Patent No. 942,700, filed 4 December 1907.

Baekeland, L. H. 1907j. "The Danger of Overspecialization," *Science* 25 (31 May 1907): 845–854.

Baekeland, L. H. 1907k. "A Family Motor Tour Through Europe." *The Horseless Age* 19–20 (between 13 March and 6 November).

Baekeland, L. H. 1909a. "The Synthesis, Constitution, and Use of Bakelite." *Journal of Industrial and Engineering Chemistry* 1: 149–161 (in German: "Bakelit, ein neues synthetischer Harz." *Chemiker Zeitung* 23: 317–318, 327–328, 347–348, 358–359).

Baekeland, L. H. 1909b. "The Use of Bakelite for Electrical and Electrochemical Purposes." *Transactions of the American Electrochemical Society* 15: 593–615.

Baekeland, L. H. 1909c. "On Soluble, Fusible, Resinous Condensation Products of Phenols and Formaldehyde." *Journal of Industrial and Engineering Chemistry* 1: 545–549.

Baekeland, L. H. 1909d. "Resinit und Bakelit." *Chemiker Zeitung* 23: 705.

Baekeland, L. H. 1909e. "Bakelit und Resit." *Zeitschrift für angewandte Chemie* 22: 2006–2007.

Baekeland, L. H. 1909f. "Method of Molding Articles." U.S. Patent No. 939,966, filed 28 January 1909.

Baekeland, L. H. 1911a. "Condensation Products of Phenols and Formaldehyde." *Journal of Industrial and Engineering Chemistry* 3: 518–520.

Baekeland, L. H. 1911b. "Recent Developments in Bakelite." *Journal of Industrial and Engineering Chemistry* 3: 932–938.

Baekeland, L. H. 1912. "Phenol-Formaldehyde Condensation Products." *Journal of Industrial and Engineering Chemistry* 4: 737–743.

Baekeland, L. H. 1913. "The Chemical Constitution of Resinous Phenolic Condensation Products. Address of Acceptance [of the Willard Gibbs Medal]." *Journal of Industrial and Engineering Chemistry* 5: 506–511.

Baekeland, L. H. 1914a. "The Invention of Celluloid." *Journal of Industrial and Engineering Chemistry* 6: 90–91.

Baekeland, L. H. 1914b. "Synthetic Resins." *Journal of Industrial and Engineering Chemistry* 6: 167–170.

Baekeland, L. H. 1916. "Practical Life as a Complement to University Education—[Perkin] Medal [Acceptance] Address." *Journal of Industrial and Engineering Chemistry* 8: 184–190.

Baekeland, L. H. 1932. "Dreams and Realities." *Journal of Chemical Education* 9: 1000–1009.

Baekeland, L. H., and Bender, H. L. 1925. "Phenol Resins and Resinoids." *Journal of Industrial and Engineering Chemistry* 17: 225–237.

Baeyer, A. 1872a. "Ueber die Verbindungen der Aldehyde mit den Phenolen." *Berichte der Deutschen Chemischen Gesellschaft* 5: 25–31.

Baeyer, A. 1872b. "Ueber die Verbindungen der Aldehyde mit den Phenolen." *Berichte der Deutschen Chemischen Gesellschaft* 5: 280–282.

Baeyer, A. 1872c. "Ueber die Verbindungen der Aldehyde mit den Phenolen und aromatischen Kohlenwasserstoffen." *Berichte der Deutschen Chemischen Gesellschaft* 5: 1094–1100.

Barnes, B. 1988. *The Nature of Power.* Cambridge: Polity Press.

Bartlett, W. E. 1890a. "Improvements in the Elastic Tyres or Rims of the Wheels of Velocipedes and Other Vehicles." British Patent No. 16384, filed 14 October 1890.

Bartlett, W. E. 1890b. "Improvements in the Tyres or Rims for Cycles and Other Vehicles." British Patent No. 16783, filed 21 October 1890.

Baudry de Saunier, L. 1936. *Histoire de la Locomotion Terrestre. Vol. 2.* Paris: L'Illustration.

Bayer, F. 1907. "Verfahren zur Herstellung von Harzartigen Kondensationsprodukten aus o-Kresol und Formaldehyd." German Patent No. 201261, filed 16 April 1907.

Becquerel, E. 1857. "Recherches sur divers effets lumineux, qui résultent de l'action de la lumière sur les corps." Paper presented on 16 November 1857 to the Académie des Sciences, published in *Annales de Chimie et de Physique*, 3rd série, Tome 55 (1859): 5–119.

Beder, S. 1991. "Controversy and Closure: Sydney's Beaches in Crisis." *Social Studies of Science* 21: 223–256.

Belt, H. van den, and Rip, A. 1987. "The Nelson-Winter-Dosi Model and Synthetic Dye Chemistry." In Bijker, Hughes, and Pinch (1987): 135–158.

Berger, P. L., and Luckmann, T. 1966. *The Social Construction of Reality. A Treatise in the Sociology of Knowledge.* Garden City, N.Y.: Doubleday.

Berthier, A. 1908. *Les nouveaux modes d'éclairage électrique.* Paris: Dunod et Pinat.

Bijker, W. E. 1985. "Images of Science. Incorporating the Results of Recent Science Studies in Science Education: An Argument and an Example." In D. Gooding (comp.), "The Uses of Experiment," abstracts. Bath: University of Bath, 62–64a.

Bijker, W. E. 1988. "Interdisciplinary Technology Studies from a Dutch Perspective." In E. Mayer, ed., *Ordnung, Rationalisierung, Kontrolle. Wechselspiel technischer und gesellschaftlicher Aspekte bei der Entwicklung technischer Grosssysteme.* Darmstadt: Technische Hochschule Darmstadt, 31–53.

Bijker, W. E. 1995. "Socio-Historical Technology Studies. Illustrated with Examples from Coastal Engineering and Hydrological Technology." In S. Jasanoff, G. E. Markle, J. C. Petersen, and T. Pinch, eds., *Handbook of Science and Technology Studies.* Newbury Park, CA, and London: Sage, 229–256.

Bijker, W. E., Hughes, T. P., and Pinch, T. J., eds. 1987a. *The Social Construction of Technological Systems: New Directions in the Sociology and History of Technology.* Cambridge, MA: MIT Press.

Bijker, W. E., Hughes, T. P., and Pinch, T. J. 1987b. "General Introduction" and "Introductions." In Bijker et al. (1987a): 1–6, 9–15, 107–110, 191–194, 307–309.

Bijker, W. E., and Law, J., eds. (1992a). *Shaping Technology/Building Society: Studies in Sociotechnical Change.* Cambridge, MA: MIT Press.

Bijker, W. E., and Law, J. (1992b) "Introduction." In Bijker and Law (1992a).

Bimber, B. 1990. "Karl Marx and the Three Faces of Technological Determinism." *Social Studies of Science* 20: 333–351.

Birr, K. 1957. *Pioneering in Industrial Research. The Story of the General Electric Research Laboratory.* Washington, D.C.: Public Affairs Press.

Blondel, C. 1985. "Industrial Science as a 'Show': A Case-Study of Georges Claude." In T. Shinn and R. Whitley, eds., *Expository Science: Forms and Functions of Popularization. Sociology of the Sciences Yearbook,* Vol. 9: 249–258.

Bloor, D. 1973. "Wittgenstein and Mannheim on the Sociology of Mathematics." *Studies in History and Philosophy of Science* 4: 173–191.

Bloor, D. 1976. *Knowledge and Social Imagery.* London: Routledge and Kegan Paul.

Bloor, D. 1981. "The Strengths of the Strong Programme." *Philosophy of the Social Sciences* 11: 199–213.

Blume, S. 1992. *Insight and Industry: On the Dynamics of Technological Change in Medicine.* Cambridge, MA: MIT Press.

Blumer, H. 1969. *Symbolic Interactionism: Prespective and Method.* Englewood Cliffs, N.J.: Prentice-Hall; republished in 1986, Berkeley, CA: University of California Press.

Blumer, L. 1902a. "Verfahren zur Herstellung eines dem Schellack ähnlichen harzartigen Kondensationsproduktes aus Phenol und Formaldehydlösung." German Patent No. 172877, filed 13 April 1902.

Blumer, L. 1902b. "Process for the Synthetical Preparation of Resinous Substances." British Patent No. 12,880, filed 5 June 1902.

Bonwitt, G. 1933. *Das Celluloid und seine Ersatzstoffe. Handbuch für Herstellung und Verarbeitung von Celluloid und seinen Ersatzstoffen.* Berlin: Union Deutsche Verlagsgesellschaft Zweigniederlassung.

Bottler, M. 1924. *Harze und Harzindustrie.* Leipzig: Max Janecke.

Bowker, G. 1992. "What's in a Patent?" In Bijker and Law (1992a), 53–74.

Bowker, G. 1994. *Science on the Run: Information Management and Industrial Geophysics at Schlumberger, 1920–1940.* Cambridge, MA: MIT Press.

Bowman Jr., W. S. 1973. *Patent and Antitrust Law. A Legal and Economic Appraisal.* Chicago: University of Chicago Press.

Brandenburger, K. 1934. "Der Kunststoff in der Tagespresse." *Kunststoffe* 24: 304–305.

Brandenburger, K. 1938. *Im Zeitalter der Kunststoffe. Allgemeinverständliche Schilderung der Entstehung und Verwendung der Kunststoffe in Wirtschaft, Industrie und im täglichen Leben.* Munich-Berlin: Lehmanns Verlag.

Braverman, H. 1974. *Labor and Monopoly Capital. The Degradation of Work in the Twentieth Century.* New York: Monthly Review Press.

Bright, A. A. 1944. "Some Broad Economic Implications of the Introduction of Hot-Cathode Fluorescent Lighting." *Transactions of the Electrochemical Society* 87: 367–378.

Bright, A. A. 1949. *The Electrical Lamp Industry: Technological Change and Economic Development from 1800 to 1947.* New York: MacMillan; republished by Arno Press, New York, 1972.

Bright, A. A., and Maclaurin, W. R. 1943. "Economic Factors Influencing the Development and Introduction of the Fluorescent Lamp." *Journal of Political Economy* 51: 429–450.

Buchanan, R. A. 1991. "Theory and Narrative in the History of Technology. *Technology and Culture* 32: 365–376.

Bugge, G. 1931. "Aus der frühen Geschichte des Formaldehyd-Herstellung." *Chemische Apparatur* 18: 157–160.

Bugge, G. 1943. "Aus der frühen Geschichte des Formaldehyds und seiner Anwendungen." *Chemische Technik* 16: 228–230.

Byck, L. C. 1952. "A Survey of the Bakelite Thermosetting Business, 1910 through 1951." Confidential internal report, Bakelite Company; one copy in the Baekeland Papers, Archives Center, National Museum of American History, Smithsonian Institution, Washington, D.C.

Callon, M. 1986. "Some Elements of a Sociology of Translation: Domestication of the Scallops and the Fishermen of St. Brieuc Bay." In J. Law, ed., *Power, Action, and Belief: A New Sociology of Knowledge?* London: Routledge and Kegan Paul: 196–233.

Callon, M., and Latour, B. 1981. "Unscrewing the Big Leviathan, or How Do Actors Macrostructure Reality?" In K. D. Knorr-Cetina and A. Cicourel, eds., *Advances in Social Theory and Methodology toward an Integration of Micro- and Macro-Sociologies.* London: Routledge and Kegan Paul: 277–303.

Callon, M., and Latour, B. 1992. "Don't Throw the Baby out with the Bath School!—A Reply to Collins and Yearley." In Pickering (1992): 343–368.

Callon, M., and Law, J. 1982. "On Interests and Their Transformation: Enrollment and Counter-Enrollment." *Social Studies of Science* 12: 615–625.

Callon, M., and Law, J. 1989. "On the Construction of Sociotechnical Networks: Content and Context Revisited." *Knowledge and Society: Studies in the Sociology of Science Past and Present* 8: 57–83.

Card, P. W. 1984. *Joseph Lucas. The First "King of the Road." Wholesale Catalogue.* Reprinted in 1984. London: John Pinkerton Reprint Catalogue Issues.

Carlson, W. B. 1991. *Innovation as a Social Process. Elihu Thomson and the Rise of General Electric, 1870–1900.* Cambridge: Cambridge University Press.

Carlson, W. B., and Gorman, M. E. 1990. "Understanding Invention as a Cognitive Process: The Case of Thomas Edison and Early Motion Pictures, 1888–91." *Social Studies of Science* 20: 387–430.

Carlson, W. B., and Gorman, M. E. 1992. "Socio-Technical Graphs and Cognitive Maps: A Response to Latour, Mauguin and Teil." *Social Studies of Scence* 22: 81–89.

Carroll, L. 1872. *Through the Looking Glass, and What Alice Found There.* Reprinted in facsimile by Avenel Books, Crown Publishers, New York.

Caunter, C. F. 1955. *The History and Development of Cycles (as Illustrated by the collection of cycles in the Science Museum). Part I: Historical Survey.* London: Science Museum and Her Majesty's Stationery Office.

Caunter, C. F. 1958. *Handbook of the Collection Illustrating Cycles. Part II: Catalogue of Exhibits with Descriptive Notes.* London: Science Museum and Her Majesty's Stationery Office.

Chandler, A. D. 1977. *The Visible Hand: The Managerial Revolution in American Business.* Cambridge, MA: Harvard University Press.

Chandler, C. F. 1914a. "Presentation Address [of the Perkin medal to J. W. Hyatt]." *Journal of Industrial and Engineering Chemistry* 6: 156–158.

Chandler, C. F. 1914b. "The Invention of Celluloid." *Journal of Industrial and Engineering Chemistry* 6: 601–602.

Chandler, C. F. 1916. "Presentation Address [of the Perkin medal to L. H. Baekeland]." *Journal of Industrial and Engineering Chemistry* 14: 148–151.

Charon, J. M. 1985. *Symbolic Interactionism: An Introduction, an Interpretation, an Integration.* Englewood Cliffs, N.J.: Prentice-Hall.

Cheney, S., and Candler Cheney, M. 1936. *Art and the Machine: An Account of Industrial Design in 20th-Century America.* New York: McGraw-Hill.

Church, R. 1955. *Over the Bridge: An Essay in Autobiography.* London: Heinemann.

Claessen, C. 1905. "Verfahren zur Herstellung celluloidartiger Massen." German Patent No. 172941, filed 7 April 1905.

Claude, A. 1936. "Gaz Rares et Lumière." *Bulletin de la Société française des Électriciens* 64: 405–432.

Claude, A. 1939. "L'éclairage par luminescence." *Bulletin de la Société Française des Électriciens* 5th série, Tome IX, no. 100, April 1939: 307–342. An English summary was published as A. Claude, "Lighting by Luminescence." *Light and Lighting* 32 (1939): 127–131.

Claude, G. 1909. *Air Liquide, Oxygène, Azote.* Paris: Dunod et Pinat.

Claude, G. 1913. *Notice sur les Travaux Scientifiques.* Paris: Dunod et Pinat.

Claude, G. 1957. *Ma vie et mes inventions.* Paris: Librairie Plon.

Clayton, R., and Algar, J. 1989. *The GEC Research Laboratories, 1919–1984.* London: Peter Peregrinus and Science Museum.

Cleaver, O. P. 1940. "Fluorescent Lighting After Two Years." *Electrical Engineering* 59: 261–266.

Clegg, S. R. 1989. *Frameworks of Power.* London: Sage.

Collins, H. M. 1981a. "The Place of the Core-Set in Modern Science: Social Contingency with Methodological Propriety in Science." *History of Science* 19: 6–19.

Collins, H. M. 1981b. "Stages in the Empirical Programme of Relativism." *Social Studies of Science* 11: 3–10.

Collins, H. M., ed. 1981c. "Knowledge and Controversy." Special issue of *Social Studies of Science* 11: 3–158.

Collins, H. M. 1985. *Changi: g Order: Replication and Induction in Scientific Practice.* Beverly Hills: Sage.

Collins, H. M. 1987. "Certainty and the Public Understanding of Science: Science on Television." *Social Studies of Science* 17: 289–313.

Collins, H. M., and Pinch, T. J. 1982. *Frames of Meaning: The Social Construction of Extraordinary Science.* London: Routledge and Kegan Paul.

Collins, H. M., and Pinch, T. J. 1993. *The Golem: What Everyone Should Know about Science.* Cambridge: Cambridge University Press.

Collins, H. M., and Yearley, S. 1992a. "Epistemological Chicken." Paper discussed at the Bath Symmetry Workshop, 10 February 1990, Bath, U.K. In A. Pickering (1992): 301–326.

Collins, H. M., and Yearley, S. 1992b. "Journey Into Space." In A. Pickering (1992): 369–389.

Committee on Patents. 1942. *Hearings before the Committee on Patents, U.S. Senate, 77th Congress, 2d Session on S.2303 and S.2491, Part 9, 18–21 August 1942: 4753–5032.* Washington, D.C.: U.S. Government Printing Office.

Constant, E. W. 1980. *The Origins of the Turbojet Revolution.* Baltimore: Johns Hopkins University Press.

Constant, E. W. 1984. "Communities and Hierarchies: Structure in the Practice of Science and Technology." In *The Nature of Technological Knowledge: Are Models of Scientific Change Relevant?* Dordrecht: Reidel, 27–46.

Cooke, M. C. 1896. *Our Social Manual.* Chicago, IL: Monarch Book Company.

Coombs, R., Saviotti, P., and Walsh, V. 1987. *Economics and Technological Change.* London: Macmillan.

Cox, J. L. 1937. "Luminescent Coating for Electric Lamps." U.S. Patent No. 2,096,693, filed 3 April 1937.

Croon, L. 1939. "Das Fahrrad und seine Entwicklung." *Deutsches Museum. Abhandlungen und Berichte* 11: 161–186.

Crozier, M., and Friedberg, E. 1977. *L'Acteur et le système.* Paris: Seuil. English translation: *Actors and Systems: The Politics of Collective Action.* Chicago: University of Chicago Press.

Cutcliffe, S. H. 1989. "The Emergence of STS as an Academic Field." *Research in Philosophy and Technology* 9: 287–301.

Daul, A. 1906. *Illustrierte Geschichte der Erfindung des Fahrrades und der Entwicklung des Motorfahrradwesens.* Dresden: R. Creutz.

Davies, L. J., Ruff, H. R., and Scott, W. J. 1942. "Fluorescent Lamps." *Transactions of the Illuminating Engineering Society (London)* 88: 447–472.

De Laire. 1905. "Procédé d'obtention de résines synthétiques transparantes, destinées être employées comme succédanés de certaines résines naturelles." French Patent No. 361539, filed 8 June 1905.

Doctorow, E. L. 1985. *World's Fair.* New York: Ballantine Books.

Doorman, G. 1947. *Het Nederlands Octrooiwezen en de techniek der 19e eeuw.* The Hague: Nijhoff.

Dosi, G. 1982. "Technological Paradigms and Technological Trajectories: A Suggested Interpretation of the Determinants and Directions of Technical Change." *Research Policy* 11: 147–162.

Dosi, G. 1984. *Technical Change and Industrial Transformation.* London: Macmillan.

Dosi, G., Freeman, C., Nelson, R., Siverberg, G., and Soete, L., eds. 1988. *Technical Change and Economic Theory.* London: Pinter.

Dubois, J. H. 1972. *Plastic Industry USA.* Boston: Cahners.

Du Cros, A. 1938. *Wheels of Fortune. A Salute to Pioneers.* London: Chapman & Hall.

Dunlop, J. B. 1888. "An Improvement in Tyres of Wheels for Bicycles, Tricycles, or Other Road Cars." British Patent No. 10607, filed 23 July 1888.

Dunlop, J. B. 1889a. "An Improved Non-return Valve." British Patent No. 4115, filed 8 March 1889.

Dunlop, J. B. 1889b. "Improvements in Wheel Tyres for Cycles and Other Vehicles, and in Means for Securing the Same to the Wheel Rims." British Patent No. 4116, filed 8 March 1889.

Elenbaas, W., ed. 1959a. *Fluorescent Lamps and Lighting.* Eindhoven: Philips' Gloeilampenfabrieken.

Elenbaas, W. 1959b. "Chapter IV: Gaseous Discharges." In Elenbaas (1959a): 67–84.

Elenbaas, W. 1959c. "Chapter V: Lamp Construction." In Elenbaas (1959a): 85–91.

Elster, J. 1983. *Explaining Technical Change: A Case Study in the Philosophy of Science.* Cambridge: Cambridge University Press.

Elzinga, A., and Jamison, A. 1994. "Changing Policy Agendas in Science and Technology." In J. C. Petersen, G. E. Markle, S. Jasanoff, and T. Pinch, eds., *Handbook of Science, Technology and Society.* Newbury Park, CA, and London: Sage.

Engelken, R. C. 1940. "Lighting the New York World's Fair." *Journal of Electrical Engineering* 59 (May 1940): 179–203.

Engineer, The. 1886. "Coventry Machinist Company's Cycle Works." *The Engineer* 62, September 10: 202.

Engineer, The. 1888a. "The Stanley Exhibition of Cycles." *The Engineer* 65, February 10: 118–119.

Engineer, The. 1888b. "The Stanley Exhibition of Cycles." *The Engineer* 65, February 17: 131.

Engineer, The. 1889. "The Stanley Exhibition of Cycles." *The Engineer* 67, 22 February: 157–158.

Engineer, The. 1890a. "The Stanley Exhibition of Cycles 1890." *The Engineer* 69, 7 February: 107–108.

Engineer, The. 1890b. "The Stanley Exhibition of Cycles 1890." *The Engineer* 69, 14 February: 138–140.

Engineer, The. 1895. "The Manufacture of Cycles. The Humber Company's Works, Beeston." *The Engineer* 79, 18 January: 53–56.

Engineer, The. 1896. "Recent Cycle Exhibitions." *The Engineer* 81, 17 January: 54–55.

Engineer, The. 1897a. "American Bicycle Tube Manufacture." *The Engineer* 83, 16 April: 403.

Engineer, The. 1897b. "Bicycle Mechanics." *The Engineer* 83, 14 May: 492.

Engineer, The. 1897c. "The National Cycle Show." *The Engineer* 84, 10 December: 569.

Engineer, The. 1898. "The Cycle Shows." *The Engineer* 86, 25 November: 514.

Fayolle, E.-H. 1903. "Procédé pour la fabrication d'une substance ayant une certaine analogie avec le caoutchouc." French Patent No. 335584, filed 26 September 1903; additions: No. 2414, filed 14 October 1903; No. 2485, filed 1 December 1903.

Fayolle, E.-H. 1904. "Procédé pour la préparation d'une substance ressemblant à la gutta-percha." French Patent No. 341013, filed 7 March 1904.

Feldhaus, F. M. 1914. *Die Technik der Vorzeit, der geschichtlichen Zeit und der Naturvölker.* Leipzig: Wilhelm Engelmann.

Ferguson, E. 1974. "Toward a Discipline of the History of Technology." *Technology and Culture* 15: 13–30.

Ferguson, E. S. 1992. *Engineering and the Mind's Eye.* Cambridge, MA: MIT Press.

Fischer, G. 1933. "Elektrische Entladungsröhre mit luminiszierender Glaswand." Austrian Patent No. 140,012, filed 11 October 1933.

Fleck, L. 1935. *Entstehung und Entwicklung einer wissenschaftlichen tatsache. Einführung in die Lehre vom Denkstil und Denkkollektiv.* Basel: Benno Schwabe. Reprinted in

1980. Frankfurt: Suhrkamp Verlag. English translation: *The Genesis and Development of a Scientific Fact.* Chicago: University of Chicago Press, 1979.

Foucault, M. 1975. *Surveillir et Punir: Naissance de la Prison.* Paris: Gallimard. English translation (1979): *Discipline and Punish: The Birth of the Prison.* Harmondsworth: Penguin.

Foucault, M. 1976. *Histoire de la sexualité, 1. La volonté de savoir.* Paris: Gallimard. English translation (1984): *The History of Sexuality: An Introduction.* Harmondsworth: Peregrine.

Frankl, P. T. 1930. *Form and Re-form: A Practical Handbook of Modern Interiors.* New York: Harper and Brothers; reprinted by Hacker Art Books, New York, 1972.

Friedel, R. 1979. "Parkesine and Celluloid: The Failure and Success of the First Modern Plastic." In *History of Technology*, A. R. Hall and N. Smith, eds. London: Mansell, vol. 4, 45–62.

Friedel, R. 1983. *Pioneer Plastic. The Making and Selling of Celluloid.* Madison: University of Wisconsin Press.

Fritze, O., and Rüttenauer, A. 1935. "Verfahren zum Anbringen von Luminophoren auf die Glaswände von elektrischen Entladungsgefäßen, insbesondere elektrischen Entladungslampen." German Patent No. 692,394, filed 29 June 1935.

Garfinkel, H. 1967. *Studies in Ethnomethodology.* Englewood Cliffs, NJ: Prentice-Hall. Paperback edition, Cambridge: Polity Press, 1984.

Giddens, A. 1979. *Central Problems in Social Theory: Action, Structure, and Contradiction in Social Analysis.* Houndmills: Macmillan.

Giddens, A. 1984. *The Constitution of Society: Outline of the Theory of Structuration.* Cambridge: Polity Press.

Gieryn, T. F., and Hirsch, R. F. 1983. "Marginality and Innovation in Science." *Social Studies of Science* 13: 87–106.

Gilfillan, S. G. 1935a. *The Sociology of Invention.* Cambridge, MA: MIT Press.

Gilfillan, S.C. 1935b. *Inventing the Ship.* Cambridge, MA: MIT Press.

Gillis, J. 1965. *Leo Hendrik Baekeland. Verzamelde oorspronkelijke documenten.* Brussels: Koninklijke Vlaamse Academie voor Wetenschappen, Letteren en Schone Kunsten van België.

Gorman, M. E., and Carlson, W. B. 1990. "Interpreting Invention as a Cognitive Process: The Case of Alexander Graham Bell, Thomas Edison, and the Telephone." *Science, Technology, and Human Values* 15: 131–164.

Grandy, D. R. 1933. "Combining Mercury Vapor and Incandescent Lamps to Produce Commercial White Light." *Transactions of the Illuminating Engineering Society* 28: 762.

Greimas, A. J. 1987. *On Meaning: Selected Writings in Semiotic Theory*. Minneapolis, MN: University of Minnesota Press.

Greimas, A. J. 1990. *The Social Sciences. A Semiotic View*. Minneapolis, MN: University of Minnesota Press.

Grew, W. F. 1921. *The Cycle Industry: Its Origins, History, and Latest Developments*. London: Pitman.

Griffin, H. H. 1877. *Bicycles of the Year 1877*. London: The Bazaar Office.

Griffin, H. H. 1886. *Bicycles & Tricycles of the Year 1886*. London: L. Upcott Gill (republished with an introduction by N. Marshman in 1971. Otley: Olicana Books Ltd.).

Griffin, H. H. 1889. *Bicycles & Tricycles of the Year 1889*. London: L.Upcott Gill (reprinted in 1985 by John Pinkerton Reprint Catalogue Issues).

Grout, W. H. J. 1870. "Improvements in the Construction and Mode of Making Wheels for Velocipedes, Carriages, and Other Vehicles." British patent No.3152 issued on 1 December 1870.

Gutting, G. 1984. "Paradigms, Revolutions, and Technology." In *The Nature of Technological Knowledge: Are Models of Scientific Change Relevant?* Dordrecht: Reidel, 47–65.

Hagendijk, R. 1990. "Structuration Theory, Constructivism, and Scientific Change." In S. E. Cozzens and T. F. Gieryn, eds., *Theories of Science in Society*. Bloomington: Indiana University Press, 43–66.

Hammond, J. W. 1941. *Men and Volts: The Story of General Electric*. Philadelphia: Lippincott.

Harrison, W., and Hibben, S. G. 1938. "Efficient Tint Lighting With Fluorescent Tubes." *Electrical World* 110 (May 1938): 1523–1530.

Haynes, W., ed. 1945a. *American Chemical Industry, Vol. 2: The World War I Period: 1912–1922*. New York: Van Nostrand.

Haynes, W., ed. 1945b. *American Chemical Industry, Vol. 3: The World War I Period: 1912–1922*. New York: Van Nostrand.

Haynes, W., ed. 1949. *American Chemical Industry, Vol. 6: The Chemical Companies*. New York: Van Nostrand.

Haynes, W., ed. 1954. *American Chemical Industry, Vol. 1: Background and Beginnings*. New York: Van Nostrand.

Helm, L. 1907. "A Process of Manufacturing from Aldehydes and Phenols Substances Adapted to Serve as Substitutes for Resins and Gums." British Patent No. 25,216, filed 13 November 1907.

Henschke, F. 1903. "Verfahren zur Darstellung eines Kondensationsproduktes aus Phenol und Formaldehyd." German Patent No. 157553, filed 22 February 1903.

Heritage, J. 1984. *Garfinkel and Ethnomethodology*. Cambridge: Polity Press.

Hindle, B. 1981. *Emulation and Invention*. New York: Norton.

Hirsch, R. F. 1989. *Technology and Transformation in the American Electric Utility Industry*. Cambridge: Cambridge University Press.

Hobbes, T. 1962. *Leviathan*. London: Collier-Macmillan.

Hogenkamp, G. J. M. 1939. *De geschiedenis van Burgers Deventer is de geschiedenis van de fiets*. Maarssen: publisher unknown.

Hooker, E. F. 1916. "An Appreciation of Dr. Baekeland." *The Journal of Industrial and Engineering Chemistry* 8: 183–184.

Hounshell, D. A. 1984. *From the American System to Mass Production, 1800–1932: The Development of Manufacturing Technology in the United States*. Baltimore: Johns Hopkins University Press.

Hounshell, D. A., and Smith Jr., J. K. 1988. *Science and Corporate Strategy: Du Pont R & D, 1902–1980*. Cambridge: Cambridge University Press.

Hughes, Th. P. 1971. *Elmer Sperry: Inventor and Engineer*. Baltimore: Johns Hopkins University Press.

Hughes, Th. P. 1983. *Networks of Power: Electrification in Western Society, 1880–1930*. Baltimore: Johns Hopkins University Press.

Hughes, Th. P. 1986. "The Seamless Web: Technology, Science, Etcetera, Etcetera." *Social Studies of Science* 16: 281–292.

Hughes, T. P. 1987. "The Evolution of Large Technological Systems." In Bijker, Hughes, and Pinch (1987a): 51–82.

Hutter, J. J. 1988. *Toepassingsgericht Onderzoek in de Industrie. De Ontwikkeling van Kwikdamplampen bij Philips, 1900–1940*. Eindhoven: Technical University of Eindhoven.

Hyatt, J. W. 1870. "Improvement in Treating and Molding Pyroxyline." U.S. Patent 105338, filed on 12 July 1870.

Hyatt, J. W. 1914. "Address of Acceptance." *Journal of Industrial and Engineering Chemistry* 6: 158–161.

Ingenieursblad, 1964. "Speciaal nummer gewijd aan de Dr. L. H. Baekeland-Herdenking." *Het Ingenieursblad* 33: 55–112.

Inman, G. E. 1954. "Fluorescent Lamps—Past, Present, and Future." *General Electric Review* 57: 34–38.

Inman, G. E., and Thayer, R. N. 1938. "Low-Voltage Fluorescent Lamps." *Journal for Electrical Engineering* 57 (June 1938): 245–48.

Institut für Wirtschaftsbeobachtung der deutschen Fertigware. 1938. "Untersuchung über Kunstharz in der Vorstellung und im Urteil des letzten Verbrauchers." Berlin: Institut für Wirtschaftsbeobachtung der deutschen Fertigware, im auftrag der Gesellschaft für Konsumforschung E.V.

Jäckel, G. 1925. "Kolben oder Glocke für Lampen und Vakuumröhren." German Patent No. 482,048, filed 12 March 1925.

Jasanoff, S., Pinch, T. J., Markle, P., and Petersen, J., eds. 1994. *Handbook of Science, Technology, and Society*. London: Sage.

Jenkins, H. G. 1935. "Improvements in Electric Discharge Lamps." British Patent No. 457,486, filed 30 May 1935.

Jenkins, H. G. 1942. "Fluorescent Lighting." *Journal of the Royal Society of Arts* 90: 282–302.

Jenkins, R. V. 1975. *Images and Enterprise: Technology and the American Photographic Industry, 1839 to 1925*. Baltimore: Johns Hopkins University Press.

Jewkes, J., Sawers, D., and Stillerman, R. 1958 (2nd ed. 1969). *The Sources of Invention*. London: Macmillan.

Kaufer, E. 1989. *The Economics of the Patent System*. London: Harwood Academic Publishers.

Kaufman, M. 1963. *The First Century of Plastics: Celluloid and Its Sequel*. London: Plastics Institute.

Kaufmann, C. B. 1968. "Grand Duke, Wizard, and Bohemian: A Biographical Profile of Leo Hendrik Baekeland (1863–1944)." Unpublished MA thesis, University of Delaware.

Kettering, Ch. F. 1946. "Biographical Memoir of Leo Hendrik Baekeland, 1863–1944." *National Academy of Sciences of the United States of America Biographical Memoirs* 24: 281–302.

Kleeberg, W. 1891. "Ueber die Einwirkung des Formaldehyds auf Phenole." *Annalen der Chemie* 263: 283–286.

Klinckowstroem, C. Graf von. 1959. *Knaurs Geschichte der Technik*. Munich, Zurich: Droemerische Verlagsanstalt Th. Knaur.

Knoll and Co. 1907. "Improvements relating to the Manufacture of Resin-like Products from Phenols and Formaldehyde." British Patent No. 28,009, filed 19 December 1907.

Knoll and Co. 1908a. "Verfahren zur Beschleunigung der Erhärtung von Kondensationsprodukten aus Phenolen und Aldehyden." German Patent No. 214194, filed 4 July 1908.

Knoll and Co. 1908b. "Verfahren zur Beschleunigung der Erhärtung von Kondensationsprodukten aus Phenolen und Aldehyden" (Zusatz zum Patente No. 214194), German Patent No. 222543, filed 21 October 1908.

Knorr-Cetina, K. D. 1981. *The Manufacture of Knowledge: An Essay on the Constructivist and Contextual Nature of Science.* Oxford: Pergamon.

Knorr-Cetina, K. D., and Mulkay, M. J., eds. 1983. *Science Observed: Perspectives on the Social Study of Science.* London and Beverly Hills: Sage.

Koch, E. 1933. "Elektrische Leuchtröhre mit einer Füllung aus Edelgasen und Quecksilber und einer die Innenwand bedeckenden Schicht aus luminiszierenden Stoffen." German Patent No. 624,758, filed 7 March 1933.

Koch, Gebrüder. 1929. "Verfahren zum Einbringen luminiszierender Stoffe in elektrische Entladungsgefäße." German Patents Nos. 536,980 and 583,305, filed 16 October 1929 and 21 June 1930 respectively.

Kras, R. J., Gras, S., Lokin, D. H. A. C., Ruhe, M. M. J. T., Zweers-van der Elst, F. J., and Tange, J. A. 1981. *Bakeliet, techniek/vormgeving/gebruik.* Rotterdam: Museum Boymans-Van Beuningen, catalogue 288.

Kruithof, A. A. 1941. "Buisvormige luminescentielampen voor algemeene verlichtingsdoeleinden." *Philips Technisch Tijdschrift.* 6, No. 3: 65–96.

Küch, R., and Retschinsky, T. 1906. "Photometrische und spektralphotometrische Messungen am Quecksilberlichtbogen bei hohem Dampfdruck." *Annalen der Physik* 20: 563–583.

Küch, R., and Retschinsky, T. 1907. Temperaturmessungen im Quecksilberlichtbogen der Quarzlampe." *Annalen der Physik* 22: 595–602.

Kuhn, Th. 1970. *The Structure of Scientific Revolutions*, 2nd ed. Chicago: Chicago University Press.

Laclau, E., and Mouffe, C. 1985. *Hegemony and Socialist Strategy: Towards a Radical Democratic Politics.* London: Verso.

Latour, B. 1984. *Les Microbes. Guerre et Paix, suivi de Irréductions.* Paris: Editions A. M. Métailié.

Latour, B. 1986b. "How to Write 'The Prince' for Machines as Well as for Machinations." Paper for seminar on Technology and Social Change, Edinburgh, June 1986.

Latour, B. 1987. *Science in Action: How to Follow Scientists and Engineers through Society.* Milton Keynes: Open University Press; and Cambridge, MA: Harvard University Press.

Latour, B. 1992a. "Where Are the Missing Masses? Sociology of a Few Mundane Artifacts." In Bijker and Law (1992a): 225–258.

Latour, B., and Woolgar, S. 1979. *Laboratory Life: The Social Construction of Scientific Facts.* Beverly Hills and London: Sage; second edition published as *Laboratory Life: The Construction of Scientific Facts.* Princeton, N.J.: Princeton University Press, 1986.

Laudan, L. 1981. "The Pseudo-Science of Science?" *Philosophy of the Social Sciences* 11: 173–198.

Laudan, R., ed. 1984a. *The Nature of Technological Knowledge: Are Models of Scientific Change Relevant?* Dordrecht: Reidel.

Laudan, R. 1984b. "Introduction." In Laudan (1984a): 1–26.

Law, J., ed. 1986. *Power, Action and Belief: A New Sociology of Knowledge?* London: Routledge and Kegan Paul.

Law, J. 1987. "Technology and Heterogeneous Engineering: The Case of Portuguese Expansion." In Bijker, Hughes, and Pinch (1987a): 111–134.

Law, J. 1991a. *A Sociology of Monsters: Essays on Power, Technology, and Domination.* London: Routledge, The Sociological Review Monograph.

Law, J. 1994. *Organizing Modernity.* Oxford: Blackwell.

Law, J., and Bijker, W. E. 1992. "Postscript: Technology, Stability, and Social Theory." In Bijker and Law (1992a), 290–308.

Law, J., and Callon, M. 1992. "The Life and Death of an Aircraft: A Network Analysis of Technical Change." In Bijker and Law (1992a), 21–52.

Lawson, H. J. 1879. "Improvements in the Construction of Bicycles and Other Velocipedes, and in Apparatus to Be Used in Connection Therewith." British Patent No. 3934, filed 30 September 1879.

Lawson, H. J. 1884. "Improvements in Velocipedes." British Patent No. 13070 issued on 2 September 1884.

Lebach, H. 1908. "Process for Hardening Condensation Products from Phenols and Aldehydes." U.S. Patent No. 965,823, filed 21 December 1908.

Lebach, H. 1909a. "Uber Resit." *Chemiker Zeitung* 23: 680, 705.

Lebach, H. 1909b. "Resinit und Bakelit." *Chemiker Zeitung* 23: 721.

Lebach, H. 1909c. "Uber Resinit." *Zeitschrift für angewandte Chemie* 22: 1598–1601, 2007–2008.

Lebach, H. 1909d. "Nachschrift." *Zeitschrift für angewandte Chemie* 22: 2007–2008.

LeBel, C. J. 1929. "Electric Lamp." U.S. Patent No. 2,126,787, filed 14 August 1929.

Lederer, L. 1894. "Eine neue Synthese von Phenolalkoholen." *Journal für Praktische Chemie. Neue Folge* 50: 223–226.

Lederer, L. 1902a. "Verfahren zur Herstellung hornartiger Produkte." German Patent No. 145106, filed 28 February 1902.

Lederer, L. 1902b. "Verfahren zur Herstellung hornartiger Produkte." German Patent No. 152111, filed 28 February 1902.

Lessing, H. E. 1990. "Karl von Drais' Two-Wheeler—What We Know." In *Proceedings of the First International Conference on Cycling History*. Glasgow: Museum of Transport, 4–18.

Leverenz, H. W. 1936. "Luminescent Material." U.S. Patent No. 2,118,091, filed 29 February 1936.

Luckiesh, M. 1938. "We have Wanted Cooler Footcandles ... Now We Have Them." *Magazine of Light* 7: 23–38.

Luckiesh, M., and Moss, F. K. 1937. *The Science of Seeing*. New York: Van Nostrand.

Luckiesh, M., Taylor, A. H., and Kerr, G. P. 1938. "Artificial White Light." *General Electric Review* 41: 89–93.

Luft, A. 1902a. "Verfahren zur Darstellung plastischer Massen." German Patent No. 140552, filed 29 April 1902.

Luft, A. 1902b. "Process of Producing Plastic Compounds." U.S. Patent No. 735,278, filed 20 September 1902.

Lukes, S. 1974. *Power: A Radical View*. London: Macmillan.

Machiavelli, N. 1958. *The Prince*. Harmondsworth: Penguin.

MacKenzie, D. 1990. *Inventing Accuracy: An Historical Sociology of Ballistic Missile Guidance*. Cambridge, MA: MIT Press.

Maher, E. 1977. "The Dynamics of Growth in the Electric Power Industry." In Sayre (1977): 149–216.

Manasse, O. 1894. "Ueber eine Synthese aromatischer Oxyalkohole." *Berichte der Deutschen Chemischen Gesellschaft* 27: 2409–2413.

Marchand, R. 1985. *Advertising the American Dream: Making Way for Modernity, 1920–1940*. Berkeley: University of California Press.

Marshman, N. R. 1971. Introduction to H. H. Griffin (1886), *Bicycles & Tricycles of the Year 1886*, republished in 1971 by Otley: Olicana Books Ltd.

Marx, K. 1867. *Das Capital. Band 1*. English edition (1976), Harmondsworth: Penguin.

Mathews, J. 1989a. *Tools of Change: New Technology and Democratisation of Work*. Sydney: Pluto Press.

Mathews, J. 1989b. *Age of Democracy: The Politics of Post-Fordism.* Oxford: Oxford University Press.

McKeag, A. H., and Randall, A. H. 1936. "Improvements in or Relating to Electric Discharge Lamps Comprising Material Adapted to Be Excited to Luminescence by an Electric Discharge." British Patent No. 480,356, filed 23 July 1936.

McKeag, A. H., and Randall, A. H. 1937. "Improvements in Combinations of a Source of Radiation with Material Adapted to Be Excited to Luminescence by the Radiation." British Patent No. 495,706, filed 16 April 1937.

McGonagle, S. 1968. *The Bicycle in Life, Love, War, and Literature.* London: Pelham Books.

McNeill, W. H. 1982. *The Pursuit of Power: Technology, Armed Force and Society since A.D. 1000.* Oxford: Blackwell.

Meikle, J. L. 1979. *Twentieth Century Limited: Industrial Design in America, 1925–1939.* Philadelphia, PA: Temple University Press.

Meikle, J. L. 1986. "Plastic, Material of a Thousand Uses." In J. J. Corn, ed., *Imagining Tomorrow: History, Technology, and the American Future.* Cambridge, MA: MIT Press, pp. 77–96.

Meinhardt, W. 1932. *Entwicklung und Aufbau der Glühlampenindustrie.* Berlin: Carl Heymanns Verlag.

Michael, A. 1883. "On the Action of Aldehydes on Phenols." *American Chemical Journal* 5: 338–349.

Micksch, K. 1918. "Kriegsersatzstoffe." *Kunststoffe* 8: 133–136.

Miller, S. 1882. "Improvements in Velocipedes." British Patent No. 4668, filed 30 September 1884.

Minck, G. H. 1968. *Fietsend door de eeuwen.* Deventer: Kluwer.

Misa, T. 1988. "How Machines Make History, and How Historians (and Others) Help Them to Do So." *Science, Technology, and Human Values* 13: 308–331.

Misa, T. 1992. "Controversy and Closure in Technological Change: Constructing Steel." In Bijker and Law (1992a), 109–139.

Moon, P. 1936. *The Scientific Basis of Illuminating Engineering.* New York: Dover Publications (revised ed. 1961).

Mueller, W. F. 1983. "The Anti-Antitrust Movement." In J. V. Craven, ed., *Industrial Organization, Antitrust, and Public Policy.* The Hague: Kluwer-Nijhoff, pp. 19–40.

Mumford, L. 1964. "Authoritarian and Democratic Technics." *Technology and Culture* 5: 1–8.

Mumford, L. 1967. *The Myth of the Machine.* Vol. 1, *Technics and Human Development.* New York: Harcourt, Brace, and World.

Mumford, L. 1970. *The Myth of the Machine.* Vol. 2, *The Pentagon of Power.* New York: Harcourt, Brace, and World.

Murdock, C. 1977. "Legal and Economic Aspects of the Electric Utility's 'Mandate to Serve.'" In Sayre (1977): 100–115.

Nelson, R. R., and Winter, S. G. 1977. "In Search of a Useful Theory of Innovation." *Research Policy* 6: 36–76.

Nelson, R. R., and Winter, S. G. 1982. *An Evolutionary Theory of Economic Change.* Cambridge, MA: Harvard University Press.

Noble, D. F. 1984. *Forces of Production: A Social History of Industrial Automation.* New York: Knopf.

Nye, D. E. 1985. *Image Worlds: Corporate Identities at General Electric, 1890–1930.* Cambridge, MA: MIT Press.

Nye, D. E. 1990. *Electrifying America: Social Meanings of a New Technology.* Cambridge, MA: MIT Press.

Oday, A. B., and Cissell, R. F. 1939. "Fluorescent Lamps and Their Applications." *Transactions of the Illuminating Engineering Society.* December 1939: 1165–1188.

Ogburn, W. F. 1933. With the assistance of S. C. Gilfillan. "The Influence of Invention and Discovery." In *Recent Social Trends in the United States: Report of the President's Research Committee on Social Trends,* 122–166. New York: McGraw-Hill Book Company.

Ogburn, W. F. 1945. *The Social Effects of Aviation.* Boston: Houghton Miffin.

Ogburn, W. F., and Meyers Nimkoff, F. 1955. *Technology and the Changing Family.* Boston: Houghton Miffin.

Oliver, S. H., and Berkebile, D. H. 1974. *Wheels and Wheeling: The Smithsonian Cycle Collection.* Washington, D.C.: Smithsonian Institution Press.

OSA Participations industrielles soc. an. 1936. "Procédé d'application de luminophores sur la paroi en verre de récipients électriques àdécharge." French Patent No. 807,991, filed 26 June 1936.

Palmer, A. J. 1958. *Riding High: The Story of the Bicycle.* London: Vision Press.

Parkes, A. 1855. "Manufacture of Elastic and Adhesive Compounds." British Patent No. 2359, filed 22 October 1855.

Parkes, A. 1864. "Preparing Compounds of Gun Cotton and Other Substances, &c." British Patent No. 2675, filed 28 October 1864.

Parkes, A. 1865a. "Manufacture of Compounds of Pyroxyline." British Patent No. 1313, filed 11 May 1865.

Parkes, A. 1865b. "On the Properties of Parkesine, and Its Application to the Arts and Manufactures." *Journal of the Society of Arts* 14, No. 683: 81–86.

Parsons, T. 1967. *Sociological Theory and Modern Society*. New York: Free Press.

Passer, H. C. 1953. *The Electrical Manufacturers, 1875–1900*. Cambridge, MA: Harvard University Press.

Pasveer, B. 1992. *Shadows of Knowledge. Making a Representing Practice in Medicine: X-ray Pictures and Pulmonary Tubercolosis, 1895–1930*. Ph.D. dissertation, University of Amsterdam.

Patent-Treuhand-Gesellschaft für elektrische Glühlampe m.b.H. 1935. "Verfahren zum Anbringen von Luminophoren auf den Glaswänden elektrischer Entladungsgefäße, insbesondere elektrische Entladungslampen." German Patent No. 692,394, filed 29 June 1935.

Pearson, H. C. 1922. *Pneumatic Tires*. New York: India Rubber Publishing Co.

Pickering, A. 1984. *Constructing Quarks—A Sociological History of Particle Physics*. Chicago and Edinburgh: University of Chicago Press and Edinburgh University Press.

Pickering, A., ed. 1992. *Science as Practice and Culture*. Chicago: University of Chicago Press.

Pinch, T. J. 1986. *Confronting Nature: The Sociology of Solar-Neutrino Detection*. Dordrecht: Reidel.

Pinch, T. J. 1993. "Turn, Turn, and Turn Again: The Woolgar Formula." *Science, Technology & Human Values* 18: 511–522.

Pinch, T. J., and Bijker, W. E. 1987. "The Social Construction of Facts and Artifacts: Or How the Sociology of Science and the Sociology of Technology Might Benefit Each Other." In Bijker, Hughes, and Pinch (1987): 17–50.

Plath, H. 1978. "Laufrad-Vélocipède-Hobbyhorse, eine typologische Untersuchung." In *Sonderdruck: Museum und Kulturgeschichte, Festschrift für Wilhelm Hansen*, Münster.

Polak, J. 1907. "Der Quecksiber-Lichtbogen und seine technische Verwendung." *Elektrotechnische Zeitschrift* 18: 599–603, 651–656, 733–738.

Prest, J. 1960. *The Industrial Revolution in Coventry*. London: Oxford University Press.

Randall, J. T. 1936. "Improvements in or Relating to Combinations of Electric Discharge Devices and Materials excited to Luminescence by the Electric Discharge." British Patent No. 469,732, filed 27 January 1936.

Randall, J. T. 1937. "Luminescence and Its Applications." *Journal of the Royal Society of Arts* 20: 354–381.

Rauck, M. J. B., Volke, G., und Paturi, F. R. 1979. *Mit dem Rad durch zwei Jahrhunderte: Das Fahrrad und seine Geschichte.* Aarau, Stuttgart: AT Verlag.

Raverat, G. 1952. *Period Piece: A Cambridge Childhood.* London: Faber and Faber.

Redman, L. V., and Mory, A. V. H. 1931. "The Bakelite Corporation." *Journal of Industrial and Engineering Chemistry* 23: 595–597.

Redman, L. V., Weith, A. J., and Brock, F. P. 1914. "Synthetic Resins." *Journal of Industrial and Engineering Chemistry* 6: 3–16.

Reich, L. S. 1985. *The Making of American Industrial Research: Science and Business at GE and Bell, 1876–1926.* Cambridge: Cambridge University Press.

Reti, L. 1974. *The Unknown Leonardo.* New York: McGraw-Hill.

Risler, J. 1922a. "Improved Process for Obtaining Phosphorescence in Luminous Paints." British Patent No. 207,786, filed 2 December 1922.

Risler, J. 1922b. "Improved Process for Obtaining Phosphorescence in Luminous Paints." British Patent No. 208,723, filed 21 December 1922.

Risler, J. 1923. "Tubes fluorescents et Radiations excitant la fluorescence." *Bulletin Scientifique des Étudiants de Paris* 12, June 1923: 18–25.

Risler, J. 1925. "Improvements in Electrical Discharge Tubes." British Patent No. 229,341, filed 14 February 1925.

Ritchie, A. 1975. *King of the Road: An Illustrated History of Cycling.* London: Wildwood House; Berkeley: Ten Speed Press.

Roberts, D. 1980. *The Years of the High Bicycle: A Compilation of Catalogues from 1877–1886.* London: John Pinkerton Reprint Catalogue Issues.

Robertson, P. 1974. *The Shell Book of Firsts.* London: Ebury Press & Michael Joseph Ltd.

Robins, N., and Aronson, S. M. L. 1985. *Savage Grace.* New York: Dell Publishing.

Rogers, R. P. 1980. *The Development and Structure of the U.S. Electric Lamp Industry: Bureau of Economics Staff Report of the Federal Trade Commission.* Washington, D.C.: U.S. Government Printing Office.

Rosenberg, N. 1976. *Perspectives on Technology.* Cambridge: Cambridge University Press.

Ryde, J. W. 1932. "Improvements in Electric Discharge Lamps." British Patent No. 401,846, filed 5 September 1932.

Ryde, J. W. 1934. "Improvements in Electric Discharge Lamps." British Patent No. 422,364, filed 4 January 1934.

Sahal, D. 1981. *Patterns of Technological Innovation.* New York: Addison-Wesley.

Schwartz Cowan, R. 1983. *More Work for Mother: The Ironies of Household Technology from the Open Hearth to the Microwave.* New York: Basic Books.

Schwartz Cowan, R. 1987. "The Consumption Junction: A Proposal for Research Strategies in the Sociology of Technology." In Bijker, Hughes, and Pinch (1987a): 261–280.

Scientific American. 1892a. "Dangers of Celluloid." *Scientific American* 66: 208.

Scientific American. 1892b. "Experiments with Celluloid." *Scientific American* 66: 261.

Scranton, Ph. 1991. "Theory and Narative in the History of Technology: Comment." *Technology and Culture* 32: 385–393.

Sewig, R., ed., 1938. *Handbuch der Lichttechnik* (two vols.). Berlin: Julius Springer.

Shrum, W. 1985. *Organized Technology: Networks and Innovation in Technical Systems.* West Lafayette, IN: Purdue University Press.

Singer, G. 1878. "Improvements in and Connected with Bicycles, Tricycles, and Other Vehicles." British Patent No. 4265, filed 24 October 1878.

Smith, A. 1899a. "A New Material Adapted for Electrical Insulation and Also for Many Purposes for Which Ebonite, Wood and Such Like Materials are Used." British Patent No. 16,247, filed 9 August 1899.

Smith, A. 1899b. "Verfahren zur Herstellung eines Ersatzmaterials für Ebonit, Holtz u. dgl." German Patent No. 112685, filed 10 October 1899.

Sotheby's. 1989. "20th Century Decorative Arts: A Private Collection of Bakelite and Other Plastics. Rietveld Furniture." Amsterdam: Sotheby's Foundation, catalogue Sale 505.

Star, S. L. 1988. "The Structure of Ill-Structured Solutions: Boundary Objects and Heterogeneous Distributed Problem Solving." Paper for the Eighth AAAI Conference on Disributed Artificial Intelligence, May 1988, Lake Arrowhead, Ca.

Starley, J., and Hillman, W. 1870. "Improvements in the Construction of Wheels Applicable Chiefly to Velocipedes, and in the Driving Gear for Such Vehicles." British Patent No. 2236, filed 11 August 1870.

Staudenmaier, J. M. 1985. *Technology's Storytellers: Reweaving the Human Fabric.* Cambridge, MA: MIT Press.

Staudenmaier, J. M. 1990. "Recent Trends in the History of Technology." *The American Historical Review* 95: 715–725.

Staudinger, H. J. 1926. "Die Chemie der hochmolekularen organischen Stoffe im Sinne der Kekuléschen Strukturlehre." *Berichte der Deutschen Chemischen Gesellschaft* 59: 3019–3043.

Staudinger, H. J. 1961. *Arbeitserinnerungen.* Heidelberg: Hüthig.

Stedelijk Museum Amsterdam. 1985. *Industrie en vormgeving.* Amsterdam: Reproduktieafdeling Stedelijk Museum.

Stocking, G. W., and Watkins, M. W. 1946. *Cartels in Action: Case Studies in International Business Diplomacy.* New York: The Twentieth Century Fund; reprinted in 1975 by Kraus Reprint Co., Millwood, NY.

Story, W. H. 1905a. "Verfahren zur Herstellung eines Ersatzmittels für Ebonit, Horn, Celluloid u. dgl. durch Kondensation von Phenolen mit Formaldehyd." German Patent No. 13990, filed 23 April 1905.

Story, W. H. 1905b. "Process of Producing Substitutes for Celluloid, Horn, Ebonite, or Similar Substances." British Patent No. 8875, filed 27 April 1905.

Strum, S., and Latour, B. 1984. "The Meaning of Social." Paper presented to the International Primatology Society Meeting, Nairobi.

Telser, L. G. 1958. "Why Should Manufacturers Want Fair Trade?" *Journal of Law and Economics* 20: 86–105.

Thayer, R. N., and Barnes, B. T. 1939. "The Basis for High Efficiency in Fluorescent Lamps." *Journal of the Society of Science and Arts* 29: 131–134.

Thinius, K. 1976. "Entwicklung der Reaktion zwischen Formaldehyd und Phenol zu einem Produktionszweig der Plastindustrie zwischen 1900 und 1930: Studien zur Geschichte der Plaste X." *Plaste und Kautschuk* 23: 746–749.

Thomas, R. E. 1955. *Salt and Water, Power and People: A Short History of Hooker Electrochemical Company.* Niagara Falls: [Hooker Electrochemical Company?].

Thompson, E. P. 1967. "Time, Work-Discipline, and Industrial Capitalism." *Past and Present* 38: 56–97.

Thompson, F. 1941. *Over to Candleford.* London: Oxford University Press.

Thomson, R. W. 1845. "An Improvement in Carriage Wheels, Which Is Also Applicable to Other Rolling Bodies." British Patent No. 10990, issued on 10 December 1845.

Thomson, R. W. 1847. *Mechanics Magazine*, Vol. XLVI: 209.

Timm, U. 1984. *Der Mann auf dem Hochrad.* Cologne: Kiepenheuer and Witsch.

Vanderpoel, F. 1914. "Personal Reminiscences." *Journal of Industrial and Engineering Chemistry* 6: 161–162.

Wallis, H. 1884. "Improvements in Bicycles." British Patent No. 15342, issued on 21 November 1884.

Weber, M. 1947. *The Theory of Social and Economic Organization*. London: Routledge and Kegan Paul.

Weber, R. J., and Perkins, D. N. 1989. "How to Invent Artifacts and Ideas." *New Ideas in Psychology* 7: 49–72.

Weingart, P. 1984. "The Structure of Technological Change: Reflections on a Sociological Analysis of Technology." In Laudan (1984a): 115–142.

Weingart, P. ed., 1989. *Technik als sozialer Prozess*. Frankfurt: Suhrkamp Verlag.

Welch, C. K. 1890. "Improvements in Rubber Tyres and Metal Rims or Felloes of Wheels for Cycles and Other Light Vehicles." British Patent No. 14563, filed 16 September 1890.

Wells, H. G. 1896. *The Wheels of Chance*. Reissued in 1984, London: Dent & Sons.

Wheatcroft, E. L. E. 1938. *Gaseous Electrical Conductors*. Oxford: Clarendon Press.

Whitt, F. R., and Wilson, D. G. 1982. *Bicycling Science*. Cambridge, MA: MIT Press.

Williamson, G. 1966. *Wheels within Wheels: The Story of the Starleys of Coventry*. London: Geoffrey Bles.

Wilson, R. G., Pilgrim, D. H., and Tashjian, D. 1986. *The Machine Age in America, 1918–1941*. New York: The Brooklyn Museum in association with Abrams Inc.

Winner, L. 1977. *Autonomous Technology: Technics-out-of-Control as a Theme in Political Thought*. Cambridge, MA: MIT Press.

Winner, L. 1980. "Do Artifacts Have Politics?" *Daedalus* 109, No. 1: 121–136; also published in Winner (1986): 19–39.

Winner, L. 1986. *The Whale and the Reactor: A Search for Limits in an Age of High Technology*. Chicago: University of Chicago Press.

Winner, L. 1991. "Upon Opening the Black Box and Finding It Empty: Social Constructivism and the Philosphy of Technology." In C. Pitt and E. Lugo, eds., *The Technology of Discovery and the Discovery of Technology: Proceedings of the 6th International Conference of the Society for Philosophy and Technology*. Blacksburg, VA: Society for Philosophy and Technology, 503–519.

Wood, R. W. 1911. *Physical Optics*. New York: Macmillan.

Woodforde, J. 1970. *The Story of the Bicycle*. London: Routledge and Kegan Paul.

Woolgar, S. 1983. "Irony in the Social Study of Science." In K. D. Knorr-Cetina and M. Mulkay, eds., *Science Observed: Perspectives on the Social Study of Science*. London: Sage: 239–266.

Woolgar, S. 1991. "The Turn to Technology in Social Studies of Science." *Science, Technology, and Human Values* 16: 20–50.

Worden, E. C. 1911. *Nitro-cellulose Industry*. New York: Van Nostrand.

Wurts, R., et al. 1977. *The New York World's Fair 1939/1940 in 155 Photographs by Richard Wurts and Others*. Selection, arrangement, and text by Stanley Appelbaum. New York: Dover.

Wynne, B. 1982. *Rationality and Ritual: The Windscale Inquiry and Nuclear Decisions in Britain*. Chalfont St.Giles Bucks: British Society for the History of Science.

Wynne, B. 1991. "Knowledges in Context." *Science, Technology, and Human Values* 16: 111–121.

Yearley, S. 1992. *The Green Case*. London: Routledge.

Yearley, S. 1994. "The Environmental Challenge to Science Studies." In J. C. Petersen, G. E. Markle, S. Jasanoff, and T. Pinch, eds., *Handbook of Science, Technology, and Society*. Newbury Park, CA, and London: Sage.

Name Index

Subject Index